A FIRST COURSE IN TURBULENCE

A FIRST COURSE IN TURBULENCE

H. Tennekes and J. L. Lumley

The MIT Press
Cambridge, Massachusetts, and London, England

This book was designed by the MIT Press Design Department.
It was set in IBM Univers Medium.

ISBN 978-0-262-53630-1 (paperback)

ISBN 978-0-262-20019-6 (hardcover)

Library of Congress catalog card number: 77–165072

CONTENTS

PREFACE

In the customary description of turbulence, there are always more unknowns than equations. This is called the closure problem; at present, the gap can be closed only with models and estimates based on intuition and experience. For a newcomer to turbulence, there is yet another closure problem: several dozen introductory texts in general fluid dynamics exist, but the gap between these and the monographs and advanced texts in turbulence is wide. This book is designed to bridge the second closure problem by introducing the reader to the tools that must be used to bridge the first.

A basic tool of turbulence theory is dimensional analysis; it is always used in conjunction with an appeal to the idea that turbulent flows should be independent of the Reynolds number if they are scaled properly. These tools are sufficient for a first study of most problems in turbulence; those requiring sophisticated mathematics have been avoided wherever possible. Of course, dimensional reasoning is incapable of actually solving the equations governing turbulent flows. A direct attack on this problem, however, is beyond the scope of this book because it requires advanced statistics and Fourier analysis. Also, even the most sophisticated studies, so far, have met with relatively little success. The purpose of this book is to introduce its readers to turbulence; it is neither a research monograph nor an advanced text.

Some understanding of viscous-flow and boundary-layer theory is a prerequisite for a successful study of much of the material presented here. On the other hand, we assume that the reader is not familiar with stochastic processes and Fourier transforms. Because the Reynolds stress is a second-rank tensor, the use of tensor notation could not be avoided; however, very little tensor analysis is needed to understand elementary operations on the equations of motion in Cartesian coordinate systems.

We use most of the material in this book in an introductory turbulence course for college seniors and first-year graduate students. We feel that this book can also serve as a supplementary text for courses in general fluid dynamics. We have attempted to avoid a bias toward any specific discipline, in the hope that the material will be useful for meteorologists, oceanographers, and astrophysicists, as well as for aerospace, mechanical, chemical, and pollution control engineers.

The scope of this book did not permit us to describe the experimental methods used in turbulence research. Also, because this is an introduction to turbulence, we have not attempted to give an exhaustive list of references. The bibliography lists the books devoted to turbulence as well as some major papers. The most comprehensive of the recent books is Monin and Yaglom's *Statistical Fluid Mechanics* (Monin and Yaglom, 1971); it contains a complete bibliography of the current journal literature.

The manuscript was read by Dr. S. Corrsin and Dr. J. A. B. Wills; they offered many valuable comments. Miss Constance Hazuda typed several drafts and the final manuscript. A preliminary set of lecture notes was compiled in 1967 by Mr. A. S. Chaplin. Several generations of students contributed to the development of the presentation of the material. While writing this book, the authors received research support from the Atmospheric Sciences Section, National Science Foundation, under grants GA-1019 and GA-18109.

HT
JLL

June 1970

Note (September 1974) In the third printing of our book, we have made revisions in the section on time spectra (pages 275-278) and added several new items to the bibliography. The changes on pages 275-278 are quite drastic. They document a striking difference between the Eulerian time spectrum and its Lagrangian counterpart. The difference was discovered recently; it invalidates the original analysis.

BRIEF GUIDE ON THE USE OF SYMBOLS

The theory of turbulence contains many, often crude, approximations. Many relations (except the equations of motion and their formal consequences) therefore do not really permit the use of the equality sign. We adopt the following usage. If the error involved in writing an equation is smaller than about 30%, we use the approximate equality sign $\cong$. For crude approximations the symbol $\sim$ is employed. This generally means that the nondimensional coefficient that would make the relation an equation is not greater than 5 and not smaller than 1/5. If the value of the coefficient is of interest (for example, if the theory is to be compared with experimental data or if a statement about the coefficient is in order), the equality sign is used and the coefficient is entered explicitly. If the problem discussed is the selection of the dominant terms in an equation of motion, the order symbol $\mathcal{O}$, which does not make any commitment on the value of the coefficient, is employed. After the dominant terms have been selected, the equality sign is used in the resulting simplified equation, with the understanding that the error involved can be made arbitrarily small by increasing the parameter in the problem (often a Reynolds number) without limit. We do not claim that we have been completely consistent, but in most cases the meaning of the symbols is made clear in the text.

Though it may sometimes seem confusing, this usage serves as a continuing reminder that relatively few accurate statements can be made about a turbulent flow without recourse to experimental evidence on that flow. If one has to study a flow for which no data are available, all one can do is to find the characteristic parameters (velocity, length, time, and other scales) and to make crude (say within a factor of two) estimates of the properties of the flow. This is no mean accomplishment; it allows one to design an experiment in a sensible way and to select the appropriate nondimensional form in which the experimental data should be presented.

1

INTRODUCTION

Most flows occurring in nature and in engineering applications are turbulent. The boundary layer in the earth's atmosphere is turbulent (except possibly in very stable conditions); jet streams in the upper troposphere are turbulent; cumulus clouds are in turbulent motion. The water currents below the surface of the oceans are turbulent; the Gulf Stream is a turbulent wall-jet kind of flow. The photosphere of the sun and the photospheres of similar stars are in turbulent motion; interstellar gas clouds (gaseous nebulae) are turbulent; the wake of the earth in the solar wind is presumably a turbulent wake. Boundary layers growing on aircraft wings are turbulent. Most combustion processes involve turbulence and often even depend on it; the flow of natural gas and oil in pipelines is turbulent. Chemical engineers use turbulence to mix and homogenize fluid mixtures and to accelerate chemical reaction rates in liquids or gases. The flow of water in rivers and canals is turbulent; the wakes of ships, cars, submarines, and aircraft are in turbulent motion. The study of turbulence clearly is an interdisciplinary activity, which has a very wide range of applications. In fluid dynamics laminar flow is the exception, not the rule: one must have small dimensions and high viscosities to encounter laminar flow. The flow of lubricating oil in a bearing is a typical example.

Many turbulent flows can be observed easily; watching cumulus clouds or the plume of a smokestack is not time wasted for a student of turbulence. In the classroom, some of the films produced by the National Committee for Fluid Dynamics Films (for example, Stewart, 1969) may be used to advantage.

1.1
The nature of turbulence

Everyone who, at one time or another, has observed the efflux from a smokestack has some idea about the nature of turbulent flow. However, it is very difficult to give a precise definition of turbulence. All one can do is list some of the characteristics of turbulent flows.

Irregularity One characteristic is the irregularity, or randomness, of all turbulent flows. This makes a deterministic approach to turbulence problems impossible; instead, one relies on statistical methods.

Diffusivity The diffusivity of turbulence, which causes rapid mixing and increased rates of momentum, heat, and mass transfer, is another important feature of all turbulent flows. If a flow pattern looks random but does not exhibit spreading of velocity fluctuations through the surrounding fluid, it is surely not turbulent. The contrails of a jet aircraft are a case in point: excluding the turbulent region just behind the aircraft, the contrails have a very nearly constant diameter for several miles. Such a flow is not turbulent, even though it was turbulent when it was generated. The diffusivity of turbulence is the single most important feature as far as applications are concerned: it prevents boundary-layer separation on airfoils at large (but not too large) angles of attack, it increases heat transfer rates in machinery of all kinds, it is the source of the resistance of flow in pipelines, and it increases momentum transfer between winds and ocean currents.

Large Reynolds numbers Turbulent flows always occur at high Reynolds numbers. Turbulence often originates as an instability of laminar flows if the Reynolds number becomes too large. The instabilities are related to the interaction of viscous terms and nonlinear inertia terms in the equations of motion. This interaction is very complex: the mathematics of nonlinear partial differential equations has not been developed to a point where general solutions can be given. Randomness and nonlinearity combine to make the equations of turbulence nearly intractable; turbulence theory suffers from the absence of sufficiently powerful mathematical methods. This lack of tools makes all theoretical approaches to problems in turbulence trial-and-error affairs. Nonlinear concepts and mathematical tools have to be developed along the way; one cannot rely on the equations alone to obtain answers to problems. This situation makes turbulence research both frustrating and challenging: it is one of the principal unsolved problems in physics today.

Three-dimensional vorticity fluctuations Turbulence is rotational and three dimensional. Turbulence is characterized by high levels of fluctuating vorticity. For this reason, vorticity dynamics plays an essential role in the description of turbulent flows. The random vorticity fluctuations that characterize turbulence could not maintain themselves if the velocity fluctuations were two dimensional, since an important vorticity-maintenance mechanism known as vortex stretching is absent in two-dimensional flow. Flows that are substantially two dimensional, such as the cyclones in the atmosphere which determine the weather, are not turbulence themselves, even though their characteristics may be influenced strongly by small-scale turbulence (generated somewhere by shear or buoyancy), which interacts with the large-scale flow. In summary, turbulent flows always exhibit high levels of fluctuating vorticity. For example, random waves on the surface of oceans are not in turbulent motion since they are essentially irrotational.

Dissipation Turbulent flows are always dissipative. Viscous shear stresses perform deformation work which increases the internal energy of the fluid at the expense of kinetic energy of the turbulence. Turbulence needs a continuous supply of energy to make up for these viscous losses. If no energy is supplied, turbulence decays rapidly. Random motions, such as gravity waves in planetary atmospheres and random sound waves (acoustic noise), have insignificant viscous losses and, therefore, are not turbulent. In other words, the major distinction between random waves and turbulence is that waves are essentially nondissipative (though they often are dispersive), while turbulence is essentially dissipative.

Continuum Turbulence is a continuum phenomenon, governed by the equations of fluid mechanics. Even the smallest scales occurring in a turbulent flow are ordinarily far larger than any molecular length scale. We return to this point in Section 1.5.

Turbulent flows are flows Turbulence is not a feature of fluids but of fluid flows. Most of the dynamics of turbulence is the same in all fluids, whether they are liquids or gases, if the Reynolds number of the turbulence is large enough; the major characteristics of turbulent flows are not controlled by the molecular properties of the fluid in which the turbulence occurs. Since the equations of motion are nonlinear, each individual flow pattern has certain unique characteristics that are associated with its initial and boundary conditions. No general solution to the Navier-Stokes equations is known; consequently, no general solutions to problems in turbulent flow are available. Since every flow is different, it follows that every turbulent flow is different, even though all turbulent flows have many characteristics in common. Students of turbulence, of course, disregard the uniqueness of any particular turbulent flow and concentrate on the discovery and formulation of laws that describe entire classes or families of turbulent flows.

The characteristics of turbulence depend on its environment. Because of this, turbulence theory does not attempt to deal with all kinds and types of flows in a general way. Instead, theoreticians concentrate on families of flows with fairly simple boundary conditions, like boundary layers, jets, and wakes.

1.2
Methods of analysis

Turbulent flows have been investigated for more than a century, but, as was remarked earlier, no general approach to the solution of problems in turbulence exists. The equations of motion have been analyzed in great detail, but it is still next to impossible to make accurate quantitative predictions without relying heavily on empirical data. Statistical studies of the equations of motion always lead to a situation in which there are more unknowns than equations. This is called the closure problem of turbulence theory: one has to make (very often ad hoc) assumptions to make the number of equations equal to the number of unknowns. Efforts to construct viable formal perturbation schemes have not been very successful so far. The success of attempts to solve problems in turbulence depends strongly on the inspiration involved in making the crucial assumption.

This book has been designed to get this point across. In turbulence, the equations do not give the entire story. One must be willing to use (and capable of using) simple physical concepts based on experience to bridge the gap between the equations and actual flows. We do not want to imply that the equations are of little use; we merely want to make it unmistakably clear that turbulence needs spirited inventors just as badly as dedicated analysts. We recognize that this is a very specific, and possibly biased, point of view. It is possible that at some time in the future, someone will succeed in developing a completely formal theory of turbulence. However, we believe that there is a far better chance of developing a physical model of turbulence in the spirit of the Rutherford model of the atom. The model need not be complete, but it would be very useful. The real challenge, it seems to us, is that no adequate model of turbulence exists today.

Turbulence theory is limited in the same way that general fluid dynamics would be if the Stokes relation between stress and rate of strain in Newtonian fluids were unknown. This illustration is not arbitrary: one approach to turbulence theory is to postulate a relation between stress and rate of strain that involves a turbulence-generated "viscosity," which then supposedly plays a role similar to that of molecular viscosity in laminar flows. This approach is based on a superficial resemblance between the way molecular motions transfer momentum and heat and the way in which turbulent velocity fluctuations transfer these quantities. Phenomenological concepts like "eddy viscosity" (to replace molecular viscosity) and "mixing length" (in analogy with the mean free path in the kinetic theory of gases) were developed by Taylor, Prandtl, and others. These concepts are studied in detail in Chapter 2.

Molecular viscosity is a property of fluids; turbulence is a characteristic of flows. Therefore, the use of an eddy viscosity to represent the effects of turbulence on a flow is liable to be misleading. However, current research seems to indicate that, in simple flows, we may, for analytical reasons, speak of a turbulent fluid rather than of a turbulent flow. Turbulent "fluids," however, are non-Newtonian: they exhibit viscoelasticity and suffer memory effects. In favorable circumstances, the memory is fading in time, so that one may be able to develop a semilocal theory relating the mean stress to the mean rate of strain.

Phenomenological theories of turbulence make crucial assumptions at a fairly early stage in the analysis. In recent years, a group of theoreticians (Kraichnan, Edwards, Orszag, Meecham, and others) have developed very formal and sophisticated statistical theories of turbulence, in the hope of finding a formalism that does not need ad hoc assumptions (see Leslie, 1973). So far, however, rather arbitrary postulates are needed in these theories, too. The mathematical complexity of this work is so overwhelming that a discussion of it has to be left out of this book.

Dimensional analysis One of the most powerful tools in the study of turbulent flows is dimensional analysis. In many circumstances it is possible to argue that some aspect of the structure of turbulence depends only on a few independent variables or parameters. If such a situation prevails, dimensional methods often dictate the relation between the dependent and independent variables, which results in a solution that is known except for a numerical coefficient. The outstanding example of this is the form of the spectrum of turbulent kinetic energy in what is called the "inertial subrange."

Asymptotic invariance Another frequently used approach is to exploit some of the asymptotic properties of turbulent flows. Turbulent flows are characterized by very high Reynolds numbers; it seems reasonable to require that any proposed descriptions of turbulence should behave properly in the limit as the Reynolds number approaches infinity. This is often a very powerful constraint, which makes fairly specific results possible. The development of the theory of turbulent boundary layers (Chapter 5) is a case in point. The limit process involved in an asymptotic approach is related to vanishingly small effects of the molecular viscosity. Turbulent flows tend to be almost independent of the viscosity (with the exception of the very smallest scales of motion); the asymptotic behavior leads to such concepts as "Reynolds-number similarity" (asymptotic invariance).

Local invariance Associated with, but distinct from, asymptotic invariance is the concept of "self-preservation" or local invariance. In simple flow geometries, the characteristics of the turbulent motion at some point in time and space appear to be controlled mainly by the immediate environment. The time and length scales of the flow may vary slowly downstream, but, if the turbulence time scales are small enough to permit adjustment to the gradually changing environment, it is often possible to assume that the turbulence is dynamically similar everywhere if nondimensionalized with local length and time scales. For example, the turbulence intensity in a wake is of order $\delta\ \partial U/\partial y$, where δ is the local width of the wake and $\partial U/\partial y$ is the average mean-velocity gradient across the wake.

Because turbulence consists of fairly large fluctuations governed by nonlinear equations, one may expect a behavior like that exhibited by simple nonlinear systems with limit cycles. Such behavior should be largely independent of initial conditions; the characteristics of the limit cycle should depend only on the dynamics of the system and the constraints imposed on it. In the same way, one expects that the structure of turbulence in a given class of shear flows might be in some state of dynamical equilibrium in which local inputs of energy should approximately balance local losses. If the energy transfer mechanisms in turbulence are sufficiently rapid, so that effects of past events do not dominate the dynamics, one may expect that this limit-cycle type of equilibrium is governed mainly by local parameters such as scale lengths and times. Simple dimensional methods and similarity arguments can be very useful in this kind of situation. Because one may want to look for local scaling laws (both in the spatial and the spectral domain), the problem of finding appropriate length and time scales becomes an important one. Indeed, scaling laws are at the heart of turbulence research.

1.3 The origin of turbulence

In flows which are originally laminar, turbulence arises from instabilities at large Reynolds numbers. Laminar pipe flow becomes turbulent at a Reynolds number (based on mean velocity and diameter) in the neighborhood of 2,000 unless great care is taken to avoid creating small disturbances that might trigger transition from laminar to turbulent flow. Boundary layers in zero pressure gradient become unstable at a Reynolds number $U\delta^*/\nu = 600$ approximately (δ^* is the displacement thickness, U is the free-stream velocity, and ν is the kinematic viscosity). Free shear flows, such as the flow in a mixing layer, become unstable at very low Reynolds numbers because of an inviscid instability mechanism that does not operate in boundary-layer and pipe flow. Early stages of transition can easily be seen in the smoke rising from a cigarette.

On the other hand, turbulence cannot maintain itself but depends on its environment to obtain energy. A common source of energy for turbulent velocity fluctuations is shear in the mean flow; other sources, such as buoyancy, exist too. Turbulent flows are generally shear flows. If turbulence arrives in an environment where there is no shear or other maintenance mechanism, it decays: the Reynolds number decreases and the flow tends to become laminar again. The classic example is turbulence produced by a grid in uniform flow in a wind tunnel.

Another way to make a turbulent flow laminar or to prevent a laminar flow from becoming turbulent is to provide for a mechanism that consumes turbulent kinetic energy. This situation prevails in turbulent flows with imposed magnetic fields at low magnetic Reynolds numbers and in atmospheric flows with a stable density stratification, to cite two examples.

Mathematically, the details of transition from laminar to turbulent flow are rather poorly understood. Much of the theory of instabilities in laminar flows is linearized theory, valid for very small disturbances; it cannot deal with the large fluctuation levels in turbulent flow. On the other hand, almost all of the theory of turbulent flow is asymptotic theory, fairly accurate at very high Reynolds numbers but inaccurate and incomplete for Reynolds numbers at which the turbulence cannot maintain itself. A noteworthy exception is the theory of the late stage of decay of wind-tunnel turbulence (Batchelor, 1953).

Experiments have shown that transition is commonly initiated by a primary instability mechanism, which in simple cases is two dimensional. The primary instability produces secondary motions, which are generally three dimensional and become unstable themselves. A sequence of this nature generates intense localized three-dimensional disturbances (turbulent "spots"), which arise at random positions at random times. These spots grow rapidly and merge with each other when they become large and numerous to form a field of developed turbulent flow. In other cases, turbulence originates from an instability that causes vortices which subsequently become unstable. Many wake flows become turbulent in this way.

1.4
Diffusivity of turbulence

The outstanding characteristic of turbulent motion is its ability to transport or mix momentum, kinetic energy, and contaminants such as heat, particles, and moisture. The rates of transfer and mixing are several orders of magnitude greater than the rates due to molecular diffusion: the heat transfer and combustion rates of turbulent combustion in an incinerator are orders of magnitude larger than the corresponding rates in the laminar flame of a candle.

Diffusion in a problem with an imposed length scale Contrasting laminar and turbulent diffusion rates is a useful exercise not only for getting acquainted with turbulence but also for recognizing the multifaceted role of the Reynolds number. Suppose one has a room (with a characteristic linear dimension L) in which a heating element (radiator) is installed. If there is no air motion in the room, heat has to be distributed by molecular diffusion. This process is governed by the diffusion equation (θ is the temperature; γ is the thermal diffusivity, assumed to be constant):

$$\frac{\partial \theta}{\partial t} = \gamma \frac{\partial^2 \theta}{\partial x_j \partial x_j}. \tag{1.4.1}$$

We are not looking for a specific solution of (1.4.1) with a given set of boundary conditions. Instead, we want to discover the gross consequences of (1.4.1) with the simple tools of dimensional analysis. Dimensionally, (1.4.1) may be interpreted as

$$\frac{\Delta \theta}{T_m} \sim \gamma \frac{\Delta \theta}{L^2}, \tag{1.4.2}$$

where $\Delta\theta$ is a characteristic temperature difference. From (1.4.2), we obtain

$$T_m \sim \frac{L^2}{\gamma}, \tag{1.4.3}$$

which relates the time scale T_m of the molecular diffusion to the independent parameters L and γ. If the characteristic linear dimension L (the *length scale*) of the room is 5 m, the time scale T_m of this diffusion process is of the order of 10^6 sec (more than 100 h). In this estimate the value of γ for air at room temperature and pressure has been used ($\gamma = 0.20\ cm^2/sec$). We conclude that molecular diffusion is rather ineffective in distributing heat through a room.

On the other hand, even fairly weak motions, such as those generated by small density differences (buoyancy), can disperse heat through the room quickly. Suppose that the turbulent motion of the air in the room may also be characterized by the length scale L (that is, motions are present of scales $\leqslant L$). This is a fair assumption, since large-scale motions are most effective in distributing heat and since the largest possible scales of motion can be no larger than the size of the room. We also need a characteristic velocity u (this u may be thought of as an rms amplitude of the velocity fluctuations in the room). For flow with a length scale L and a velocity scale u, the characteristic time is

$$T_t \sim \frac{L}{u}. \tag{1.4.4}$$

Apparently, T_t can be determined only if u can be estimated. Suppose the radiator heats the air in its vicinity by $\Delta\theta$ degrees Kelvin. This causes a buoyant acceleration $g\,\Delta\theta/\theta$, which is of order 0.3 m/sec^2 if $\Delta\theta = 10°K$. This acceleration probably occurs only near the surface of the radiator. If it has a height $h = 0.1$ m, the kinetic energy of the air above the radiator is $gh\Delta\theta/\theta$, which is of order 0.03 (m/sec)^2 per unit mass. This corresponds to a velocity of 24 cm/sec. Much of the kinetic energy, however, is lost because of the stable vertical temperature gradient in the room (the air near the ceiling tends to be hotter than the air near the floor). A characteristic velocity u of order 5 cm/sec may be a reasonable average throughout the room. With $u = 5$ cm/sec and $L = 5$ m, T_t becomes 100 sec, or about 2 min. Of course, we still have to rely on molecular diffusion to even out small-scale irregularities in the temperature distribution. However, the turbulence generates eddies as small as about 1 cm (this estimate can be obtained with simple equations based on the dissipation of kinetic energy; those are discussed in Section 1.5). The temperature gradients associated with these small eddies are smeared out by molecular diffusion in a time of order ℓ^2/γ (see Section 7.3), which is only a few seconds if $\ell = 1$ cm.

Diffusion by random motion apparently is very rapid compared to molecular diffusion. The ratio of the turbulent time scale T_t to the molecular time scale T_m is the inverse of the Péclet number:

$$\frac{T_t}{T_m} \sim \frac{L}{u}\frac{\gamma}{L^2} = \frac{\gamma}{uL}. \tag{1.4.5}$$

Since for gases the heat conductivity γ is of the same order of magnitude as the kinematic viscosity ν (for air $\nu/\gamma = 0.73$; this ratio is known as the Prandtl number), and since we are discussing only orders of magnitude, we may write without compromise,

$$\frac{T_t}{T_m} \sim \frac{\nu}{uL} = \frac{1}{R}. \tag{1.4.6}$$

In our example, the Reynolds number R is about 15,000.

This exercise shows that the Reynolds number of a turbulent flow may be interpreted as a ratio of a turbulence time scale to a molecular time scale that would prevail in the absence of turbulence in a problem with the same length scale. This point of view is often more reliable than thinking of R as a ratio of inertia terms to viscous terms in the governing equations. The latter point of view tends to be misleading because at high Reynolds numbers viscous and other diffusion effects tend to operate on smaller length scales than inertia effects.

Eddy diffusivity Since the equations governing turbulent flow are very complicated, it is tempting to treat the diffusive nature of turbulence by means of a properly chosen effective diffusivity. In doing so, the idea of trying to understand the turbulence itself is partly discarded. If we use an effective diffusivity, we tend to treat turbulence as a property of a fluid rather than as a property of a flow. Conceptually, this is a very dangerous approach. However, it often makes the mathematics a good deal easier.

If the effects of turbulence could be represented by a simple, constant scalar diffusivity, one should be able to write for the diffusion of heat by turbulent motions,

$$\frac{\partial \theta}{\partial t} = K \frac{\partial^2 \theta}{\partial x_i \partial x_i}, \tag{1.4.7}$$

in which K is the representative diffusivity (often called "eddy" diffusivity but sometimes called the "exchange coefficient" for heat). In order to make this equation at least a crude representation of reality, one must insist that the value of K be chosen such that the time scale of the hypothetical turbulent diffusion process is equal to that of the actual mixing process. The time scale associated with (1.4.7) is roughly

$$T \sim \frac{L^2}{K}, \tag{1.4.8}$$

and the actual time scale is T_t, given by (1.4.4). Equating T with T_t, one finds

$$K \sim uL. \tag{1.4.9}$$

It should be noted that this is a dimensional estimate, which cannot predict the numerical values of coefficients that may be needed. Expressions like (1.4.9), with experimentally determined coefficients, are used frequently in practical applications.

The eddy diffusivity (or viscosity) K may be compared with the kinematic viscosity ν and the thermal diffusivity γ:

$$\frac{K}{\gamma} \cong \frac{K}{\nu} \sim \frac{uL}{\nu} = R. \tag{1.4.10}$$

One concludes that this particular Reynolds number may also be interpreted as a ratio of apparent (or turbulent) viscosity to molecular viscosity. A note of warning is in order, though. In most flow problems, many different length scales exist, so that the interpretation of Reynolds numbers based on these length scales may not always be as straightforward as in the example used here.

It cannot be stressed too strongly that the eddy diffusivity K is an artifice which may or may not represent the effects of turbulence faithfully. We investigate this question carefully in Chapter 2.

Diffusion in a problem with an imposed time scale As another example of the diffusivity of turbulence, we look at boundary layers in the atmosphere. The boundary layer in the atmosphere is exposed to the rotation of the earth.

In a rotating frame of reference, flows are accelerated by the Coriolis force, which is twice the vector product of the flow velocity and the rotation rate. If the angular velocity of the frame of reference is $f/2$, it follows that atmospheric flows have an imposed time scale of order $1/f$. At a latitude of 40 degrees, the value of f for a Cartesian coordinate system whose z axis is parallel to the local vertical is about 10^{-4} $\sec^{-1}$ (f is called the Coriolis parameter).

If the boundary layer in the atmosphere were laminar, it would be governed by a diffusion equation like (1.4.1), so that its length and time scales would be related by

$$L_{\mathrm{m}}^2 \sim \nu T. \tag{1.4.11}$$

With $\nu = 0.15$ cm^2 sec^{-1} and $T = f^{-1} = 10^4$ sec, this gives $L_{\mathrm{m}} = 40$ cm.

In reality, however, the atmospheric boundary layer is nearly always turbulent; a typical thickness is about 10^3 m (1 km). One can obtain some appreciation for this by replacing ν by K in (1.4.11) and substituting for K with (1.4.9). This yields

$$L_{\mathrm{t}} \sim uT, \tag{1.4.12}$$

which, of course, merely rephrases (1.4.4). In turbulent boundary-layer flows, the characteristic velocity of the turbulence is typically about $\frac{1}{30}$ of the mean wind speed. For a wind speed of 10 m/sec, we thus estimate that $u \sim 0.3$ m/sec. With $T = 1/f = 10^4$ sec, (1.4.12) then yields $L_{\mathrm{t}} \sim 3 \times 10^3$ m (3 km), which is indeed of the same order as the observed thickness.

From a somewhat different point of view, we may argue that turbulent eddies with a characteristic velocity u, exposed to a Coriolis acceleration which imposes a time scale $1/f$, must have a size (length scale) of order u/f. It should be noted that we can equate eddy size and boundary-layer thickness only because in most turbulent flows the larger eddies seem to have sizes comparable to the characteristic size of the flow in a direction normal to the mean flow field (Figure 1.1). In estimates of diffusion or mixing, the large eddies are relevant because they perform most of the mixing ($K \sim u\ell$ increases with eddy size).

Arguments of this nature are often supplemented by experiments to determine the numerical coefficient in formulas like (1.4.12), because this coefficient cannot be found by dimensional reasoning. In the case of the atmospheric boundary layer,

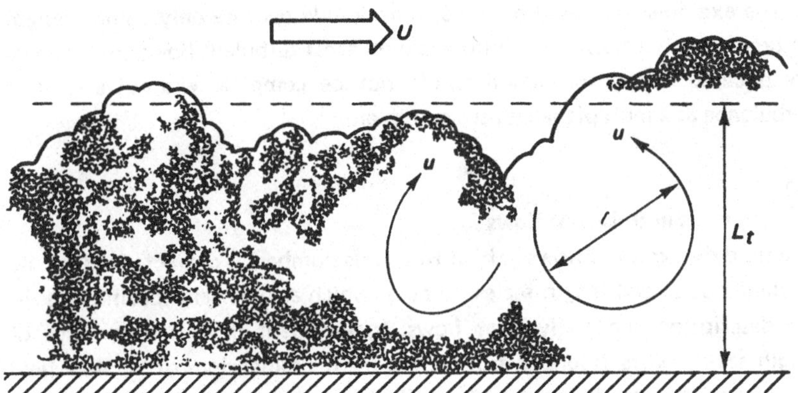

Figure 1.1. Large eddies in a turbulent boundary layer. The flow above the boundary layer has a velocity U; the eddies have velocities u. The largest eddy size (ℓ) is comparable to the boundary-layer thickness (L_t). The interface between the turbulence and the flow above the boundary layer is quite sharp (Corrsin and Kistler, 1954).

$$L_t = \tfrac{1}{4}\, u/f \qquad (1.4.13)$$

would give very close agreement between "theory" and experimental evidence.

Using (1.4.11), (1.4.12), and $T = 1/f$, we find the ratio between the thicknesses of the laminar and turbulent atmospheric boundary layers to be

$$\frac{L_t}{L_m} \sim \frac{u}{f}\left(\frac{f}{\nu}\right)^{1/2} = \left(\frac{u^2}{f\nu}\right)^{1/2} = R^{1/2}. \qquad (1.4.14)$$

This is the square root of the Reynolds number associated with the turbulent boundary layer in the atmosphere, since u/f is proportional to the actual length scale L_t. In this example, the Reynolds number R is clearly associated with the ratio of the turbulent and molecular diffusion length scales: turbulent flow penetrates much deeper into the atmosphere than laminar flow. In our example, $R \sim 10^7$.

The results obtained here concerning the different aspects of the Reynolds number may be summarized by stating that in flows with imposed length scales the Reynolds number is proportional to the ratio of time scales, while in flows with imposed time scales the Reynolds number is proportional to the square of a ratio of length scales. Since the Reynolds numbers of most flows are large, these relations clearly show that turbulence is a far more effective diffusion agent than molecular motion.

The examples discussed here are rather crude because only a single length or time scale has been taken into account. Most turbulent flows are far more complicated; this introduction would not be complete without a look at turbulence as a multiple length-scale problem.

1.5
Length scales in turbulent flows

The fluid dynamics of flows at high Reynolds numbers is characterized by the existence of several length scales, some of which assume very specific roles in the description and analysis of flows. In turbulent flows a wide range of length scales exists, bounded from above by the dimensions of the flow field and bounded from below by the diffusive action of molecular viscosity. Incidentally, this is the reason why spectral analysis of turbulent motion is useful.

Laminar boundary layers Let us take a look at the problem of multiple scales in laminar shear flows. For steady flow of an incompressible fluid with constant viscosity, the Navier-Stokes equations are

$$u_j \frac{\partial u_i}{\partial x_j} = -\frac{1}{\rho}\frac{\partial p}{\partial x_i} + \nu \frac{\partial^2 u_i}{\partial x_j \partial x_j}. \tag{1.5.1}$$

One would be tempted to estimate the inertia terms as U^2/L (U being a characteristic velocity and L a characteristic length) and to estimate the viscous terms as $\nu U/L^2$. The ratio of these terms is $UL/\nu = R$, indicating that viscous terms should become negligible at large Reynolds numbers. However, boundary conditions or initial conditions may make it impossible to neglect viscous terms everywhere in the flow field. For example, a boundary layer has to exist in the flow along a solid surface to satisfy the no-slip condition. This can be understood by allowing for the possibility that viscous effects may be associated with small length scales. The viscous terms can survive at high Reynolds numbers only by choosing a new length scale ℓ such that the viscous terms are of the same order of magnitude as the inertia terms. Formally,

$$U^2/L \sim \nu U/\ell^2. \tag{1.5.2}$$

The viscous length ℓ is thus related to the scale L of the flow field as

$$\frac{\ell}{L} \sim \left(\frac{\nu}{UL}\right)^{1/2} = R^{-1/2}. \tag{1.5.3}$$

The viscous length ℓ is a *transverse* length scale: it represents the width (thickness) of the boundary layer, because it relates to the molecular diffusion of momentum deficit across the flow, away from the surface. Molecular diffusion along the flow, of course, is negligible compared to the downstream transport of momentum by the flow itself. Figure 1.2 illustrates this situation.

Diffusive and convective length scales As (1.5.3) indicates, the boundary-layer thickness may be considerably smaller than the scale L of the flow field in which the boundary layer (or other laminar shear flow) develops. The distinction between a "diffusive" length scale across the flow and a "convective" length scale along the flow is essential to the understanding of all shear flows, both laminar and turbulent. Many shear flows are very slender: their width is much smaller than their "length" (that is, the distance from some suitably defined origin). The wide separation between lateral and longitudinal length scales in shear flows leads to very attractive simplifying approximations in the equations of motion; without this feature, analysis would be next to impossible.

The most powerful of the asymptotic approximations associated with $\ell/L \to 0$ is that the shear flow becomes independent of most of its environment, except for the boundary conditions imposed by the overall flow. The use of words like boundary layers, wakes, fronts in weather systems, jetstreams, and the Gulf Stream is not a semantic accident. Because of the wide difference in length scales, these shear flows are identifiable as distinct regions in flow fields. These regions have distinct dynamics and distinct

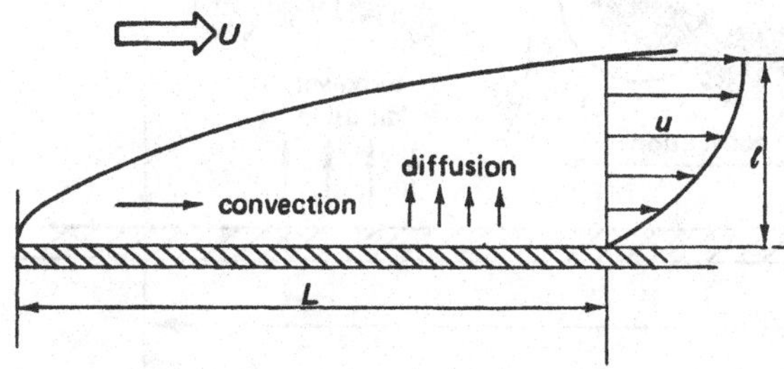

Figure 1.2. Length scales, diffusion, and convection in a laminar boundary layer over a flat plate.

characteristics; they are governed by specific equations of motion, which, in the asymptotic approximation $\ell/L \to 0$, may be substantially simpler than the equations governing other parts of the flow field.

Turbulent boundary layers It is useful to compare turbulent shear flows to laminar ones, even though we can do so at this moment only in a very rudimentary way. The relevant length and velocity scales in a turbulent boundary layer are illustrated in Figure 1.3. The turbulent eddies transfer momentum deficit away from the surface. With characteristic velocity fluctuations of order u, the boundary-layer thickness ℓ presumably increases roughly as $d\ell/dt \sim u$ (see Section 5.5). The time interval elapsed between the origin of the boundary layer and the downstream position L is of order L/U (convective time scale), so that we may estimate $\ell \sim ut \sim uL/U$. In effect, we are equating the turbulent "diffusion" time scale ℓ/u to the convective time scale L/U. This procedure could also have been used for laminar boundary layers. In laminar boundary layers, the diffusion distance ℓ increases as $(\nu t)^{1/2}$; with $t = L/U$, the result (1.5.2, 1.5.3) is retrieved.

In analogy to (1.4.4) and (1.4.12), we thus can write the scale relations for turbulent boundary layers as

$$\ell/L \sim u/U, \tag{1.5.4}$$

$$\ell/u \sim L/U. \tag{1.5.5}$$

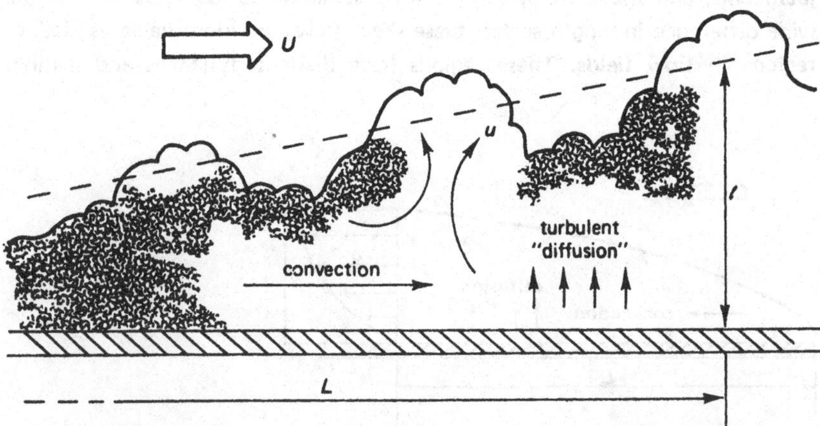

Figure 1.3. Length and velocity scales in a turbulent boundary layer. The time passed since the fluid at L passed the origin of the boundary layer is of order L/U.

These relations merely relate characteristic lengths and velocities; they should not be used as formulas to compute the rate of spreading of a turbulent boundary layer. The relation between the time scales, (1.5.5), rephrases the fundamental assumption we implicitly encountered earlier, that is, that in a situation with an imposed external flow the turbulence, being part of the flow, must have a time scale commensurate with the time scale of the flow. As we will see later, this assumption conflicts with eddy-viscosity concepts. Fortunately, not all of the turbulence has such a large time scale: the small eddies in turbulence have very short time scales, which tend to make them statistically independent of the mean flow.

Laminar and turbulent friction If we compare (1.5.3) and (1.5.4) and introduce experimental data, which suggest that u/U is of the order of 10^{-2} over a wide range of Reynolds numbers, we again get some appreciation for the relatively rapid growth of turbulent shear flows. This rapid growth should correspond to a larger drag coefficient.

For a steady laminar boundary layer in two-dimensional flow on a plate with length L, the drag D per unit span is equal to the total rate of loss of momentum. Estimating the momentum loss as $\rho U^2 \ell$, where ℓ is a boundary-layer thickness at the end of the plate, we may put

$$D \sim \rho U^2 \ell. \tag{1.5.6}$$

The drag coefficient (or friction coefficient) c_d is defined by

$$c_d \equiv \frac{D}{\frac{1}{2}\rho U^2 L}. \tag{1.5.7}$$

Substituting (1.5.6) into (1.5.7) and using the relation for ℓ/L given by (1.5.3), we obtain

$$c_d \sim 2\frac{\ell}{L} = 2R^{-1/2}. \tag{1.5.8}$$

For a turbulent boundary layer, on the other hand, the mass flow deficit at the end of the plate is proportional to $\rho u \ell$ (see Chapter 5), so that the rate of loss of momentum is proportional to $(\rho u \ell)U$. Consequently,

$$D \sim \rho u U \ell. \tag{1.5.9}$$

The drag coefficient then becomes, if we use the definition (1.5.7) and the scale relation (1.5.4),

$$c_d \sim 2\frac{u}{U}\frac{\ell}{L} \sim 2\left(\frac{u}{U}\right)^2. \tag{1.5.10}$$

Experimental evidence shows that the turbulence level u/U varies very slowly with Reynolds number, so that the drag coefficient of a turbulent boundary layer, given by (1.5.10), should be very much greater than the drag coefficient of a laminar boundary layer (1.5.8). Figure 1.4 illustrates this point. Similar conclusions are valid for heat- and mass-transfer coefficients.

Equation (1.5.4) has another interesting implication. In boundary layers and wakes u/U and ℓ/L tend to zero as L increases beyond limit. In jets entering fluid at rest and shear layers, on the other hand, u/U and ℓ/L approach finite asymptotic values as $L \to \infty$. This distinction is the origin of some important differences in the asymptotic treatment of the two different types of flow. In particular, jets and mixing layers spread linearly, while wakes and boundary layers grow slower the farther downstream they travel. Even so, most turbulent shear flows spread slowly enough to make $\ell/L \to 0$ a useful approximation.

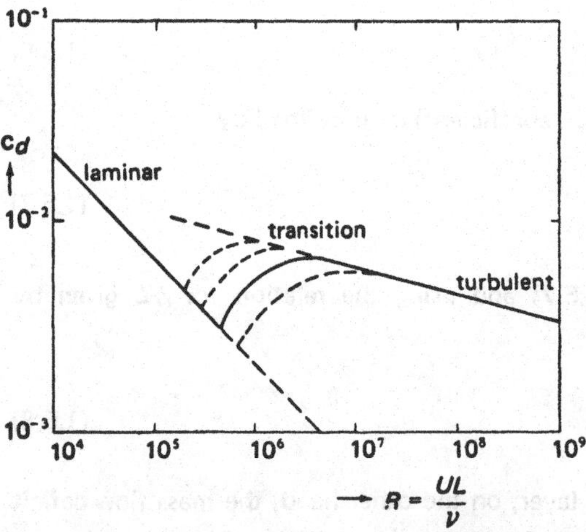

Figure 1.4. The drag coefficient of a flat plate. The several curves drawn in the transition range (partially laminar, partially turbulent flow over the plate) illustrate that transition is very sensitive to small disturbances.

Small scales in turbulence So far only the largest eddy sizes in turbulent flows have been considered, because the large eddies do most of the transport of momentum and contaminants. We have suggested that large eddies are as big as the width of the flow and that the latter is the relevant length scale in the analysis of the interaction of the turbulence with the mean flow. For some of the other aspects of the dynamics of turbulence, however, other length scales are needed.

We shall attempt to find the smallest length scales in turbulent flows. At very small length scales, viscosity can be effective in smoothing out velocity fluctuations. The generation of small-scale fluctuations is due to the nonlinear terms in the equations of motion; the viscous terms prevent the generation of infinitely small scales of motion by dissipating small-scale energy into heat. This is characteristic of a small parameter like ν (more properly $1/R$) with a singular behavior. One might expect that at large Reynolds numbers the relative magnitude of viscosity is so small that viscous effects in a flow tend to become vanishingly small. The nonlinear terms in the Navier-Stokes equation counteract this threat by generating motion at scales small enough to be affected by viscosity. The smallest scale of motion automatically adjusts itself to the value of the viscosity. There seems to be no way of doing away with viscosity: as soon as the scale of the flow field becomes so large that viscosity effects could conceivably be neglected, the flow creates small-scale motion, thus keeping viscosity effects (in particular dissipation rates) at a finite level.

Since small-scale motions tend to have small time scales, one may assume that these motions are statistically independent of the relatively slow large-scale turbulence and of the mean flow. If this assumption makes sense, the small-scale motion should depend only on the rate at which it is supplied with energy by the large-scale motion and on the kinematic viscosity. It is fair to assume that the rate of energy supply should be equal to the rate of dissipation, because the net rate of change of small-scale energy is related to the time scale of the flow as a whole. The net rate of change, therefore, should be small compared to the rate at which energy is dissipated. This is the basis for what is called Kolmogorov's *universal equilibrium theory* of the small-scale structure (Chapter 8).

This discussion suggests that the parameters governing the small-scale motion include at least the dissipation rate per unit mass ϵ ($m^2\ sec^{-3}$) and the kinematic viscosity ν (m^2 sec^{-1}). With these parameters, one can form length, time, and velocity scales as follows:

$$\eta \equiv (\nu^3/\epsilon)^{1/4}, \quad \tau \equiv (\nu/\epsilon)^{1/2}, \quad \upsilon \equiv (\nu\epsilon)^{1/4}. \tag{1.5.11}$$

These scales are referred to as the *Kolmogorov microscales* of length, time, and velocity (see Friedlander and Topper, 1962). In the Russian literature, these scales are called "inner" scales.

The Reynolds number formed with η and υ is equal to one

$$\eta\upsilon/\nu = 1, \tag{1.5.12}$$

which illustrates that the small-scale motion is quite viscous and that the viscous dissipation adjusts itself to the energy supply by adjusting length scales.

An inviscid estimate for the dissipation rate One can form an impression of the differences between the large-scale and small-scale aspects of turbulence if the dissipation rate ϵ can be related to the length and velocity scales of the large-scale turbulence. A plausible assumption is to take the rate at which large eddies supply energy to small eddies to be proportional to the reciprocal of the time scale of the large eddies. The amount of kinetic energy per unit mass in the large-scale turbulence is proportional to u^2; the rate of transfer of energy is assumed to be proportional to u/ℓ, where ℓ represents the size of the largest eddies or the width of the flow. We shall see later that ℓ relates to the "integral" scales of turbulence, which can be measured by statistical methods. To avoid confusion, we identify ℓ from here on as the "integral scale," leaving a more precise definition for Chapter 2. Russian scientists speak of "outer" scales rather than of integral scales.

The rate of energy supply to the small-scale eddies is thus of order $u^2 \cdot u/\ell = u^3/\ell$. This energy is dissipated at a rate ϵ, which should be equal to the supply rate. Hence (Taylor, 1935),

$$\epsilon \sim u^3/\ell, \tag{1.5.13}$$

which states that viscous dissipation of energy can be estimated from the large-scale dynamics, which do not involve viscosity. In this sense, dissipation again is clearly seen as a passive process in the sense that it proceeds at a rate dictated by the inviscid inertial behavior of the large eddies.

The estimate (1.5.13) should not be passed over lightly. It is one of the

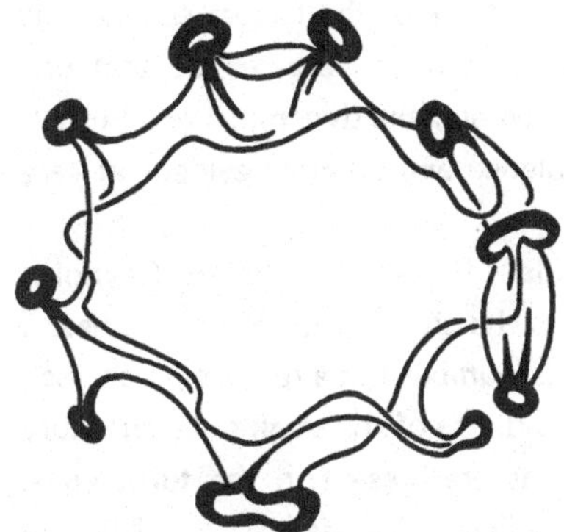

Figure 1.5. Sketch of the nonlinear breakdown of a drop of ink in water.

cornerstone assumptions of turbulence theory; it claims that large eddies lose a significant fraction of their kinetic energy $\frac{1}{2}u^2$ within one "turnover" time ℓ/u. This implies that the nonlinear mechanism that makes small eddies out of larger ones is as "dissipative" as its characteristic time permits. In other words, turbulence is a strongly damped nonlinear stochastic system. Some researchers believe that this feature may be related to the entropy production concept embodied in the second law of thermodynamics. It should be kept in mind, however, that large eddies lose a negligible fraction of their energy to direct viscous dissipation effects. The time scale of their decay is ℓ^2/ν, so that their viscous energy loss proceeds at a rate $\nu u^2/\ell^2$, which is small compared to u^3/ℓ if the Reynolds number $u\ell/\nu$ is large. The nonlinear mechanism is dissipative because it creates smaller and smaller eddies until the eddy sizes become so small that viscous dissipation of their kinetic energy is almost immediate. The reader may gain some appreciation for the vigor of this process by observing drops of ink or milk that are put in a glass of water (Figure 1.5).

Scale relations Substituting (1.5.13) into (1.5.11), we obtain

$$\eta/\ell \sim (u\ell/\nu)^{-3/4} = R^{-3/4}. \tag{1.5.14}$$

$$\tau u/\ell \sim \tau/t = (u\ell/\nu)^{-1/2} = R^{-1/2}, \tag{1.5.15}$$

$$v/u \sim (u\ell/\nu)^{-1/4} = R^{-1/4}. \tag{1.5.16}$$

These relations indicate that the length, time, and velocity scales of the smallest eddies are very much smaller than those of the largest eddies. The separation in scales widens as the Reynolds number increases, so that one may suspect that the statistical independence and the dynamical equilibrium state of the small-scale structure of turbulence will be most evident at very large Reynolds numbers.

The main difference between two turbulent flows with different Reynolds numbers but with the same integral scale is the size of the smallest eddies: a turbulent flow at a relatively low Reynolds number has a relatively "coarse" small-scale structure (Figure 1.6). Visual evidence of the small-scale structure can be obtained if temperature fluctuations are present in the turbulence. Temperature and index of refraction gradients are steepest if they are associated with the smallest eddies; any optical system that is sensitive to such fluctuating gradients "sees" the small-scale structure of turbulence. The trembling, jittery horizon seen on a very hot day and the random pattern of

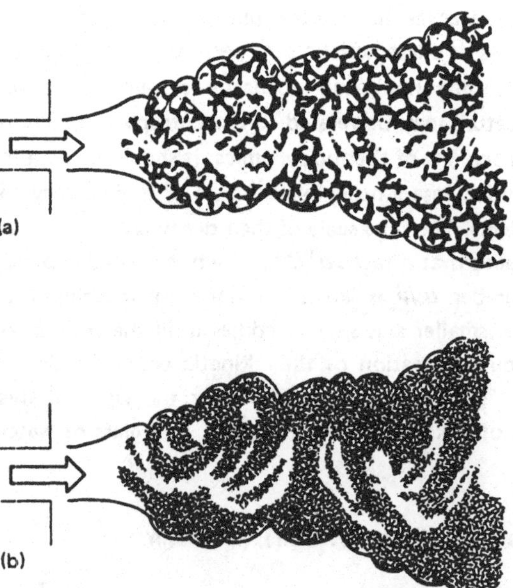

Figure 1.6. Turbulent jets at different Reynolds numbers: (a) relatively low Reynolds number, (b) relatively high Reynolds number (adapted from a film sequence by R. W. Stewart, 1969). The shading pattern used closely resembles the small-scale structure of turbulence seen in shadowgraph pictures.

light and dark seen on the wall next to a heating element in sunlight are good illustrations.

Vorticity has the dimensions of a frequency (sec^{-1}). The vorticity of the small-scale eddies should be proportional to the reciprocal of the time scale τ. From (1.5.15) we conclude that the vorticity of the small-scale eddies is very much larger than that of the large-scale motion. On the other hand, (1.5.16) indicates that the small-scale energy is small compared to the large-scale energy. This is typical of all turbulence: most of the energy is associated with large-scale motions, most of the vorticity is associated with small-scale motions.

Molecular and turbulent scales The Kolmogorov length and time scales are the smallest scales occurring in turbulent motion. At this point, it is convenient to demonstrate that most turbulent flows are indeed continuum phenomena. The Kolmogorov scales of length and time decrease with increasing dissipation rates. High dissipation rates are associated with large values of u. In gases, large values of u are more likely to occur than in liquids. Therefore, it is sufficient to show that in gases the smallest turbulent scales of motion are normally very much larger than molecular scales of motion. The relevant molecular length scale is the mean free path ξ. The velocity scale of molecular motion in a gas is proportional to the speed of sound a in the gas. Kinetic theory of gases shows that the product $a\xi$ is proportional to the kinematic viscosity of the gas:

$$\nu \sim a\xi. \tag{1.5.17}$$

The ratio of the mean free path ξ to the Kolmogorov length scale η (this might be called a microstructure Knudsen number) becomes (Corrsin, 1959)

$$\xi/\eta \sim M/R^{1/4}, \tag{1.5.18}$$

where we have used (1.5.14) and (1.5.17). In (1.5.18) the turbulence Reynolds number $R = u\ell/\nu$ and the turbulence Mach number $M = u/a$ are used as independent variables. It is seen that turbulence might interfere with molecular motion at high Mach numbers and low Reynolds numbers. This kind of situation is unlikely to occur, because M is seldom large, but R is typically very large. A pertinent illustration is the situation in gaseous nebulae (cosmic gas clouds) (Spitzer, 1968). In clouds that consist mainly of neutral hydrogen, the turbulent Mach number is of order 10 ($u \sim 10$ km/sec, $a \sim 1$ km/sec), while the Reynolds number is of order 10^7 ($\ell \sim 10^{17}$ m, $\xi \sim 10^{11}$ m). With (1.5.18), we compute that $\xi/\eta \sim 1/6$. In this extreme case, it seems doubtful that the smallest eddies perceive a continuum. In clouds that consist mainly of ionized hydrogen, temperatures are quite high, increasing a to about 10 km/sec and decreasing M to about 1. The mean free path ξ remains roughly the same (the density in ionized clouds is not appreciably different from that in neutral clouds), so that R reduces to about 10^6. In this case, $\xi/\eta \sim \frac{1}{32}$, which may be small enough for the smallest eddies to operate in a continuum.

The ratio of the time scale τ to the collision time scale ξ/a associated with molecular motion is, in terms of R and M,

$$\tau a/\xi \sim R^{1/2} M^{-2}. \tag{1.5.19}$$

For $M = 10$ and $R = 10^7$, the smallest time scale of turbulence is 32 times as large as the collision time scale of the gas molecules; for $M = 1$ and $R = 10^6$ the ratio is 1000. It should be recognized that in ionized gases other length and time scales are associated with the motion of the microscopic particles and with the several other dynamical processes (radiation, cosmic rays, magnetic fields) that may be present, so that η may not always be a relevant length scale.

Because the smallest time scales in turbulent motion tend to be much larger than molecular time scales, the motion of the gas molecules is in approximate statistical equilibrium, so that molecular transport effects may indeed be represented by transport coefficients such as viscosity and heat conductivity. These representations would become invalid if the departures from equilibrium were large; the case $\xi/\eta \sim \frac{1}{6}$, $\tau a/\xi \sim 32$ would probably require treatment with the methods of statistical mechanics.

1.6
Outline of the material

The bird's-eye view of turbulence dynamics given in the preceding sections sets the stage for a brief outline of this book. In Chapter 2, we deal with eddy-viscosity and mixing-length theories. The dimensional framework of these theories is useful in the analysis of typical shear flows. In Chapter 3, the energy and vorticity equations of turbulent flow are derived. In Chapter 4, some free shear flows like wakes and jets are discussed. In Chapter 5, boundary layers are analyzed. To prepare a formal basis for the study of diffusion and spectral dynamics, an introduction to statistics is given in Chapter 6. In Chapter 7, turbulent diffusion and mixing are studied.

The study of the spatial dynamics of turbulent flows precedes that of the spectral dynamics. There exist many similarities and analogies between spatial and spectral dynamics of turbulence. Also, spatial dynamics can be visualized more easily by those new to the subject. Once some of the subtle features of turbulent shear flow are understood, the dynamics of turbulence in wave-number space should not be too perplexing. Spectral dynamics is studied in Chapter 8.

Problems

1.1 Estimate the energy dissipation rate in a cumulus cloud, both per unit mass and for the entire cloud. Base your estimates on velocity and length scales typical of cumulus clouds. Compute the total dissipation rate in kilowatts. Also estimate the Kolmogorov microscale η. Use $\rho = 1.25$ kg/m^3 and $\nu = 15 \times 10^{-6}$ m^2/sec.

1.2 A cubical box of volume L^3 is filled with fluid in turbulent motion. No source of energy is present, so that the turbulence decays. Because the turbulence is confined to the box, its length scale may be assumed to be equal to L at all times. Derive an expression for the decay of the kinetic energy $\frac{3}{2}u^2$ as a function of time. As the turbulence decays, its Reynolds number decreases. If the Reynolds number uL/ν becomes smaller than 10, say, the inviscid estimate $\epsilon = u^3/L$ should be replaced by an estimate of the type $\epsilon = c\nu u^2/L^2$, because the weak eddies remaining at low Reynolds numbers lose their energy directly to viscous dissipation. Compute c by requiring that the dissipation rate is continuous at $uL/\nu = 10$. Derive an expression tor the decay of the kinetic energy when $uL/\nu < 10$ (this is called the "final" period of decay). If $L = 1$ m, $\nu = 15 \times 10^{-6}$ m^2/sec and $u = 1$m/sec at time $t = 0$, how long does it take before the turbulence enters the final period of decay? Assume that the effects of the walls of the box on the decay of the turbulence may be ignored. Can you support this assumption in any way?

1.3 The large eddies in a turbulent flow have a length scale ℓ, a velocity scale $v(\ell) = u$, and a time scale $t(\ell) = \ell/u$. The smallest eddies have a length scale η, a velocity scale υ, and a time scale τ. Estimate the characteristic velocity $v(r)$ and the characteristic time $t(r)$ of eddies of size r, where r is any length in the range $\eta < r < \ell$. Do this by assuming that $v(r)$ and $t(r)$ are determined by ϵ and r only. Show that your results agree with the known velocity and time scales at $r = \ell$ and $r = \eta$. The energy spectrum of turbulence is a plot of $E(\kappa) = \kappa^{-1}\ v^2(\kappa)$, where $\kappa = 1/r$ is the "wave number" associated with eddies of size r. Find an expression for $E(\kappa)$ and compare your result with the data in Chapter 8.

1.4 An airplane with a hot-wire anemometer mounted on its wing tip is to fly through the turbulent boundary layer of the atmosphere at a speed of 50 m/sec. The velocity fluctuations in the atmosphere are of order 0.5 m/sec, the length scale of the large eddies is about 100 m. The hot-wire anemometer is to be designed so that it will register the motion of the smallest eddies. What is the highest frequency the anemometer will encounter? What should the length of the hot-wire sensor be? If the noise in the electronic circuitry is expressed in terms of equivalent turbulence intensity, what is the permissible noise level?

2

TURBULENT TRANSPORT OF MOMENTUM AND HEAT

Turbulence consists of random velocity fluctuations, so that it must be treated with statistical methods. The statistical analysis does not need to be sophisticated at this stage; a simple decomposition of all quantities into mean values and fluctuations with zero mean will suffice for the next few chapters. We shall find that turbulent velocity fluctuations can generate large momentum fluxes between different parts of a flow. A momentum flux can be thought of as a stress; turbulent momentum fluxes are commonly called Reynolds stresses. The momentum exchange mechanism superficially resembles molecular transport of momentum. The latter gives rise to the viscosity of a fluid; by analogy, the turbulent momentum exchange is often represented by an eddy viscosity. This analogy will be explored in great detail.

2.1
The Reynolds equations

In turbulence, a description of the flow at all points in time and space is not feasible. Instead, following Reynolds (1895), we develop equations governing mean quantities, such as the mean velocity. The equations of motion of an incompressible fluid are

$$\frac{\partial \tilde{u}_i}{\partial t} + \tilde{u}_j \frac{\partial \tilde{u}_i}{\partial x_j} = \frac{1}{\rho} \frac{\partial}{\partial x_j} \tilde{\sigma}_{ij}, \tag{2.1.1}$$

$$\frac{\partial \tilde{u}_i}{\partial x_i} = 0. \tag{2.1.2}$$

Here, $\tilde{\sigma}_{ij}$ is the *stress tensor*. Repeated indices in any term indicate a summation over all three values of the index; a tilde denotes the instantaneous value at (x_i, t) of a variable on which no Reynolds decomposition into a mean value and fluctuations (see next section) has been performed.

If the fluid is Newtonian, the stress tensor $\tilde{\sigma}_{ij}$ is given by

$$\tilde{\sigma}_{ij} = -\tilde{p}\delta_{ij} + 2\mu\tilde{s}_{ij}. \tag{2.1.3}$$

In (2.1.3), δ_{ij} is the Kronecker delta, which is equal to one if $i = j$ and zero otherwise; $\tilde{p}$ is the hydrodynamic pressure and μ is the dynamic viscosity (which will be assumed to be constant). The *rate of strain* $\tilde{s}_{ij}$ is defined by

$$\tilde{s}_{ij} = \frac{1}{2}\left(\frac{\partial \tilde{u}_i}{\partial x_j} + \frac{\partial \tilde{u}_j}{\partial x_i}\right). \tag{2.1.4}$$

If (2.1.3) is substituted into (2.1.1) and if the continuity equation (2.1.2) is invoked, the *Navier-Stokes equations* are obtained:

$$\frac{\partial \tilde{u}_i}{\partial t} + \tilde{u}_j \frac{\partial \tilde{u}_i}{\partial x_j} = -\frac{1}{\rho}\frac{\partial \tilde{p}}{\partial x_i} + \nu \frac{\partial^2 \tilde{u}_i}{\partial x_j \partial x_j} . \tag{2.1.5}$$

Here, ν is the kinematic viscosity ($\nu = \mu/\rho$).

The Reynolds decomposition The velocity $\tilde{u}_i$ is decomposed into a mean flow U_i and velocity fluctuations u_i, such that

$$\tilde{u}_i = U_i + u_i . \tag{2.1.6}$$

We interpret U_i as a time average, defined by

$$U_i = \lim_{T\to\infty} \frac{1}{T} \int_{t_0}^{t_0+T} \tilde{u}_i \, dt. \tag{2.1.7}$$

Time averages (mean values) of fluctuations (which are denoted by lowercase letters) and of their derivatives, products, and other combinations are denoted by an overbar. The mean value of a fluctuating quantity itself is zero by definition; for example,

$$\overline{u_i} = \lim_{T\to\infty} \frac{1}{T} \int_{t_0}^{t_0+T} (\tilde{u}_i - U_i) \, dt \equiv 0. \tag{2.1.8}$$

The use of time averages corresponds to the typical laboratory situation, in which measurements are taken at fixed locations in a statistically steady, but often inhomogeneous, flow field. In an inhomogeneous flow, a time average like U_i is a function of position, so that the use of a spatial average would be inappropriate for most purposes. For a time average to make sense, the integrals in (2.1.7) and (2.1.8) have to be independent of t_0. In other words, the mean flow has to be steady:

$$\frac{\partial U_i}{\partial t} = 0. \tag{2.1.9}$$

Without this constraint (2.1.7) and (2.1.8) would be meaningless. The averaging time T needed to measure mean values depends on the accuracy desired; this problem is discussed in Section 6.4.

The mean value of a spatial derivative of a variable is equal to the corresponding spatial derivative of the mean value of that variable; for example,

$$\frac{\partial \bar{\tilde{u}}_i}{\partial x_j} = \frac{\partial U_i}{\partial x_j}, \quad \overline{\frac{\partial u_i}{\partial x_j}} = \frac{\partial}{\partial x_j}\overline{u_i} = 0. \tag{2.1.10}$$

These operations can be performed because averaging is carried out by integrating over a long period of time, which commutes with differentiation with respect to another independent variable.

The pressure $\tilde{p}$ and the stress $\tilde{\sigma}_{ij}$ are also decomposed into mean and fluctuating components. Again, capital letters are used for mean values and lowercase letters for fluctuations with zero mean. Specifically,

$$\tilde{p} = P + p, \quad \bar{p} \equiv 0, \tag{2.1.11}$$

$$\tilde{\sigma}_{ij} = \Sigma_{ij} + \sigma_{ij}, \quad \bar{\sigma}_{ij} \equiv 0. \tag{2.1.12}$$

Like U_i, P and Σ_{ij} are independent of time. The mean stress tensor Σ_{ij} is given by

$$\Sigma_{ij} = -P\delta_{ij} + 2\mu S_{ij}, \tag{2.1.13}$$

and the stress fluctuations σ_{ij} are given by

$$\sigma_{ij} = -p\delta_{ij} + 2\mu s_{ij}. \tag{2.1.14}$$

Here, the mean strain rate S_{ij} and the strain-rate fluctuations s_{ij} are defined by

$$S_{ij} \equiv \frac{1}{2}\left(\frac{\partial U_i}{\partial x_j} + \frac{\partial U_j}{\partial x_i}\right), \quad s_{ij} \equiv \frac{1}{2}\left(\frac{\partial u_i}{\partial x_j} + \frac{\partial u_j}{\partial x_i}\right). \tag{2.1.15}$$

The commutation between averaging and spatial differentiation involved here is based on (2.1.10).

Correlated variables Averages of products are computed in the following way:

$$\begin{aligned} \overline{\tilde{u}_i\tilde{u}_j} &= \overline{(U_i + u_i)(U_j + u_j)} \\ &= U_iU_j + \overline{u_iu_j} + \overline{U_iu_j} + \overline{U_ju_i} \\ &= U_iU_j + \overline{u_iu_j}. \end{aligned} \tag{2.1.16}$$

The terms consisting of a product of a mean value and a fluctuation vanish if they are averaged, because the mean value is a mere coefficient as far as the averaging is concerned, and the average of a fluctuating quantity is zero.

If $\overline{u_iu_j} \neq 0$, u_i and u_j are said to be *correlated*; if $\overline{u_iu_j} = 0$, the two are *uncorrelated*. Figure 2.1 illustrates the concept of correlated fluctuating variables. A measure for the degree of correlation between the two variables u_i and u_j is obtained by dividing $\overline{u_iu_j}$ by the square root of the product of the *variances* $\overline{u_i^2}$ and $\overline{u_j^2}$; this gives a *correlation coefficient* c_{ij}, which is defined by

$$c_{ij} \equiv \overline{u_iu_j}/(\overline{u_i^2} \cdot \overline{u_j^2})^{1/2}, \tag{2.1.17}$$

with the understanding that the summation convention does not apply in this case. If $c_{ij} = \pm 1$, the correlation is said to be *perfect*. Each variable, of course, is perfectly correlated with itself ($c_{\alpha\alpha} = 1$ if $i = j = \alpha$).

The square root of a variance is called a *standard deviation* or root-mean-square (rms) amplitude; it is denoted by a prime (for example, $u_i' = (\overline{u_i^2})^{1/2}$). A characteristic velocity, or "velocity scale," of turbulence at some downstream position in a boundary layer might be defined as the mean rms velocity taken across the boundary layer at that position; in this way velocity scales used in dimensional analysis could be given a precise definition whenever desired.

Equations for the mean flow If we apply the decomposition rule (2.1.6) to the continuity equation (2.1.2), we obtain

$$\frac{\partial \tilde{u}_i}{\partial x_i} = \frac{\partial}{\partial x_i}(U_i + u_i) = \frac{\partial U_i}{\partial x_i} + \frac{\partial u_i}{\partial x_i} = 0. \tag{2.1.18}$$

If the average of all terms in this equation is taken, the last term vanishes because of (2.1.8, 2.1.10). Hence, the mean flow is incompressible:

$$\partial U_i/\partial x_i = 0. \tag{2.1.19}$$

Subtracting (2.1.19) from (2.1.18), we find that the turbulent velocity fluctuations are also incompressible:

$$\partial u_i/\partial x_i = 0. \tag{2.1.20}$$

The equations of motion for the mean flow U_i are obtained by substituting (2.1.6) and (2.1.12) into (2.1.1) and taking the average of all terms in the resulting equation. This yields, if all rules on averaging are observed (in particular, recall that $\partial U_i/\partial t = 0$),

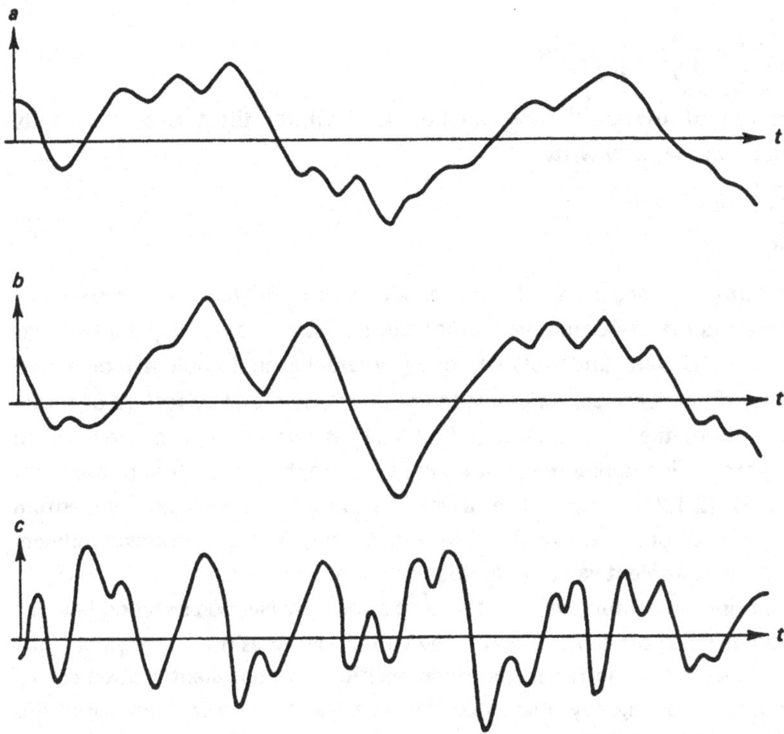

Figure 2.1. Correlated and uncorrelated fluctuations. The fluctuating variable a has the same sign as the variable b for most of the time; this makes $\overline{ab} > 0$. The variable c, on the other hand, is uncorrelated with a and b, so that $\overline{ac} = 0$ and $\overline{bc} = 0$ (note that $\overline{ab} \neq 0$, $\overline{ac} \neq 0$ does not necessarily imply that $\overline{bc} \neq 0$).

$$U_j \frac{\partial U_i}{\partial x_j} + \overline{u_j \frac{\partial u_i}{\partial x_j}} = \frac{1}{\rho} \frac{\partial}{\partial x_j} \Sigma_{ij}. \tag{2.1.21}$$

With use of the continuity equation (2.1.20) for the turbulent velocity fluctuations, we may write

$$\overline{u_j \frac{\partial u_i}{\partial x_j}} = \frac{\partial}{\partial x_j} \overline{u_i u_j}. \tag{2.1.22}$$

This term is analogous to the convection term $U_j \, \partial U_i / \partial x_j$; it represents the mean transport of fluctuating momentum by turbulent velocity fluctuations. If u_i and u_j were uncorrelated, there would be no turbulent momentum transfer. Experience shows that momentum transfer is a key feature of turbulent motion; the term (2.1.22) of (2.1.21) is not likely to be zero. Mean transport of fluctuating momentum may change the momentum of the mean flow, as (2.1.21) shows. The term (2.1.22) thus exchanges momentum between the turbulence and the mean flow, even though the mean momentum of the turbulent velocity fluctuations is zero ($\rho \overline{u_i} = 0$).

Because momentum flux is related to a force by Newton's second law, the turbulent transport term (2.1.22) may be thought of as the "divergence" of a stress. Because of the Reynolds decomposition, the turbulent motion can be perceived as an agency that produces stresses in the mean flow. For this reason, (2.1.21, 2.1.22) are rearranged, so that all stresses can be put together. This yields the *Reynolds momentum equation*:

$$U_j \frac{\partial U_i}{\partial x_j} = \frac{1}{\rho} \frac{\partial}{\partial x_j} (\Sigma_{ij} - \rho \overline{u_i u_j}). \tag{2.1.23}$$

If we recall that Σ_{ij} is given by (2.1.13), the total mean stress T_{ij} in a turbulent flow may be written as

$$T_{ij} = \Sigma_{ij} - \rho \overline{u_i u_j} = -P \, \delta_{ij} + 2\mu \, S_{ij} - \rho \overline{u_i u_j}. \tag{2.1.24}$$

The Reynolds stress The contribution of the turbulent motion to the mean stress tensor is designated by the symbol τ_{ij}:

$$\tau_{ij} \equiv - \rho \overline{u_i u_j}. \tag{2.1.25}$$

In honor of the original developer of this part of the theory, τ_{ij} is called the *Reynolds stress tensor*. The Reynolds stress is symmetric: $\tau_{ij} = \tau_{ji}$, as can be seen by inspection of (2.1.25). The diagonal components of τ_{ij} are normal stresses (pressures); their values are $\overline{\rho u_1^2}$, $\overline{\rho u_2^2}$, and $\overline{\rho u_3^2}$. In many flows, these normal stresses contribute little to the transport of mean momentum. The off-diagonal components of τ_{ij} are shear stresses; they play a dominant role in the theory of mean momentum transfer by turbulent motion.

The decomposition of the flow into a mean flow and turbulent velocity fluctuations has isolated the effects of fluctuations on the mean flow. However, the equations for the mean flow (2.1.23, 2.1.24) contain the nine components of τ_{ij} (of which only six are independent of each other) as unknowns additional to P and the three components of U_i. This illustrates the closure problem of turbulence. Indeed, if one obtains additional equations for τ_{ij} from the original Navier-Stokes equations, unknowns like $\overline{u_i u_j u_j}$ are generated by the nonlinear inertia terms. This problem is characteristic of all nonlinear stochastic systems.

This is a frustrating prospect. Therefore, many investigators have attempted to guess at a relation between τ_{ij} and S_{ij}. This is a tempting approach because the function of the Reynolds stress in the equations of motion seems to be similar to that of the viscous stress $2\mu S_{ij}$. We investigate the nature of possible relations between τ_{ij} and S_{ij} in Section 2.3; before this is done, some background material on the viscous stress is given in Section 2.2.

Turbulent transport of heat Turbulence transports passive contaminants such as heat, chemical species, and particles in much the same way as momentum. For later use, we develop the equation governing heat transfer in turbulent flow of a constant-density fluid. The density is approximately constant if temperature differences remain relatively small, if gravity-induced density stratification may be neglected, and if the Mach number of the flow is small.

The starting point is the diffusion equation for heat in a flow:

$$\frac{\partial \tilde{\theta}}{\partial t} + \tilde{u}_j \frac{\partial \tilde{\theta}}{\partial x_j} = \gamma \frac{\partial^2 \tilde{\theta}}{\partial x_j \partial x_j} . \tag{2.1.26}$$

The thermal diffusivity γ is assumed to be constant; its dimensions are $m^2 \sec^{-1}$. The ratio ν/γ is called the Prandtl number.

The temperature $\tilde{\theta}$ at (x_i, t) is decomposed in a mean value Θ and temperature fluctuations θ, such that

$$\tilde{\theta} = \Theta + \theta, \tag{2.1.27}$$

$$\bar{\tilde{\theta}} \equiv \Theta = \lim_{T\to\infty} \frac{1}{T}\int_{t_0}^{t_0+T} \tilde{\theta}\, dt, \tag{2.1.28}$$

$$\bar{\theta} \equiv 0, \quad \partial\Theta/\partial t = 0. \tag{2.1.29}$$

The last condition has been imposed because time averages would not make sense in an unsteady situation.

Substituting (2.1.27) into (2.1.26) and taking the average of all terms in the resulting equation, we obtain

$$U_j \frac{\partial \Theta}{\partial x_j} = \frac{\partial}{\partial x_j}\left(-\overline{\theta u_j} + \gamma \frac{\partial \Theta}{\partial x_j}\right). \tag{2.1.30}$$

The mean heat flux Q_j per unit area and unit time in a turbulent flow then becomes (c_p is the specific heat at constant pressure)

$$Q_j = c_p\, \rho(\overline{\theta u_j} - \gamma\, \partial\Theta/\partial x_j). \tag{2.1.31}$$

The heat flux is thus a sum of the contributions of the molecular motion and of the turbulent motion. The analogy between (2.1.24) and (2.1.31) is striking; it is the analytical foundation for the belief that turbulence may transport heat in much the same way as momentum.

2.2
Elements of the kinetic theory of gases

In this section we discuss the molecular background of the viscosity and other molecular transport coefficients in dilute perfect gases (Jeans, 1940). For gases, the rudiments of kinetic theory are straightforward, but the kinetic theory of liquids is not nearly as well developed.

Pure shear flow Let us take a steady pure shear flow, homogeneous in the x_1, x_3 plane. The only nonvanishing velocity component is taken to be U_1; it is a function of x_2 only. If the flow is laminar, the only nonvanishing components of the viscous shear stress are

$$\sigma_{12} = \sigma_{21} = \mu\, \partial U_1/\partial x_2. \tag{2.2.1}$$

The flow situation corresponding to (2.2.1) is sketched in Figure 2.2.

The shear stress σ_{12} must result from molecular transport of momentum in the x_2 direction. Let v_1 and v_2 be the x_1 and x_2 components of the instantaneous velocity of a molecule relative to the mean flow. The x_1

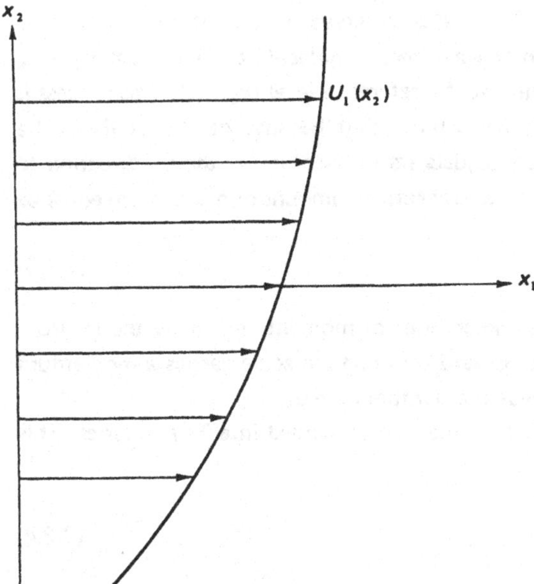

Figure 2.2. Pure shear flow. $U_2 = U_3 = 0$ and all derivatives with respect to x_1 and x_3 vanish.

momentum mv_1 of a molecule with mass m is transported in the x_2 direction if v_2 is correlated with v_1. The momentum transport per molecule is proportional to $m\, v_1 v_2$. If there are N molecules per unit volume, the transport of x_1 momentum in the x_2 direction is $Nm\, \overline{v_1 v_2}$ per unit time and area. Here, the overbar represents an average taken over a large number of molecules. Now, Nm is the mass per unit volume, which is the density ρ, and momentum flux per unit area and time may be equated with a stress. Hence,

$$\sigma_{12} = -\overline{\rho v_1 v_2}. \tag{2.2.2}$$

The minus sign in (2.2.2) is needed because positive values of v_2 should carry momentum deficit in a flow with positive σ_{12} and $\partial U_1/\partial x_2$. The analogy between (2.2.2) and the definition of the Reynolds stress given in (2.1.25) is intentional: a stress that is generated as a momentum flux can always be written as (2.2.2), no matter what mechanism causes the momentum flux.

Molecular collisions Kinetic theory of transport coefficients in gases estimates the right-hand side of (2.2.2) as follows. Suppose the mean free path (the average distance between collisions of molecules) is ξ. The unusual notation is selected because λ has to be reserved for one of the length scales occurring in turbulence. On the average, a molecule coming from $x_2 = -\xi$ collides with another molecule at the reference level ($x_2 = 0$). This process is illustrated in Figure 2.3. If we assume that because of this collision the molecule coming from below adjusts its momentum in the x_1 direction to that of its new environment, it has to absorb an amount of momentum equal to

$$M = m[U_1(0) - U_1(-\xi)]. \tag{2.2.3}$$

The quantity M is equal to the amount of momentum lost by the environment at $x_2 = 0$, because the upward-traveling molecule carries a momentum deficit with respect to the mean momentum $x_2 = 0$.

The right-hand side of (2.2.3) may be expanded in a Taylor series. This yields

$$M = m\xi \frac{\partial U_1}{\partial x_2} + \frac{1}{2} m\xi^2 \frac{\partial^2 U_1}{\partial x_2^2} + \ldots . \tag{2.2.4}$$

The second and higher terms in the expansion may be neglected if

$$\frac{\partial U_1}{\partial x_2} >> \frac{1}{2} \xi \frac{\partial^2 U_1}{\partial x_2^2}. \tag{2.2.5}$$

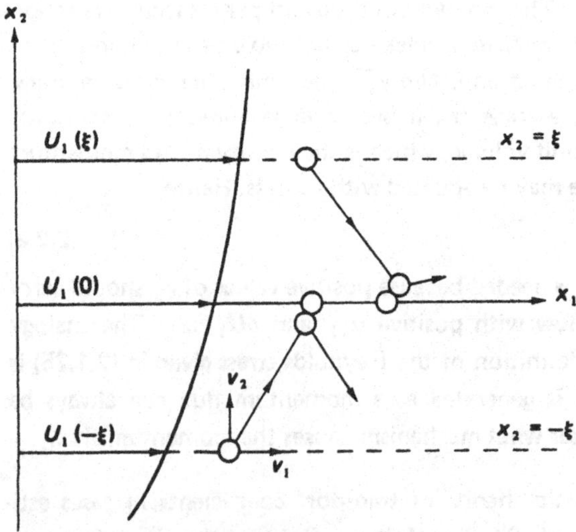

Figure 2.3. Molecular motion in a shear flow.

A local length scale ℓ of the flow $U_1(x_2)$ is defined as

$$\ell \equiv \frac{\partial U_1/\partial x_2}{\partial^2 U_1/\partial x_2^2}. \tag{2.2.6}$$

Hence, (2.2.5) may be written as

$$\ell >> \tfrac{1}{2}\xi. \tag{2.2.7}$$

For air at room temperature and density, $\xi = 7 \times 10^{-6}$ cm, so that for almost all flows the condition (2.2.7) is indeed satisfied. This implies that (2.2.4) may be approximated by

$$M = m\xi\, \partial U_1/\partial x_2. \tag{2.2.8}$$

In this simplified model, the quantity $\xi\, \partial U_1/\partial x_2$ is the part of v_1 that is correlated with v_2, apart from a minus sign needed due to the sign convention for σ_{12}. The number of collisions occurring at the reference level $x_2 = 0$ per unit area and time may be estimated as Na, where N again is the number of molecules per unit volume and a is the speed of sound (which is a good representative for the rms molecular velocity). If the momentum transfer per collision is M, the momentum transfer per unit area and time must be proportional to MNa. Using (2.2.8), we thus can write

$$\sigma_{12} = \alpha MNa = \alpha Nma\xi\, \partial U_1/\partial x_2. \tag{2.2.9}$$

Here, α is an unknown coefficient, which should be of order one. In air at ordinary temperatures and pressure, α is approximately $\frac{2}{3}$; we shall use this value for convenience.

Because $Nm = \rho$, (2.2.10) becomes

$$\sigma_{12} = \tfrac{2}{3}\rho a\xi\, \partial U_1/\partial x_2. \tag{2.2.10}$$

If we compare this with (2.2.1) and use $\mu = \rho\nu$, we obtain

$$\nu = \tfrac{2}{3}a\xi. \tag{2.2.11}$$

The Reynolds number formed with these variables is

$$\frac{a\xi}{\nu} = \frac{3}{2}. \tag{2.2.12}$$

That this Reynolds number turns out to be of the order one is no accident, because the viscosity is defined on the basis of molecular motion with velocity scale a and length scale ξ. The Reynolds number (2.2.12), however, is not a dynamically significant number because at length scales of order ξ the gas is not a continuum. For air at room temperature and pressure, $\xi = 7 \times 10^{-8}$ m, $a = 3.4 \times 10^2$ m/sec, so that $\nu = 15 \times 10^{-6}$ m^2/sec. It should be noted that elementary kinetic theory as given here cannot predict ratios of diffusivities (such as Prandtl number ν/γ).

Characteristic times and lengths The ratio of ξ to the local length scale ℓ of the flow is called the *Knudsen number K*. With (2.2.12), we obtain

$$K = \frac{\xi}{\ell} = \frac{3}{2}\frac{\nu}{a\ell} = \frac{3}{2}\frac{U}{a}\frac{\nu}{U\ell} = \frac{3}{2}\frac{M}{R}. \tag{2.2.13}$$

The Knudsen number is thus proportional to the ratio of the Mach number M and the Reynolds number R. In most flows $M \ll R$, so that the condition (2.2.7) is easily satisfied.

The Knudsen number is a ratio of length scales. The time scales involved in molecular transport of momentum are of interest, too. The molecular time scale is the time interval ξ/a between collisions; this is typically of the order of 10^{-10} sec. The time scale of the flow is the reciprocal of the velocity gradient $\partial U_1/\partial x_2$. If the velocity gradient is 10^4 sec^{-1}, corresponding to quite rapid shearing, the time scale of the flow is 10^{-4} sec. It is seen that changes in the flow are slow compared to the time scale representing molecular motion. This suggests that the thermal motion of the molecules should not be disturbed very much by the flow: molecules collide many thousands of times before the flow has advanced appreciably.

The correlation between v_1 and v_2 For future reference, it is useful to obtain some idea of how well the molecular velocity components v_1 and v_2 are correlated. The part of v_1 correlated with v_2 is proportional to $\xi\, \partial U_1/\partial x_2$, as shown by (2.2.8). Taking representative values for a rapid shearing flow in air ($\xi = 7 \times 10^{-8}$ m, $\partial U_1/\partial x_2 = 10^4$ sec^{-1}), we find that $\xi\, \partial U_1/\partial x_2 = 7 \times 10^{-4}$ m/sec. A correlation coefficient c between v_1 and v_2 may be defined as

$$c \equiv -\frac{\overline{v_1 v_2}}{(v_2')^2}. \tag{2.2.14}$$

Here, v_2' is the rms value of the x_2 component of the molecular velocity. As a comparison of (2.2.14) and (2.1.17) shows, we have used $v_1{}' = v_2{}'$. If we use the results previously given, we may estimate that

$$c \sim \frac{\xi\, \partial U_1/\partial x_2}{v_2{}'} . \tag{2.2.15}$$

Since $v_2{}'$ is of the same order of magnitude as the speed of sound a, which is 3.4×10^2 m/sec for air at room conditions, we find that c is approximately 2×10^{-6}, indicating that v_1 and v_2 are very poorly correlated. If $\partial U_1/\partial x_2$ is estimated as U/ℓ, we find that the correlation coefficient is of the order of M^2/R, a parameter which indeed tends to be extremely small in most flows. We may conclude that the state of the gas is hardly disturbed by molecular momentum transfer. In other words, the dynamical equilibrium of the thermal motion of the molecules in shear flow of gases is, to a very close approximation, the same as the equilibrium state in a gas at rest. This implies that shear flow is not likely to upset the equation of state of the gas, unless M^2/R is large.

In anticipation of results that are obtained in Section 2.3, we note that the correlation coefficient of turbulent velocity fluctuations, defined in a manner similar to (2.2.14), is not small in turbulent shear flow. Consequently, the "state" of the turbulence is not independent of the mean flow field; on the contrary, the interaction between the mean flow and the turbulence tends to be quite strong.

Thermal diffusivity Molecular transport of scalar quantities is similar to the transport of momentum. The heat transfer rate is given by the second term of (2.1.31); in the model flow used here, the only nonvanishing component is

$$Q_2 = -\, \rho c_p \gamma\, \partial\Theta/\partial x_2 . \tag{2.2.16}$$

In terms of molecular parameters, this is

$$Q_2 = -0.93 c_p \rho a \xi\, \partial\Theta/\partial x_2 . \tag{2.2.17}$$

In this equation we have used (2.2.11) and $\nu/\gamma = 0.73$ (air at room conditions). The thermal diffusivity is larger than the diffusivity for momentum because molecules that travel faster than average carry more thermal energy with them and make more collisions per unit time. Energetic molecules thus do more than a proportional share in transporting heat.

2.3
Estimates of the Reynolds stress

We have seen that molecular transport can be interpreted fairly easily in terms of the parameters of molecular motion. It is very tempting to apply a similar heuristic treatment to turbulent transport. We again use a pure shear flow as a basis for our discussion. This flow is illustrated in Figure 2.4. Using (2.1.25) and (2.1.31), we find the rates of turbulent momentum transfer and heat transfer to be

$$\tau_{12} = -\rho \overline{u_1 u_2}, \tag{2.3.1}$$

$$H_2 = \rho c_p \overline{\theta u_2}. \tag{2.3.2}$$

The symbol H_2 is used to avoid confusion with the total rate of heat transfer Q_2.

Reynolds stress and vortex stretching Let us consider the Reynolds stress only. The existence of a Reynolds stress requires that the velocity fluctuations u_1 and u_2 be correlated. In a shear flow with $\partial U_1/\partial x_2 > 0$, negative values of u_1 should occur more frequently than positive ones when u_2 is positive, and vice versa. This is a rather intricate problem: the energy of the eddies has to be maintained by the shear flow, because they are continuously losing energy to smaller eddies. Molecules do not depend on the flow for their energy because the collisions between molecules are elastic. Eddies, on the

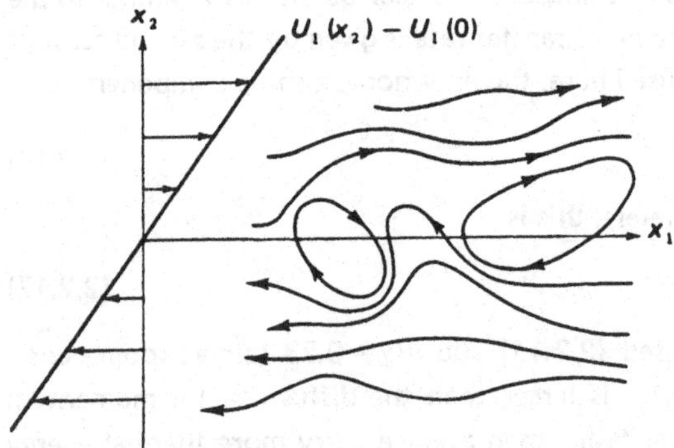

Figure 2.4. Turbulent pure shear flow. The mean velocity is steady: $U_2 = U_3 = 0$ and $U_1 = U_1(x_2)$. The instantaneous streamline pattern sketched refers to a coordinate system that moves with a velocity $U_1(0)$.

other hand, need shear to maintain their energy; the most powerful eddies thus are those that can absorb energy from the shear flow more effectively than others. Evidence (for example, Townsend, 1956, Bakewell and Lumley, 1967) suggests that the eddies that are more effective than most in maintaining the desired correlation between u_1 and u_2 and in extracting energy from the mean flow are vortices whose principal axis is roughly aligned with that of the mean strain rate. Such eddies are illustrated in Figure 2.5. The energy transfer mechanism for eddies of this kind is believed to be associated with vortex stretching: as the eddies in Figure 2.5 are being strained by the shear, conservation of angular momentum tends to maintain the good correlation between u_1 and u_2, thus allowing (as we discuss in more detail in Chapter 3) efficient energy transfer.

The interaction between eddies and the mean flow described here is essentially three dimensional. Two-dimensional eddies (velocity fluctuations without a component normal to the x_1, x_2 plane) may on occasion have appreciable Reynolds stress, but the mean shear tends to rotate and strain them in such a way that they would lose their capacity for extracting energy from the mean flow rather quickly.

These considerations suggest that a simple transport theory patterned after kinetic theory of gases is at best a very crude representation of reality. The dynamic interaction between the mean flow and the turbulence is too strong

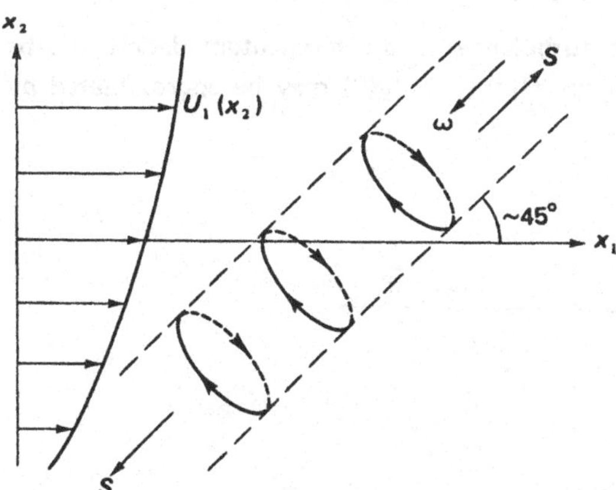

Figure 2.5. Three-dimensional eddies (vortices with vorticity ω) being stretched by the rate of strain S. The fluctuating velocity has strong components in the plane normal to the vorticity vector. Note that the shape of these eddies may differ widely from flow to flow.

to allow for a simple transport model. Also, a more detailed analysis of the energy and vorticity dynamics of the eddies (Chapter 3) is essential to the understanding of turbulence.

It should be noted that this discussion applies only to shear flows. If eddies receive energy in other ways (from buoyancy or a magnetic field, say), the picture may be entirely different.

The mixing-length model An estimate for the turbulent momentum flux can be obtained by analyzing the random motion of moving points ("fluid particles") in a turbulent shear flow. A formal treatment of the statistics of the motion of wandering points is given in Chapter 7; the less rigorous analysis presented here is more than adequate for a first look at turbulent transport.

Suppose a moving point starts from a level $x_2 = 0$ (see Figure 2.6) at time $t = 0$. Its x_1 momentum per unit volume is $\rho\tilde{u}_1(0, 0)$, where $\tilde{u}_1(0, 0)$ stands for the instantaneous velocity at $x_2 = 0$, $t = 0$. If we assume that the moving point does not lose its momentum as it travels upward, it has a momentum deficit $\Delta M = \rho\tilde{u}_1(x_2, t) - \rho\tilde{u}_1(0, 0)$ when it passes an arbitrary level x_2 at time t. Using the Reynolds decomposition of velocities, we can write the momentum deficit as

$$\Delta M = \rho[U_1(x_2) - U_1(0)] + \rho[u_1(x_2, t) - u_1(0, 0)]. \qquad (2.3.3)$$

If the contribution of the turbulence to the momentum deficit can be neglected and if the difference $U_1(x_2) - U_1(0)$ may be approximated by

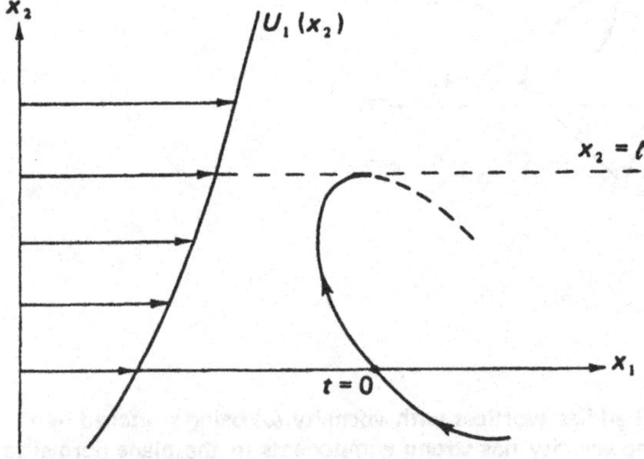

Figure 2.6. Transport of momentum by turbulent motion.

$x_2\, \partial U_1/\partial x_2$, where the gradient is taken at $x_2 = 0$, ΔM may be approximated by

$$\Delta M = \rho x_2\, \partial U_1/\partial x_2. \tag{2.3.4}$$

The volume transported per unit area and unit time in the x_2 direction is $\tilde{u}_2$ of the moving point. Now, $\tilde{u}_2 = dx_2/dt$, so that the average momentum flux at $x_2 = 0$ may be written as

$$\tau_{12} = \tfrac{1}{2}\rho \frac{\partial U_1}{\partial x_2} \frac{d}{dt}(\overline{x_2^2}). \tag{2.3.5}$$

The overbar here denotes an average over all moving points that start from $x_2 = 0$.

The dispersion rate $d(\overline{x_2^2})/dt$ may be written as (see also Section 7.1)

$$\frac{d}{dt}(\overline{x_2^2}) = 2\overline{x_2 \frac{dx_2}{dt}} = 2\overline{x_2 u_2}. \tag{2.3.6}$$

If the fluid at any point did not continually exchange momentum with its environment, u_2 would remain constant for any given moving point, and $\overline{x_2 u_2}$ would continue to increase in time as x_2 increased. This is not realistic; instead, we expect that the correlation between u_2 and x_2 of a moving point decreases as the distance traveled increases. If we assume that u_2 and x_2 become essentially uncorrelated at values of x_2 comparable to some transverse length scale ℓ (see Figure 2.6), we may estimate that $\overline{x_2 u_2}$ is of order $u_2'\ell$. Here, u_2' is the rms velocity in the x_2 direction; the dispersion length scale ℓ is called the *mixing length*. Of course, this very estimate of $\overline{x_2 u_2}$ implies that momentum is not conserved when the moving point travels in the x_2 direction, so that this estimate makes the expression for the momentum deficit ΔM given in (2.3.4) very dubious, to say the least.

With $2\overline{x_2 u_2} = 2c_1 u_2'\ell$, (2.3.5) becomes

$$\tau_{12} = c_1 \rho u_2'\ell\, \partial U_1/\partial x_2. \tag{2.3.7}$$

The numerical coefficient c_1 is unknown.

We define the *eddy viscosity* ν_T (or turbulent exchange coefficient for momentum), in analogy with (2.2.1), by the equation

$$\tau_{12} \equiv \rho \nu_T\, \partial U_1/\partial x_2. \tag{2.3.8}$$

Comparing (2.3.7) and (2.3.8), we find that the eddy viscosity is given by

$$\nu_T = c_1 u_2' \ell. \tag{2.3.9}$$

If the mixing length ℓ and the velocity u_2' were known everywhere in the flow field and if the mixing-length model were accurate, the closure problem would be solved. The unknown Reynolds stress would be related to known variables and to the mean velocity gradient, making a solution of the equations of motion possible. However, the situation is not quite that simple. Even if we were willing to accept (2.3.7) as a model, u_2' and ℓ are not properties of the fluid but properties of the flow. This implies that u_2' and ℓ may vary throughout the flow field, making the eddy viscosity variable, dependent on the position in the flow. This is not a very promising prospect. Consequently, applications of (2.3.7) are usually restricted to flows for which it can be argued that u_2' is approximately constant (at least in the cross-stream direction) and for which ℓ is either constant or depends in a simple way on the geometry of the shear flow concerned.

In reality, turbulence consists of fluctuating motion in a broad spectrum of length scales. However, in view of the way ℓ occurs in (2.2.7), one may argue that large eddies contribute more to the momentum transfer than small eddies. The mixing-length model therefore favors large-scale motions; for simplicity, ℓ may be taken to be proportional to the size of the larger eddies.

The length-scale problem The approximations involved in the estimate (2.3.4) of the momentum deficit carried by a moving point need to be carefully considered. Because the distance over which momentum is transported is of order ℓ, the approximation (2.3.4) of (2.3.3) should be accurate over transverse distances of order ℓ. Let us define a local length scale $\mathscr{L}$ of the mean flow by (von Kármán, 1930)

$$\mathscr{L} \equiv \frac{\partial U_1/\partial x_2}{\partial^2 U_1/\partial x_2^2}. \tag{2.3.10}$$

The approximation $U_1(x_2) - U_1(0) = x_2\, \partial U_1/\partial x_2$ for all values of x_2 of order ℓ is valid only if

$$\mathscr{L} \gg \tfrac{1}{2}\ell. \tag{2.3.11}$$

In turbulent flows, however, the largest eddies tend to have sizes comparable to the width of the flow, as we have seen in Chapter 1. Consequently, ℓ is usually of the same order as the local length scale $\mathscr{L}$. This makes the "turbulent Knudsen number" $\ell/\mathscr{L}$ of order one. Note that both ℓ and $\mathscr{L}$ are transverse length scales: they are associated with the x_2 direction, which is normal to the mean flow.

We have to conclude that the truncation of the Taylor series expansion involved in (2.3.4) is not justified. Therefore, a *gradient-transport model*, which links the stress to the rate of strain at the same point in time and space, cannot be used for turbulent flow. It should be emphasized again that turbulence is an irreducible part of the flow, not a mere property of the fluid. Turbulence interacts strongly with its environment; the "state" of turbulence depends strongly on the flow in which it finds itself.

A neglected transport term The approximation (2.3.4) to the momentum deficit ΔM given by (2.3.3) neglected the contribution $\rho[u_1(x_2, t) - u_1(0, 0)]$. Let us call this $\rho\, \Delta u_1$. The momentum flux associated with this term is $\rho\, \overline{u_2 \cdot \Delta u_1}$, where the overbar again denotes an average over many moving points. The velocity difference Δu_1 should be very small for transverse distances small compared to ℓ, but it could be appreciable for values of x_2 of order ℓ, so that there is no a priori reason why this term can be neglected. However, in view of all of the other dubious assumptions involved in the mixing-length model, it does not seem useful to pursue this issue.

The mixing length as an integral scale In the derivation of (2.3.7), we used

$$\frac{1}{2}\frac{d}{dt}\overline{x_2^2} = \overline{x_2 u_2} = c_1\, u_2{'}\ell. \tag{2.3.12}$$

It is worthwhile to investigate how ℓ could be defined. For this purpose, consider how the value of x_2 increases as the moving point travels away from the reference level $x_2 = 0$. We can write

$$x_2(t) = \int_0^t u_2(t')\, dt'. \tag{2.3.13}$$

This implies that (2.3.12) may be written as (Taylor, 1921; see Friedlander and Topper, 1962)

$$\frac{1}{2}\frac{d}{dt}\overline{x_2^2} = \int_0^t \overline{u_2(t)u_2(t')}\, dt'. \tag{2.3.14}$$

The velocity $u_2(t)$ can be taken inside the integral because it is independent of t'; the averaging process can be performed on the integrand because it is done over many moving points, not over time.

In a statistically steady situation like the flow considered in this chapter, the origin of time is irrelevant, so that the correlation between $u_2(t)$ and $u_2(t')$ should depend only on the time difference $t - t' = \tau$. Let us define a correlation coefficient $c(\tau)$ by

$$c(\tau) \equiv \frac{\overline{u_2(t)u_2(t-\tau)}}{\overline{u_2^2}}. \tag{2.3.15}$$

Substituting (2.3.15) into (2.3.14), we obtain

$$\frac{1}{2}\frac{d}{dt}\overline{x_2^2} = \overline{u_2^2}\int_0^t c(\tau)\,d\tau. \tag{2.3.16}$$

The correlation coefficient $c(\tau)$ decreases as the time interval τ increases; at large values of τ the velocities $u_2(t)$ and $u_2(t')$ are uncorrelated. A sketch of $c(\tau)$ is shown in Figure 2.7.

The area under the curve in Figure 2.7 is given by

$$\mathcal{T} = \int_0^\infty c(\tau)\,d\tau; \tag{2.3.17}$$

it is assumed that $c(\tau)$ decreases rapidly enough at large τ to make $\mathcal{T}$ finite. The time $\mathcal{T}$ is called the *Lagrangian integral scale*. The adjective "Lagrangian" is used to indicate that it relates to moving points ("fluid particles"). The adjective "Eulerian" is used whenever correlations between two fixed points in a fixed frame of reference are considered. A more detailed discussion is given in Chapter 7.

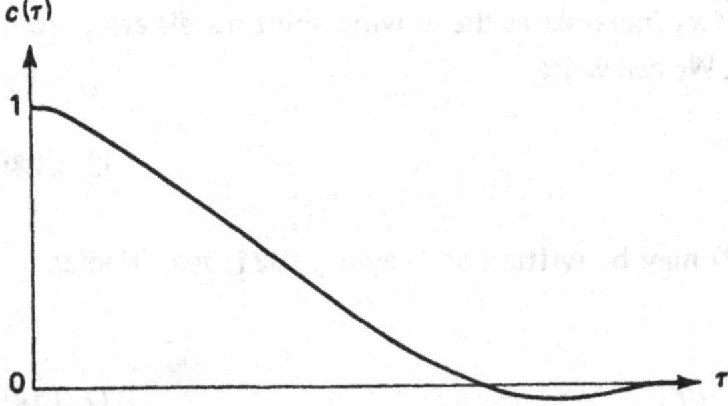

Figure 2.7. The Lagrangian correlation curve. Some correlation curves have negative tails, many do not.

Moving fluid loses its capability of transporting momentum when the correlation between x_2 and u_2 becomes zero. The time interval t involved in (2.3.16) should thus be large enough to make $c(t)$ zero. The dispersion rate then becomes (see also Section 7.1)

$$\frac{1}{2}\frac{d}{dt}\overline{x_2^2} = \overline{u_2^2}\mathcal{T}. \tag{2.3.18}$$

If we define (Taylor, 1921) a Lagrangian integral length scale ℓ_L by

$$\ell_L \equiv u_2'\mathcal{T}, \tag{2.3.19}$$

we can write (2.3.18) as

$$\frac{1}{2}\frac{d}{dt}(\overline{x_2^2}) = u_2'\ell_L. \tag{2.3.20}$$

The time scale $\mathcal{T}$ is hard to determine experimentally, because it requires that the motion of many tagged fluid particles be followed, say with photographic or radioactive tracer methods. In most turbulent flows, however, the length scale ℓ_L is believed to be comparable to the transverse Eulerian integral scale ℓ, which is defined by

$$\overline{u_2^2}\ell \equiv \int_0^\infty \overline{u_2(x_2)u_2(0)}\, dx_2. \tag{2.3.21}$$

The averaging process used in (2.3.21) is performed over a long period of time, with a fixed transverse separation x_2 and zero time delay between the two velocities. Experimental determination of ℓ is relatively simple.

If ℓ_L and ℓ are of the same order of magnitude, we thus may estimate $\overline{x_2 u_2}$ as $c_1 u_2'\ell$, where ℓ is defined by (2.3.21) (see also Sections 7.1 and 8.5).

The gradient-transport fallacy The mixing-length model has been discussed in great detail because of its ubiquitous use in much of turbulence theory. Let us now demonstrate that (2.3.7) is merely a dimensional necessity in a turbulent shear flow dominated by a single velocity scale u_2' and a single length scale ℓ.

The correlation coefficient c_{12} between u_1 and u_2 is defined as

$$c_{12} \equiv \overline{u_1 u_2}/(u_1' u_2'). \tag{2.3.22}$$

Hence, we may write

$$\tau_{12} = -c_{12}\rho u_1' u_2'. \tag{2.3.23}$$

In all turbulent flows, u_1' and u_2' are of the same order of magnitude so that

(2.3.23) may be written as

$$\tau_{12} = c_2 \rho (u_2')^2. \tag{2.3.24}$$

In turbulent flows driven by shear, the unknown coefficients c_{12} and c_2 are always of order one: u_1 and u_2 are well correlated in eddies that can absorb energy from the mean flow by vortex stretching (Figure 2.5). Note, however, that in turbulence maintained in other ways, say by buoyancy, c_{12} and c_2 may be quite small.

The eddies involved in momentum transfer have characteristic vorticities of order u_2'/ℓ; they maintain their vorticity because of their interaction with the mean shear $\partial U_1/\partial x_2$. Let us write

$$u_2'/\ell = c_3 \, \partial U_1/\partial x_2, \tag{2.3.25}$$

so that c_3 is a nondimensional coefficient. If the straining of eddies is the effective mechanism that Figure 2.5 suggests it is, c_3 should be of order one. In effect, we are merely saying that the characteristic time of eddies (ℓ/u_2') and the characteristic time of the mean flow $(\partial U_1/\partial x_2)^{-1}$ should be of the same order if no other characteristic times or lengths are present, because turbulence is the fluctuating part of the flow. In particular, it is implied that ℓ and the differential length scale $\mathscr{L}$ defined in (2.3.10) are of the same order and that the mixing length is of the same order as the length scale of large eddies. The statement about time scales made here may be transposed into a statement about vorticities or strain rates if so desired: if $c_3 \sim 1$, (2.3.25) states that the vorticity found in the larger eddies is of the same order as the vorticity of the mean flow, and that the respective strain rates are also comparable.

If we use (2.3.25) to substitute for one of the u_2' occurring in (2.3.24), we find

$$\tau_{12} = c_2 c_3 \rho u_2' \ell \, \partial U_1/\partial x_2, \tag{2.3.26}$$

which, of course, is equivalent to (2.3.7). We see that we can relate the stress at $x_2 = 0$ to the mean velocity gradient at $x_2 = 0$ because the correlation between u_1 and u_2 is good and because the time-scale ratio is of order one. No conservation of momentum needs to be assumed; the mean-velocity gradient $\partial U_1/\partial x_2$ at $x_2 = 0$ may be used because it is a convenient representative of $\partial U_1/\partial x_2$ throughout an environment of scale ℓ. Indeed, (2.3.16) is only one member of a class of expressions

$$\tau_{12}(x_2 = 0) \sim \rho u_2' \ell \frac{\partial U_1}{\partial x_2} \qquad (|x_2| \leq \ell), \tag{2.3.27}$$

all of which are implied by (2.3.24) and (2.3.25). The localized estimate (2.3.26) merely is the most convenient member of this class. In other words, we may treat the local stress as if it were determined by the local rate of strain because there is only one characteristic length and one characteristic time. In short, (2.3.26) is a dimensional necessity that does not imply conservation of momentum or "localness" of the mechanism that produces the stress; (2.3.26) should not be mistaken for a gradient-transport postulate.

Further estimates Comparing (2.3.23) and (2.3.26), we see that the part of u_1 that is correlated with u_2 is of order $\ell\,\partial U_1/\partial x_2$. If the correlation between u_1 and u_2 is good and if u_1' and u_2' are of the same order, we may write

$$\tau_{12} = c_4 \rho \ell^2 \frac{\partial U_1}{\partial x_2}\left|\frac{\partial U_1}{\partial x_2}\right|. \tag{2.3.28}$$

In (2.3.28), c_4 is a coefficient of order one; the modulus of $\partial U_1/\partial x_2$ is used to make τ_{12} switch signs with $\partial U_1/\partial x_2$. This expression is the one originally proposed by Prandtl (see Hinze, 1959).

The eddy viscosity is of order $u_2'\ell$. The ratio of the Reynolds stress to the viscous stress is thus

$$\frac{\tau_{12}}{\mu\,\partial U_1/\partial x_2} = \frac{\nu_T}{\nu} = c_1 \frac{u_2'\ell}{\nu} = c_1 R_\ell. \tag{2.3.29}$$

This substantiates one of the results obtained in Chapter 1: the Reynolds number $u_2'\ell/\nu$ of the turbulent eddies may be interpreted as a ratio of diffusivities. In most flows, R_ℓ is very large, which implies that the Reynolds stress is much larger than the viscous stress. In other words, turbulent transport of momentum tends to be much more effective than molecular transport. If this is the case, the viscous terms in the equations for the mean flow may be neglected. The dependence of the mean flow on the Reynolds number is thus small, except in regions where ℓ and ν/u_2' are of the same order of magnitude.

Recapitulation We have found that, in a shear flow with one characteristic velocity and one characteristic length, the time scale of the turbulence is proportional to the time scale of the mean flow. Under certain circumstances, ℓ/u_2' may be as small as one-tenth of the reciprocal of $\partial U_1/\partial x_2$, but the general conclusion must be that turbulence in a shear flow cannot possibly be in a state of equilibrium which is independent of the flow field involved. The turbulence is continually trying to adjust to its environment, without ever succeeding. This conclusion is substantiated by the good correlation between u_1 and u_2. In all turbulent shear flows $|-\overline{u_1 u_2}| \sim 0.4 u_1' u_2'$; the value of 0.4 should be contrasted to the correlation coefficient for molecular motion, which was seen to be of order 10^{-6}. A theory for the Reynolds stress thus cannot be patterned after the kinetic theory of gases; the mixing-length model must be rejected, even though a mixing-length expression like (2.3.26) makes good dimensional sense in a situation where only one length scale and only one time scale are relevant.

In situations where more than one characteristic length and time are involved, the problem of the relation between stress and rate of strain generally becomes nearly intractable. If, for instance, the turbulence is mainly generated by buoyancy (as in an atmospheric boundary layer with an unstable temperature gradient), there is no need for the vorticity $\partial U_1/\partial x_2$ of the mean flow to be of order u_2'/ℓ, so that nothing can be said a priori about the value of the coefficient c_1 in (2.3.7). Problems such as this require a very careful study of the kinetic energy budget of turbulent motion.

In the model problem considered in this chapter, downstream variation in the flow was suppressed by virtue of the assumption that U_1 is only a function of x_2. In most flows, however, downstream changes do occur, introducing time scales such as the reciprocal of $\partial U_1/\partial x_1$ and length scales such as the distance x_1 from some suitably defined origin. These parameters would have to be taken into account were it not for the fact that in many flows of practical interest

$$\frac{\partial U_1}{\partial x_1} \ll \frac{\partial U_1}{\partial x_2}, \quad \ell \ll x_1. \tag{2.3.30}$$

If these inequalities hold almost everywhere in the flow, the downstream changes in the flow field are slow compared to the time scale of the turbulence, so that the turbulence may be in approximate equilibrium with respect to its environment at all values of the downstream distance x_1. This concept is vital to the theory of turbulent shear flows (Chapters 4 and 5).

2.4
Turbulent heat transfer

Passive contaminants are transported by turbulent motions in much the same way as momentum. The transfer of heat in the pure shear flow considered in this chapter is a good example. We assume here that the heat flux does not cause significant buoyancy effects.

Reynolds' analogy The vertical heat flux H_2 is given by (2.3.2):

$$H_2 = \rho c_p \overline{u_2 \theta}.$$

An eddy diffusivity for heat, γ_T, is defined by

$$H_2 \equiv -\rho c_p \gamma_T \, \partial\Theta/\partial x_2. \tag{2.4.1}$$

This is a mere definition, which does not assume anything about the nature of γ_T. In most turbulent flows, the "turbulent Prandtl number" ν_T/γ_T is close to one: turbulence transports heat just as rapidly as momentum (Hinze, 1959). Recall that τ_{12} may be expressed as (2.3.8):

$$\tau_{12} = \rho \nu_T \, \partial U_1/\partial x_2.$$

If ν_T/γ_T is equal to one, heat and momentum transfer are related by

$$\frac{H_2}{c_p \tau_{12}} = -\frac{\partial\Theta/\partial x_2}{\partial U_1/\partial x_2}. \tag{2.4.2}$$

This is called *Reynolds' analogy*. It is used to estimate the turbulent heat flux if the stress and the mean velocity and temperature fields are known. The analogy avoids an explicit statement on the magnitudes of the eddy diffusivities for heat and momentum, so that it can be applied even if ν_T and γ_T cannot be determined.

The mixing-length model Mixing-length theory (Taylor, 1915) estimates the heat flux as

$$H_2 = -\rho c_p c_5 u_2' \ell \, \partial\Theta/\partial x_2, \tag{2.4.3}$$

where c_5 is a coefficient of order one. The mixing-length model of turbulent heat transfer is not as misleading as the model of momentum transfer, because the temperature of a fluid particle is more nearly conserved than its momentum. Even so, (2.4.3), like its stress counterpart, does not need to be defended with a mixing-length model in order to justify its use in situations with a single characteristic length and velocity. If the correlation between u_2 and θ is good and if

$$\theta'/\ell \sim \partial\Theta/\partial x_2, \tag{2.4.4}$$

the heat transfer can be expressed as (2.4.3).

The assertion (2.4.4) may be understood as follows. Consider turbulent motion between $x_2 = 0$ and $x_2 = \mathscr{L}$, where $\mathscr{L}$ is the local length scale of the flow field, defined by (2.3.10). Let us assume that the mean temperature difference between $x_2 = 0$ and $x_2 = \mathscr{L}$ is $\Delta\Theta$. In turbulent flows, $\mathscr{L}$ and ℓ are of the same order of magnitude, so that the eddies, in attempting to mix the temperature field, create temperature fluctuations of order $\Delta\Theta$. This implies that $\theta' \sim \Delta\Theta$ if $\ell \sim \mathscr{L}$, which is expressed most concisely in the differential form (2.4.4). Strictly speaking, an average value of $\partial\Theta/\partial x_2$ between $x_2 = 0$ and $x_2 = \mathscr{L}$ should be used, but the definition of $\mathscr{L}$ implies that $\partial\Theta/\partial x_2$ is of the same order of magnitude everywhere between $x_2 = 0$ and $x_2 = \mathscr{L}$, so that a local value may be used to represent the average. It should be kept in mind, however, that a local interpretation of (2.4.3), though often convenient, is more restrictive than it needs to be.

The expression $\theta'/\ell \sim \partial\Theta/\partial x_2$ often is more reliable than its momentum counterpart $u_2'/\ell \sim \partial U_1/\partial x_2$, because the former merely expresses that turbulence mixes passive scalar contaminants over scales of order ℓ, whereas the latter is valid only if the turbulent motion is maintained by a mean strain rate. Momentum is not a passive contaminant; "mixing" of mean momentum relates to the dynamics of turbulence, not merely to its kinematics.

2.5 Turbulent shear flow near a rigid wall

Let us apply the concepts developed in this chapter to a pure shear flow in the vicinity of a rigid, but porous wall. The flow geometry is sketched in Figure 2.8. If there is no mass transfer (blowing or suction) through the wall, we shall find that there is only one velocity scale. In that case, mixing-length models may be used. However, if the mass-transfer velocity is different from zero, there are two velocity scales. We shall see that mixing-length theory cannot cope with that problem.

We take the mean flow to be steady and homogeneous in the x_1, x_3 plane. We take $U_3 = 0$ and $\partial P/\partial x_i = 0$ for $i = 1,2,3$. The flow may be thought of as occurring in a very wide channel, with the upper wall at $x_2 \to \infty$ moving at a certain velocity to maintain the momentum of the flow. The entire half-space $x_2 > 0$ is supposed to be filled with turbulent flow.

The equations of motion are

$$\frac{\partial U_2}{\partial x_2} = 0, \tag{2.5.1}$$

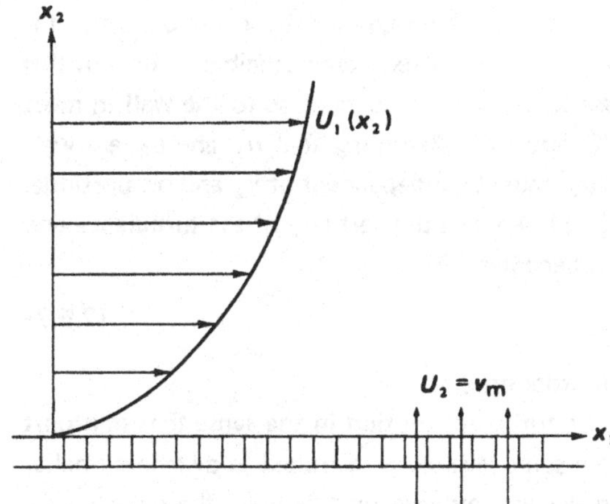

Figure 2.8. Turbulent flow near a rigid surface with mass transfer. The surface is at rest ($U_1(0) = 0$).

$$U_2 \frac{\partial U_1}{\partial x_2} = \frac{1}{\rho} \frac{\partial}{\partial x_2} T_{12}. \qquad (2.5.2)$$

Equation (2.5.1) can be solved at once; because U_2 has to be independent of x_1 by virtue of downstream homogeneity, U_2 is uniform:

$$U_2 = v_m. \qquad (2.5.3)$$

The mass-transfer velocity v_m is independent of x_1 and x_2, but it does not need to be zero as in the flow considered in Section 2.3.

With (2.5.3), (2.5.2) can be integrated to yield

$$\rho v_m U_1 = T_{12} - T_{12}(0). \qquad (2.5.4)$$

The boundary condition $U_1(0) = 0$ is implied in (2.5.4). Let us define a *friction velocity* u_* by

$$T_{12}(0) = \rho u_*^2. \qquad (2.5.5)$$

If the analysis is restricted to values of x_2 where $x_2 U_1/\nu >> 1$, the viscous contribution to the total shear stress T_{12} should be negligible, so that we may write

$$v_m U_1 = -\overline{u_1 u_2} - u_*^2. \qquad (2.5.6)$$

A flow with constant stress If $V_m = 0$, the Reynolds stress $-\overline{u_1 u_2}$ is equal to u_*^2 at all values of x_2 for which viscous effects are negligible. A flow of this kind is called a *constant-stress layer*; it also occurs close to the wall in most turbulent boundary layers (Chapter 5). Assuming that u_1 and u_2 are well correlated, we conclude that u_2' must be independent of x_2 and proportional to u_*. The scale relation (2.3.25) between the vorticity of the turbulence and the vorticity of the mean flow becomes

$$u_*/\ell = \alpha_1 \, \partial U_1/\partial x_2, \tag{2.5.7}$$

in which α_1 is a coefficient of order one.

The rigid wall constrains the turbulent motion in the sense that transport of momentum downward from some level x_2 is restricted to distances smaller than x_2 itself. If no length scales are *imposed* on this flow, the only dimensionally correct choice for ℓ is

$$\ell = \alpha_2 x_2. \tag{2.5.8}$$

A comprehensive study of the implications of (2.5.8) is deferred until Chapter 5. With (2.5.8), (2.5.7) becomes

$$\partial U_1/\partial x_2 = u_*/\kappa x_2, \tag{2.5.9}$$

which readily integrates to

$$\frac{U_1}{u_*} = \frac{1}{\kappa} \ln x_2 + \text{const.} \tag{2.5.10}$$

The coefficient κ is known as the constant of von Kármán (Kármán constant, for short). Experiments have shown that κ is approximately equal to 0.4 (Hinze, 1959).

The additive constant in (2.5.10) is presumably determined by the no-slip condition ($U_1 = 0$ at $x_2 = 0$). However, this condition cannot be enforced because (2.5.10) is not valid at values of x_2 which are so small that the Reynolds number $x_2 U_1/\nu$ is of order unity.

In this flow without mass transfer through the surface, mixing-length models can be used because there is only one length scale (x_2) and one velocity scale (u_*), so that no ambiguity can arise. Specifically, (2.3.7) becomes

$$-\overline{u_1 u_2} = \kappa u_* x_2 \, \partial U_1/\partial x_2. \tag{2.5.11}$$

Because $-\overline{u_1 u_2}$ is equal to u_*^2 if $v_m = 0$, (2.5.11) produces (2.5.10) upon integration. Prandtl's version of the mixing-length formula can be applied with equal success.

Nonzero mass transfer If $v_m \neq 0$, the problem has two characteristic velocities, u_* and v_m. The length scale, however, remains proportional to x_2. This problem cannot be solved without making further assumptions. The least restrictive assumption we can make is that $\partial U_1/\partial x_2$ should be proportional to w/x_2, where w is an undetermined velocity scale that depends on u_* and v_m. Let us write

$$\partial U_1/\partial x_2 = w/x_2. \tag{2.5.12}$$

The numerical coefficient needed in (2.5.12) has been absorbed in the unknown velocity scale w.

Integration of (2.5.12) yields

$$U_1/w = \ln x_2 + \text{const.} \tag{2.5.13}$$

This equation is not a solution to the equations of motion; it is merely a consequence of the differential similarity law (2.5.12). Because w is unknown, it has to be determined experimentally. In this flow, v_m and u_* are the only two velocity scales, so that we may write

$$w/u_* = f(v_m/u_*). \tag{2.5.14}$$

Experimental results on w/u_* are given in Figure 2.9. In the case of blowing ($v_m > 0$), the Reynolds stress is larger than u_*^2; this results in an increase of w/u_*. If $v_m \gg u_*$, the friction velocity becomes relatively unimportant, so that w should be proportional to v_m. In the case of suction ($v_m < 0$), the Reynolds stress is smaller than u_*^2, so that w/u_* decreases. If the suction rate is large, the Reynolds stress becomes so small that turbulence cannot be maintained; this causes reverse transition from turbulent to laminar flow. If $v_m < 0$, the situation is further complicated by the fact that the suction imports not only mean momentum toward the wall but also turbulent kinetic energy.

The mixing-length approach The preceding analysis was based on the assumption expressed by (2.5.12). If the resulting velocity profile (2.5.13) is substituted into the equation of motion (2.5.6), there results

$$-\overline{u_1 u_2} = u_*^2 + v_m w(\ln x_2 + c). \tag{2.5.15}$$

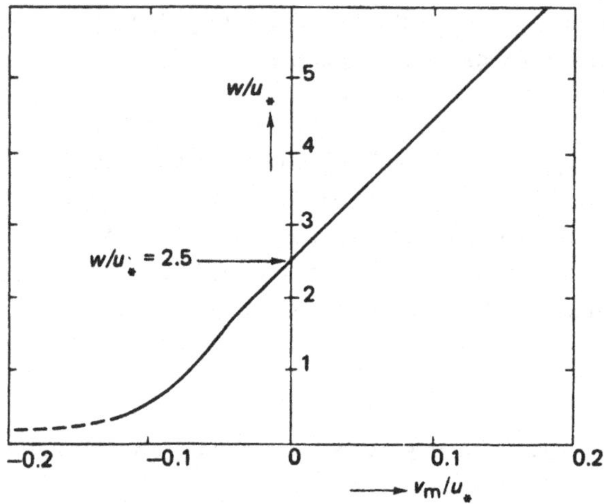

Figure 2.9. The velocity scale of flow near a rigid wall with mass transfer (based on data collected by Tennekes, 1965).

However, if we insist on using a mixing-length model and if we continue to use w as a characteristic velocity, we should write

$$-\overline{u_1 u_2} = \alpha_3 w x_2 \, \partial U_1 / \partial x_2, \tag{2.5.16}$$

where α_3 is an unknown coefficient. If we substitute (2.5.12) into (2.5.16), we obtain

$$-\overline{u_1 u_2} = \alpha_3 w^2. \tag{2.5.17}$$

A stress that is independent of x_2 is clearly not a correct solution: (2.5.6) states that the stress depends on x_2 because U_1 presumably depends on x_2. However, the difference between (2.5.15) and (2.5.17) is not as large as it seems. For $v_m = 0$, $w = 2.5\, u_*$ (Figure 2.9), so that $\alpha_3 = 0.16$. For small values of v_m/u_*, Figure 2.9 shows that $w/u_* = 2.5\ (1 + 9\ v_m/u_*)$, so that $\alpha_3 w^2$ may be approximated by $u_*^2 + 18\ v_m u_*$ if v_m/u_* is small. This is very much like (2.5.15) except for the suppressed dependence on x_2.

A third approach would be to substitute (2.5.16) into (2.5.6) without making a further substitution based on (2.5.12). Upon integration, this yields

$$-\overline{u_1 u_2} = v_m (\alpha_4 x_2)^{v_m/\alpha_3 w}. \tag{2.5.18}$$

This expression agrees neither with (2.5.15) nor with (2.5.17).

A fourth approach would be to use (2.5.12) to remove w from the mixing-length formula (2.5.16). This results in Prandtl's version of the mixing-length formula; after integration of the equation of motion (2.5.6) there results

$$-\overline{u_1 u_2} = [\alpha_5 v_m (\ln \alpha_6 x_2)]^2. \tag{2.5.19}$$

The corresponding velocity profile is obtained by substitution of (2.5.19) into (2.5.6). The proponents of (2.5.19) claim that it agrees with their experimental data. However, (2.5.19) contains two adjustable coefficients (α_5 and α_6), both of which may depend on v_m/u_*. Like (2.5.15), (2.5.17), and (2.5.18), (2.5.19) is not a solution to the equations of motion.

The limitations of mixing-length theory At this point it has become abundantly clear that mixing-length models are incapable of describing turbulent flows containing more than one characteristic velocity with any degree of consistency. None of the versions that were tried gives a clear picture of the roles of the two velocity scales; the effects of v_m/u_* on the integration constants remain altogether unresolved. Let us recall that mixing-length expressions can be understood as the combination of a statement about the stress ($-\overline{u_1 u_2} \sim w^2$) and a statement about the mean-velocity gradient ($\partial U_1/\partial x_2 \sim w/x_2$). These statements do not give rise to inconsistencies if there is only one characteristic velocity, but they cannot be used to obtain solutions to the equations of motion if there are two or more characteristic velocities that contribute to w in unknown ways. In other words, mixing-length theory is useless because it cannot predict anything substantial; it is often confusing because no two versions of it can be made to agree with each other. Mixing-length and eddy-viscosity models should be used only to generate analytical expressions for the Reynolds stress and the mean-velocity profile if those are desired for curve-fitting purposes in turbulent flows characterized by a single length scale and a single velocity scale. The use of mixing-length theory in turbulent flows whose scaling laws are not known beforehand should be avoided.

Problems

2.1 Consider a fully developed turbulent Couette flow in a channel between two infinitely long and wide parallel plane walls. The distance between the walls is $2h$, the lower wall is at rest and the upper wall moves with a velocity U_0 in its own plane. Assume that the flow consists of two wall layers (Section

2.5) which match at the center line of the channel. Find an expression for the friction coefficient at the lower wall ($c_f = 2u_*^2/U_c^2$, where U_c is the mean velocity at the center line) in terms of an appropriate Reynolds number. Estimate the additive constant in the logarithmic velocity profile (2.5.10) by assuming that near the walls there exist "viscous sublayers" in which the Reynolds number is so small that the Reynolds stress is negligible. The thickness of these sublayers is equal to $10\nu/u_*$. Sketch the velocity profile in the channel.

2.2 Experimental evidence obtained in pipe flow (Hinze, 1959) suggests that a more accurate representation of the velocity profile in turbulent Couette flow is obtained if it is assumed that the eddy viscosity is nowhere larger than $0.07hu_*$. Repeat the analysis of Problem 2.1 on this basis.

2.3 A certain amount of hot fluid is released in a turbulent flow with characteristic velocity u and characteristic length ℓ. The temperature of the patch is higher than the ambient temperature, but the density difference and the effects of buoyancy may be neglected. Estimate the rate of spreading of the patch of hot fluid and the rate at which the maximum temperature difference decreases. Assume that the size of the patch at the time of release is much smaller than ℓ and much larger than the Kolmogorov microscale η. The use of an eddy diffusivity is appropriate, but the choice of the velocity and length scales that are needed to form an eddy diffusivity requires careful thought, in particular as long as the size of the patch remains smaller than the length scale ℓ. In this context, a review of Problem 1.3 will be helpful.

2.4 A vortex generator in the shape of a low aspect-ratio wing is located on the wing of a Boeing 707. The height of the vortex generator is comparable to the thickness of the turbulent boundary layer over the wing. Give a qualitative description of the effect of the vortex generator on the momentum transfer in the boundary layer.

3

THE DYNAMICS OF TURBULENCE

In Chapter 2, we studied the effects of the turbulent velocity fluctuations on the mean flow. We now turn to the other side of the issue. Two major questions arise. First, how is the kinetic energy of the turbulence maintained? Second, why are vorticity and vortex stretching so important to the study of turbulence? To help answer these questions, we shall proceed as follows. We first derive equations for the kinetic energy of the mean flow and that of the turbulence. We shall see that turbulence extracts energy from the mean flow at large scales and that this gain is approximately balanced by viscous dissipation of energy at very small scales. Realizing that dissipation of energy at small scales occurs only if there exists a dynamical mechanism that transfers energy from large scales to small scales, we then turn to a study of vorticity. In order to gain an appreciation of the role of vorticity fluctuations, we first analyze how they are involved in the generation of Reynolds stresses. It turns out to be convenient to associate the Reynolds shear stress with transport and stretching of vorticity. With the understanding obtained that way, the vorticity equations can be studied. We shall discover that energy is transferred to small scales by vortex stretching and that the dissipation rate of energy is proportional to the mean-square vorticity fluctuations if the Reynolds number is large enough. The analysis of the interaction between the vorticity and the strain rate demonstrates the dynamical role of strain-rate fluctuations; this gives us the opportunity to discuss some other problems in which the strain-rate fluctuations play a role.

3.1
Kinetic energy of the mean flow

We found in Section 2.1 that the equations of motion for steady mean flow in an incompressible fluid are

$$U_j \frac{\partial U_i}{\partial x_j} = \frac{\partial}{\partial x_j}\left(\frac{T_{ij}}{\rho}\right), \tag{3.1.1}$$

$$\frac{\partial U_i}{\partial x_i} = 0. \tag{3.1.2}$$

The stress tensor T_{ij} is

$$T_{ij} = -P\delta_{ij} + 2\mu S_{ij} - \rho\,\overline{u_i u_j}. \tag{3.1.3}$$

The mean rate of strain S_{ij} is defined by

$$S_{ij} \equiv \frac{1}{2}\left(\frac{\partial U_i}{\partial x_j} + \frac{\partial U_j}{\partial x_i}\right). \tag{3.1.4}$$

Since the mean momentum $\overline{u_i}$ of the turbulent velocity fluctuations is zero, we cannot discuss the effects of the mean flow on the turbulence very well in terms of mean momentum. We shall study the equations for the kinetic energy of the mean flow and of the turbulence instead. The equation governing the dynamics of the mean-flow energy $\frac{1}{2}U_iU_i$ is obtained by multiplying (3.1.1) by U_i. It is useful to split the stress term in the resulting equation into two components. The energy equation becomes

$$\rho U_j \frac{\partial}{\partial x_j}(\tfrac{1}{2}\, U_iU_i) = \frac{\partial}{\partial x_j}(T_{ij}U_i) - T_{ij}\frac{\partial U_i}{\partial x_j}. \tag{3.1.5}$$

Because T_{ij} is a symmetric tensor, the product $T_{ij}\, \partial U_i/\partial x_j$ is equal to the product of T_{ij} and the symmetric part S_{ij} of the deformation rate $\partial U_i/\partial x_j$; (3.1.5) thus becomes

$$\rho U_j \frac{\partial}{\partial x_j}(\tfrac{1}{2}\, U_iU_i) = \frac{\partial}{\partial x_j}(T_{ij}U_i) - T_{ij}S_{ij}. \tag{3.1.6}$$

The first term on the right-hand side of (3.1.6) represents transport of mean-flow energy by the stress T_{ij}. This term integrates to zero if the integration refers to a control volume on whose surface either T_{ij} or U_i vanishes. According to the divergence theorem,

$$\int_V \frac{\partial}{\partial x_j}(T_{ij}U_i)\, dV = \int_S n_j T_{ij} U_i\, ds. \tag{3.1.7}$$

The vector n_j is a unit vector normal to the surface element ds. If the work performed by the stress on the surface S of the control volume V is zero, only the volume integral of $T_{ij}\, S_{ij}$ can change the total amount of kinetic energy. The term $T_{ij}\, S_{ij}$ is called *deformation work*; by virtue of conservation of energy, it represents kinetic energy of the mean flow that is lost to or retrieved from the agency that generates the stress. The distinction between spatial energy transfer and deformation work is crucial to the understanding of the dynamics of turbulence.

Pure shear flow As an illustration, let us take a pure shear flow in which all variables depend on x_2 only and in which the only nonzero component of U_i

is U_1. For this turbulent Couette flow, which is sketched in Figure 3.1, the energy equation reads

$$0 = \frac{\partial}{\partial x_2}(T_{12}U_1) - T_{12}\frac{\partial U_1}{\partial x_2}. \tag{3.1.8}$$

Figure 3.1 illustrates that the rate of work done by the stresses per unit volume is equal to the first term in (3.1.8). The average value of the stress is T_{12}; the work done by the average stress is equal to the second term in (3.1.8). Because the left-hand side of (3.1.8) is zero, the work $\partial(T_{12}U_1)/\partial x_2$ performed by the stresses does not result in a change of the kinetic energy of this flow; instead, it is all traded for deformation work. This is consistent with (3.1.8), because this equation implies that T_{ij} is constant. A constant stress field does not accelerate a flow; the tendency to change $\frac{1}{2}U_iU_i$ by $\partial(T_{12}U_1)/\partial x_2$ is balanced exactly by the deformation work $T_{12}\ \partial U_1/\partial x_2$. Work is performed, but $\frac{1}{2}U_iU_i$ does not change. We expect that deformation work generally will be an input term for the energy of the agency that generates the stress and that the kinetic energy $\frac{1}{2}U_iU_i$ will decrease because of the deformation work unless this loss is balanced by a net input of energy. However, no specific conclusions can be made without a study of the individual contributions of the various stresses to the deformation work.

The deformation work is caused by the stresses that contribute to T_{ij}. Substitution of (3.1.3) into $T_{ij}S_{ij}$ yields

$$T_{ij}S_{ij} = 2\mu\, S_{ij}S_{ij} - \rho\, \overline{u_iu_j}\, S_{ij}. \tag{3.1.9}$$

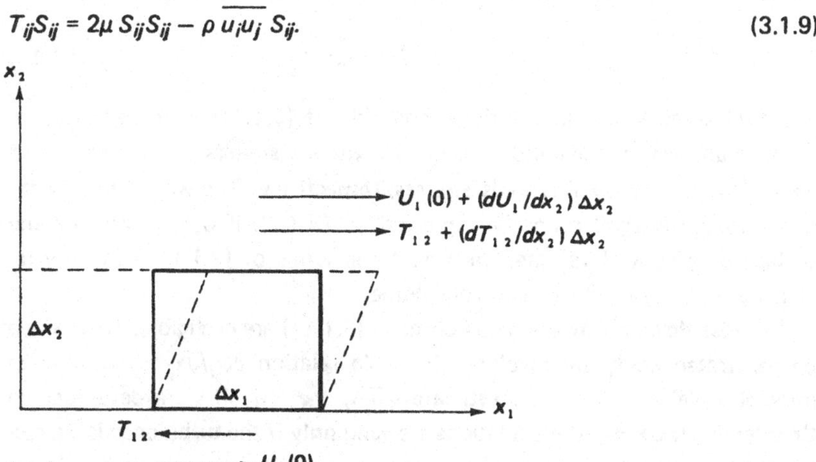

Figure 3.1. Stresses on a small volume element in a pure shear flow.

The contribution of the pressure to deformation work in an incompressible fluid is zero:

$$-P\delta_{ij}S_{ij} = -PS_{ii} = -P\frac{\partial U_i}{\partial x_i} = 0. \tag{3.1.10}$$

The contribution of viscous stresses to the deformation work is always negative; consequently, viscous deformation work always represents a loss of kinetic energy. For this reason, the term $2\mu S_{ij}S_{ij}$ is called *viscous dissipation.* Note that the dissipation is related to the strain rate, not to the vorticity (the vorticity is related to the skew-symmetric part of $\partial U_i/\partial x_j$).

The contribution of Reynolds stresses to the deformation work is also dissipative in most flows: negative values of $\overline{u_iu_j}$ tend to occur in situations with positive S_{ij}, as we have seen in Chapter 2. Positive values of $\overline{u_iu_j}\,S_{ij}$ can occur in unusual situations; even then the region in which $\overline{u_iu_j}\,S_{ij} > 0$ is a small fraction of the entire flow. Since turbulent stresses perform the deformation work, the kinetic energy of the turbulence benefits from this work. For this reason $-\rho\,\overline{u_iu_j}\,S_{ij}$ is known as *turbulent energy production.*

The effects of viscosity If (3.1.3) is substituted into (3.1.5), the energy equation for the mean flow becomes

$$U_j\frac{\partial}{\partial x_j}(\tfrac{1}{2}U_iU_i) = \frac{\partial}{\partial x_j}\left(-\frac{P}{\rho}U_j + 2\nu U_iS_{ij} - \overline{u_iu_j}\,U_i\right)$$

$$- 2\nu S_{ij}S_{ij} + \overline{u_iu_j}\,S_{ij}. \tag{3.1.11}$$

The first three terms on the right-hand side of (3.1.11) are called pressure work, transport of mean-flow energy by viscous stresses, and transport of mean-flow energy by Reynolds stresses, respectively. The word "transport" refers to the integral property expressed by (3.1.7): if U_iT_{ij} is zero on the surface of a control volume, the first three terms of (3.1.11) can only redistribute energy inside the control volume.

In most flows the two viscous terms in (3.1.11) are negligible. This can be demonstrated easily by invoking the scale relation $\partial U_i/\partial x_j \sim u/\ell$ (ℓ is an integral scale) and the stress estimate $-\overline{u_iu_j} \sim u^2$ which were developed in Chapter 2. Of course, these relations are valid only if the turbulence is characterized by u and ℓ and if no other characteristic scales are present. We define the representative velocity u by

$$\mathscr{u}^2 \equiv \tfrac{1}{3}\,\overline{u_i u_i}. \tag{3.1.12}$$

With $S_{ij} \sim \mathscr{u}/\ell$ and $-\overline{u_i u_j} \sim \mathscr{u}^2$, turbulence production is estimated as

$$\overline{u_i u_j}\,S_{ij} = C_1\,\mathscr{u}\ell S_{ij}S_{ij}; \tag{3.1.13}$$

in the same way, energy transport by turbulent motion is estimated as

$$-\overline{u_i u_j}\,U_i = C_2\,\mathscr{u}\ell U_i S_{ij}. \tag{3.1.14}$$

In most simple shear flows, the undetermined coefficients C_1 and C_2 are of order one. Comparing (3.1.13) and (3.1.14) with the corresponding viscous terms (3.1.11), we see that the turbulence terms are $\mathscr{u}\ell/\nu$ times as large as the viscous terms. This Reynolds number tends to be very large (except in situations very close to smooth surfaces), so that the viscous terms in (3.1.11) can ordinarily be neglected. This conclusion again illustrates that the gross structure of turbulent flows tends to be virtually independent of viscosity. Viscosity makes itself felt only indirectly.

Although the equation for the energy of the mean flow is helpful in obtaining additional insight into the dynamics of turbulent motion, it does not contain any more information than the momentum equation for the mean flow since the former is obtained from the latter by mere manipulation.

3.2 Kinetic energy of the turbulence

The equation governing the mean kinetic energy $\frac{1}{2}\,\overline{u_i u_i}$ of the turbulent velocity fluctuations is obtained by multiplying the Navier-Stokes equations (2.1.1) by $\tilde{u}_i$, taking the time average of all terms, and subtracting (3.1.11), which governs the kinetic energy of the mean flow. This is a fairly tedious exercise, which is left to the reader. The final equation, the *turbulent energy budget*, reads

$$U_j \frac{\partial}{\partial x_j}\left(\tfrac{1}{2}\,\overline{u_i u_i}\right) = -\frac{\partial}{\partial x_j}\left(\frac{1}{\rho}\overline{u_j p} + \tfrac{1}{2}\,\overline{u_i u_i u_j} - 2\nu\,\overline{u_i s_{ij}}\right)$$

$$-\overline{u_i u_j}\,S_{ij} - 2\nu\,\overline{s_{ij}s_{ij}}. \tag{3.2.1}$$

The quantity s_{ij} is the fluctuating rate of strain, defined by

$$s_{ij} \equiv \frac{1}{2}\left(\frac{\partial u_i}{\partial x_j} + \frac{\partial u_j}{\partial x_i}\right). \tag{3.2.2}$$

The rate of change of $\frac{1}{2}\overline{u_i u_i}$ is thus due to pressure-gradient work, transport by turbulent velocity fluctuations, transport by viscous stresses, and two kinds of deformation work. The transport terms, like those in (3.1.11), are divergences of energy flux. If the energy flux out of or into a closed control volume is zero, these terms merely redistribute energy from one point in the flow to another.

The deformation-work terms are more important. The turbulence production $-\overline{u_i u_j}\, S_{ij}$ occurs in (3.1.11) and in (3.2.1) with opposite signs. As we had anticipated, this term apparently serves to exchange kinetic energy between the mean flow and the turbulence. Normally, the energy exchange involves a loss to the mean flow and a profit to the turbulence.

The last term in (3.2.1) is the rate at which viscous stresses perform deformation work against the fluctuating strain rate. This always is a drain of energy, since the term is quadratic in s_{ij}. The term is called *viscous dissipation*; unlike the dissipation term in (3.1.11), it is essential to the dynamics of turbulence and cannot ordinarily be neglected.

Production equals dissipation In a steady, homogeneous, pure shear flow (in which all averaged quantities except U_i are independent of position and in which S_{ij} is a constant), (3.2.1) reduces to

$$-\overline{u_i u_j}\; S_{ij} = 2\nu\, \overline{s_{ij} s_{ij}}. \tag{3.2.3}$$

This equation states that in this flow the rate of production of turbulent energy by Reynolds stresses equals the rate of viscous dissipation. It should be noted that in most shear flows production and dissipation do not balance, though they are nearly always of the same order of magnitude. Keeping this in mind, we may use (3.2.3) as an aid in understanding those features of turbulence that are not directly related to spatial transport. For this reason, (3.2.3) is often written in symbolical form. If we define

$$\mathscr{P} \equiv -\overline{u_i u_j}\; S_{ij}, \tag{3.2.4}$$

$$\epsilon \equiv 2\nu\, \overline{s_{ij} s_{ij}}, \tag{3.2.5}$$

(3.2.3) reads simply

$$\mathscr{P} = \epsilon. \tag{3.2.6}$$

In order to interpret (3.2.6), we again employ the scale relation $S_{ij} \sim u/\ell$ and the stress estimate $-\overline{u_i u_j} \sim u^2$, keeping in mind that these estimates are valid only in shear-generated turbulence with one length scale and one velocity scale.

With this provision, we use (3.1.13) as an estimate for the Reynolds stress. The energy budget (3.2.3) becomes

$$C_1 u\ell S_{ij} S_{ij} = 2\nu \overline{s_{ij} s_{ij}}. \tag{3.2.7}$$

Since the Reynolds number $u\ell/\nu$ is generally very large, we conclude that

$$\overline{s_{ij} s_{ij}} \gg S_{ij} S_{ij}. \tag{3.2.8}$$

The fluctuating strain rate s_{ij} is thus very much larger than the mean rate of strain S_{ij} when the Reynolds number is large. Since strain rates have the dimension of $\sec^{-1}$, this implies that the eddies contributing most to the dissipation of energy have very small convective time scales compared to the time scale of the flow. This suggests that there should be very little direct interaction between the strain-rate fluctuations and the mean flow if the Reynolds number is large. In other words, S_{ij} and s_{ij} do not interact strongly, because they are not tuned to the same frequency band. Therefore, the small-scale structure of turbulence tends to be independent of any orientation effects introduced by the mean shear, so that all averages relating to the small eddies do not change under rotations or reflections of the coordinate system. If this is the case, the small-scale structure is called *isotropic* (Figure 3.2). Isotropy at small scales is called *local isotropy* (see Chapter 8).

Taylor microscale The preceding considerations suggest that any length scale involved in estimates of s_{ij} must be very much smaller than ℓ if a balance between production and dissipation is to be obtained. The situation is similar to the one in laminar boundary-layer theory (Section 1.5). In laminar boundary layers, we had to select the thickness δ in such a way that the essential viscous term in the equation of motion could be retained; this yielded $\delta/L \sim R^{-1/2}$ (1.5.3). Here, we should be able to proceed in a similar way. The dissipation of energy is proportional to $\overline{s_{ij} s_{ij}}$; this consists of several terms like $\overline{(\partial u_i/\partial x_j)^2}$, most of which cannot be measured conveniently. However, as we mentioned, the small-scale structure of turbulence tends to be

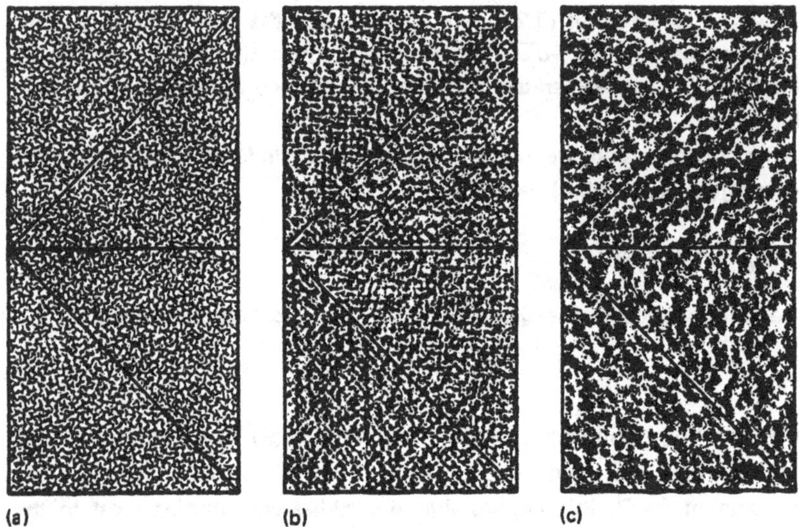

Figure 3.2. The shading pattern used in this book: (a) was selected because it is an isotropic random field, like the small-scale structure of turbulence. The other patterns, (b) and (c), have preferred directions; they are not isotropic.

isotropic. In isotropic turbulence, the dissipation rate is equal to

$$\epsilon = 2\nu \overline{s_{ij}s_{ij}} = 15\nu\, \overline{(\partial u_1/\partial x_1)^2}. \tag{3.2.9}$$

The derivation of (3.2.9) is not given here; it involves bookkeeping with terms like $\overline{(\partial u_1/\partial x_1)^2}$ that contribute to $\overline{s_{ij}s_{ij}}$ (Hinze, 1959). The coefficient 15 in (3.2.9) is considerably larger than one because so many components are involved. In many flows, $\overline{(\partial u_1/\partial x_1)^2}$ can be measured relatively easily.

Let us define a new length scale λ by

$$\overline{(\partial u_1/\partial x_1)^2} \equiv \overline{u_1^2}/\lambda^2 = u^2/\lambda^2. \tag{3.2.10}$$

The length scale λ is called the *Taylor microscale* in honor of G. I. Taylor who first defined (3.2.10). The Taylor microscale is also associated with the curvature of spatial velocity autocorrelations; this is discussed in Section 6.4. The substitution $\overline{u_1^2} = u^2$ can be made because in isotropic turbulence $\overline{u_1^2} = \overline{u_2^2} = \overline{u_3^2}$, so that u^2, which was defined as $\frac{1}{3}\overline{u_iu_i}$, is equal to $\overline{u_1^2}$. Since the small-scale structure of turbulence at large Reynolds numbers is always approximately isotropic (see Section 8.3), we use

$$\epsilon = 15\nu\, u^2/\lambda^2, \tag{3.2.11}$$

with λ defined by (3.2.10), as a convenient estimate of ϵ.

A relation between λ and ℓ can be obtained from the simplified energy budget (3.2.3). If S_{ij} is of order u/ℓ and if $-\overline{u_i u_j}$ is of order u^2, we obtain

$$A u^3/\ell = 15\nu u^2/\lambda^2. \tag{3.2.12}$$

The ratio λ/ℓ is then given by

$$\frac{\lambda}{\ell} = \left(\frac{15}{A}\right)^{1/2} \left(\frac{u\ell}{\nu}\right)^{-1/2} = \left(\frac{15}{A}\right)^{1/2} R_\ell^{-1/2}. \tag{3.2.13}$$

In (3.2.12, 3.2.13), A is an undetermined constant, which is presumably of order one. Because in all turbulent flows $R_\ell >> 1$, the Taylor microscale λ is always much smaller than the integral scale ℓ. Again we see that dissipation of energy is due to the small eddies of turbulence.

Scale relations The Taylor microscale λ is not the smallest length scale occurring in turbulence. The smallest scale is the Kolmogorov microscale η, which was introduced in Chapter 1:

$$\eta = (\nu^3/\epsilon)^{1/4}. \tag{3.2.14}$$

The difference between λ and η can be understood if we return to the definition (3.2.5) and the estimate (3.2.11) of the dissipation rate ϵ. The strain-rate fluctuations s_{ij} have the dimension of a frequency (sec^{-1}); the definition of ϵ thus defines a time scale associated with the dissipative structure of turbulence. Calling this time scale τ, we find that

$$\tau = (\nu/\epsilon)^{1/2}. \tag{3.2.15}$$

This time scale is identical to the one discovered by elementary considerations in Chapter 1. This is no coincidence. The dimensions of s_{ij} are such that the length scale λ was found by taking u as a velocity scale. There is no physical reason at all for this choice of characteristic velocity; the only scale that can be determined unambiguously is the time scale τ. The Taylor microscale should thus be used only in the combination (3.2.11):

$$u/\lambda = 0.26\,\tau^{-1} = 0.26\,(\epsilon/\nu)^{1/2}. \tag{3.2.16}$$

The Taylor microscale is thus not a characteristic length of the strain-rate field and does not represent any group of eddy sizes in which dissipative effects are strong. It is not a dissipation scale, because it is defined with the assistance of a velocity scale which is not relevant for the dissipative eddies. Even so, λ is used frequently because the estimate $s_{ij} \sim u/\lambda$ is often convenient. For future use, expressions relating ℓ, λ, and η are given:

$$\frac{\lambda}{\ell} = \left(\frac{15}{A}\right)^{1/2} R_\ell^{-1/2} = \frac{15}{A} R_\lambda^{-1}, \tag{3.2.17}$$

$$\frac{\lambda}{\eta} = \left(\frac{225}{A}\right)^{1/4} R_\ell^{1/4} = 15^{1/4} R_\lambda^{1/2}. \tag{3.2.18}$$

The undetermined constant A is the same as the one used in (3.2.12) and (3.2.13). The parameter R_λ is the *microscale Reynolds number*, which is defined by

$$R_\lambda \equiv u\lambda/\nu. \tag{3.2.19}$$

This Reynolds number may be interpreted as the ratio of the large-eddy time scale ℓ/u (which is proportional to λ^2/ν by virtue of (3.2.13)) and the time scale λ/u of the strain-rate fluctuations (Corrsin, 1959).

Spectral energy transfer The energy exchange between the mean flow and the turbulence is governed by the dynamics of the large eddies. This is clear from (3.2.7): large eddies contribute most to the turbulence production $\mathscr{P}$ because $\mathscr{P}$ increases with eddy size. The energy extracted by the turbulence from the mean flow thus enters the turbulence mainly at scales comparable to the integral scale ℓ.

The viscous dissipation of turbulent energy, on the other hand, occurs mainly at scales comparable to the Kolmogorov microscale η. As we found in Chapter 1, this implies that the internal dynamics of turbulence must transfer energy from large scales to small scales. All of the available experimental evidence suggests that this spectral energy transfer proceeds at a rate dictated by the energy of the large eddies (which is of order u^2) and their time scale (which is of order ℓ/u). Thus, the dissipation rate may always be estimated as

$$\epsilon = A\, u^3/\ell, \tag{3.2.20}$$

provided there exists only one characteristic length ℓ (Taylor, 1935). The estimate (3.2.20) is independent of the presence of turbulence production; (3.2.12) is thus a valid statement about the dissipation rate even if production and dissipation do not balance.

Of course, turbulence can maintain itself only if it receives a continuous supply of energy. If $-\overline{u_i u_j} S_{ij}$ is the only production term and if ϵ is estimated with (3.2.20), the approximate balance between $\mathscr{P}$ and ϵ which occurs in many turbulent shear flows may be written as

$$-\overline{u_i u_j}\, S_{ij} \sim A\, \mathscr{u}^3/\ell. \qquad (3.2.21)$$

This equation needs careful interpretation. It states that $\mathscr{P}$ must be of order $\mathscr{u}^3/\ell$ if $\mathscr{P} = \epsilon$ and if ϵ is estimated by (3.2.20). This is distinct from the original interpretation of (3.2.12), which stated that $\mathscr{P}$ must be of order $\mathscr{u}^3/\ell$ because $-\overline{u_i u_j}$ is of order $\mathscr{u}^2$ and S_{ij} is of order $\mathscr{u}/\ell$, so that ϵ must be of order $\mathscr{u}^3/\ell$ if $\mathscr{P} = \epsilon$. This discrepancy arises because the estimate $-\overline{u_i u_j} \sim \mathscr{u}^2$ was introduced in Chapter 2 as an empirical statement without theoretical justification. This estimate now receives support from (3.2.21). With ϵ of order $\mathscr{u}^3/\ell$ because spectral energy transfer is of that order and with S_{ij} of order $\mathscr{u}/\ell$ because the vorticity of the large eddies is maintained by the vorticity and the strain rate of the mean flow, we conclude from (3.2.21) that $-\overline{u_i u_j}$ has to be of order $\mathscr{u}^2$ if a balance between $\mathscr{P}$ and ϵ, however approximate, is to be obtained. Conversely, (3.2.21) states that a good correlation between u_i and u_j can be obtained only if S_{ij} and $\mathscr{u}/\ell$ occur in the same range of frequencies.

Further estimates The orders of magnitude of the other terms of the original energy budget (3.2.1) need to be established. We shall use $s_{ij} \sim \mathscr{u}/\lambda$ and $\lambda/\ell \sim R_\ell^{-1/2}$ wherever needed.

The pressure-work term in (3.2.1) is estimated as

$$-\frac{\partial}{\partial x_j}\left(\frac{1}{\rho}\overline{u_j p}\right) \sim \frac{\mathscr{u}^3}{\ell}, \qquad (3.2.22)$$

because the pressure fluctuations p should be of order $\rho\mathscr{u}^2$ and because the local length scale of the flow, which determines the gradients of averaged quantities, should be of the same order as the large-eddy size ℓ.

Mean transport of turbulent energy by turbulent motion is estimated as

$$-\frac{\partial}{\partial x_j}\left(\tfrac{1}{2}\,\overline{u_i u_i u_j}\right) \sim \frac{\mathscr{u}^3}{\ell}. \qquad (3.2.23)$$

It is tempting to estimate transport by viscous stresses in the following way:

$$2\nu \frac{\partial}{\partial x_j} (\overline{u_i s_{ij}}) \sim \frac{\nu u^2}{\ell\lambda} \sim \frac{u^3}{\ell} R_\ell^{-1/2}. \tag{3.2.24}$$

This estimate, however, is too large because it assumes that u_i and s_{ij} are well correlated. This is not likely because the time scale of the eddies contributing most to s_{ij} is much smaller than the time scale of the eddies contributing most to u_i. The problem can easily be resolved by substituting the definition (3.2.2) of s_{ij} into (3.2.24) and employing $\partial u_i/\partial x_i = 0$:

$$2\nu \frac{\partial}{\partial x_j} (\overline{u_i s_{ij}}) = \nu \frac{\partial^2}{\partial x_j \partial x_j} (\tfrac{1}{2} \overline{u_i u_i}) + \nu \frac{\partial^2}{\partial x_i \partial x_j} \overline{u_i u_j}. \tag{3.2.25}$$

Both terms on the right-hand side of (3.2.25) are of order $\nu u^2/\ell^2$, so that the correct estimate for the viscous transport term is

$$2\nu \frac{\partial}{\partial x_j} (\overline{u_i s_{ij}}) \sim \nu \frac{u^2}{\ell^2} \sim \frac{u^3}{\ell} R_\ell^{-1}. \tag{3.2.26}$$

Comparing (3.2.24) and (3.2.26), we see that the correlation coefficient between u_i and s_{ij} must be of order $R_\ell^{-1/2}$. The time scale of the large eddies is of order ℓ/u and the time scale of the dissipative eddies is of order λ/u. The ratio of these time scales is λ/ℓ, which is of order $R_\ell^{-1/2}$ by virtue of (3.2.17). The correlation coefficient thus scales with the ratio of the time scales involved. One might say that u_i and s_{ij} cannot interact strongly at large Reynolds numbers because they are not tuned to the same frequency range.

The estimates (3.2.22) through (3.2.26) show that only the viscous transport of turbulent energy can be neglected if the Reynolds number is large. The other transport terms are of the same order of magnitude as the production and dissipation rates, so that they need to be retained in most flows. The pressure-work term is sometimes neglected, partly because it cannot be measured and partly because p tends to be rather poorly correlated with u_i, except near a wall (Townsend, 1956). A possible explanation is that the pressure is a weighted integral of $u_i u_j$, so that its fluctuations tend to have scales that are larger than those of the velocity fluctuations.

Wind-tunnel turbulence As an application of the equations and estimates developed here, we discuss the decay of nearly homogeneous turbulence in a low-speed wind tunnel. Wind-tunnel turbulence is commonly generated by a grid or screen in a uniform flow without shear. The flow geometry is illustrated in Figure 3.3. If S_{ij} is zero, there is no turbulence production. The turbulence should then decay through viscous dissipation. This serves as a reminder that the approximation $\mathscr{P} \sim \epsilon$ is not always relevant.

If the frame of reference is chosen such that U_1 (a constant) is the only nonzero component of the mean velocity, the energy budget (3.2.1) becomes

$$U_1 \frac{\partial}{\partial x_1} (\tfrac{1}{2} \overline{u_i u_i}) = -\frac{\partial}{\partial x_1} \left(\frac{1}{\rho} \overline{u_1 p} + \tfrac{1}{2} \overline{u_i u_i u_1} \right) - \epsilon. \tag{3.2.27}$$

It has been assumed that the Reynolds number R_ℓ is so large that the viscous transport term can be neglected. The orders of magnitude of the various terms in (3.2.27) may be estimated as follows:

$$U_1 \frac{\partial}{\partial x_1} (\tfrac{1}{2} \overline{u_i u_i}) = \mathcal{O}\left(\frac{U_1}{x_1} u^2 \right), \tag{3.2.28}$$

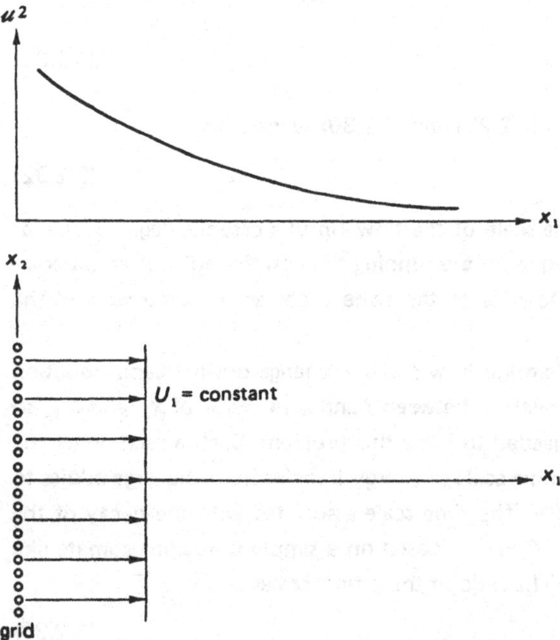

Figure 3.3. Geometry of wind-tunnel turbulence. The mean flow velocity U_1 is independent of x_1, but u^2 decreases downstream because of viscous dissipation.

$$-\frac{\partial}{\partial x_1}\left(\frac{1}{\rho}\overline{u_1 p}+\tfrac{1}{2}\overline{u_i u_i u_1}\right)=\mathcal{O}\left(\frac{u^3}{x_1}\right), \tag{3.2.29}$$

$$\epsilon=\mathcal{O}\left(\frac{u^3}{\ell}\right). \tag{3.2.30}$$

The distance x_1 is measured from a virtual origin which is presumably in the immediate vicinity of the turbulence-producing grid. The downstream distance x_1 is the appropriate length scale in the estimate of the downstream decay of $\frac{1}{2}\overline{u_i u_i}$ and $\overline{u_1(p/\rho+\frac{1}{2}u_i u_i)}$: the integral scale ℓ is not a measure for the downstream inhomogeneity of the turbulence and no characteristic length in the downstream direction is imposed, so that $\partial/\partial x_1$ can scale only with x_1 itself. More specifically, if $u^2 \sim x_1^{\alpha}$ then $\partial u^2/dx_1 \sim \alpha u^2/x_1$, so that $\partial/\partial x_1 \sim x_1^{-1}$.

In grid turbulence, the velocity fluctuations are small: $u \ll U$. The turbulent transport terms in (3.2.27) then should be negligible compared to the transport by the mean flow, so that the energy equation reduces to

$$U_1\frac{\partial}{\partial x_1}(\tfrac{1}{2}\overline{u_i u_i})=-\epsilon. \tag{3.2.31}$$

The dimensional estimates (3.2.28) and (3.2.30) suggest that

$$U_1/x_1=Cu/\ell, \tag{3.2.32}$$

which states that the time scale of the flow (in this case the "age" x_1/U_1 of the mean flow, which is equal to the running time on the clock of an observer moving with the mean flow) is of the same order as the time scale of the turbulence.

We would like to determine how ℓ and u change downstream. Equation (3.2.32) gives only one relation between ℓ and u in terms of x_1 and U_1, so that another relation is needed to solve this problem. Such a relation can be obtained as follows. The time scale of energy transfer from the large eddies to the small eddies is $\tau \sim \ell/u$. The time scale associated with the decay of the large eddies themselves is $T \sim \ell^2/\nu$ (based on a simple diffusion estimate like those used in Chapter 1). The ratio of these time scales is

$$T/\tau \sim \ell u/\nu, \tag{3.2.33}$$

which suggests that at large values of $R_\ell = \ell u/\nu$ the large eddies are affected very little by direct dissipation. We now assume that these time scales, together with the running time x_1/U_1, are the only independent variables of the problem. This is a fair assumption, since this flow has no time scales imposed from outside. A relation between the independent and dependent variables should exist; in nondimensional form it may be written as

$$\frac{u\ell}{\nu} = f\left(\frac{x_1}{U_1\tau}, \frac{\tau}{T}\right). \tag{3.2.34}$$

Since T/τ is proportional to $u\ell/\nu$, this can be rewritten

$$\frac{u\ell}{\nu} = g\left(\frac{x_1}{U_1\tau}\right) = g\left(\frac{x_1 u}{U_1\ell}\right). \tag{3.2.35}$$

Now, the only way in which $u\ell$ can be a function of u/ℓ is by requiring that g be a constant. This is supported by the fact that the argument of g should be a constant, as predicted by (3.2.32). Hence, wind-tunnel turbulence in its initial period of decay (where $R_\ell >> 1$) should have an approximately constant Reynolds number. Keeping in mind that R_ℓ should be independent of x_1, we find from (3.2.32)

$$u^2 = \frac{1}{3}\overline{u_i u_i} = \mathscr{C}\frac{\nu U_1 R_\ell}{x_1}, \tag{3.2.36}$$

$$\ell = \mathscr{C}\left(\frac{x_1}{U_1}\right)^{1/2} (R_\ell \nu)^{1/2}. \tag{3.2.37}$$

The constants $\mathscr{C}$ are undetermined. Because R_ℓ is a constant, the ratios ℓ/λ and ℓ/η are constant by virtue of (3.2.17, 3.2.18). Hence λ and η also are proportional to $x_1{}^{1/2}$.

We conclude that the turbulent energy decays as $x_1{}^{-1}$, while all length scales grow as $x_1{}^{1/2}$. These results are expected to be rather crude approximations, because they are based on the assumption that only a small number of nondimensional groups is relevant. Experimental evidence indicates that the predicted exponents are within 30% of the observed values (Comte-Bellot and Corrsin, 1965). More realistic results can be obtained by spectral analysis (Problem 8.3).

At large distances from the grid, the turbulence decays much faster than indicated in the preceding analysis. The final period of decay, as this is called, cannot be understood with simple dimensional estimates, since the asymptotic behavior of the largest eddies (much larger than ℓ, but with little energy) is very complicated. The largest eddies are the ones that survive in the end; spectral analysis (Problem 8.4) is needed to resolve their decay.

Pure shear flow The energy budget of steady pure shear flow is also of interest if only because it relates to the situation discussed in Chapter 2. We adopt the notation used in that chapter: $U_1 = U_1(x_2)$, $U_2 = U_3 = 0$, $U_j\, \partial/\partial x_j = 0$, $\partial/\partial x_1 = \partial/\partial x_3 = 0$. In this flow, the only nonzero component of $\partial U_i/\partial x_j$ is $\partial U_1/\partial x_2$, so that the only nonzero components of S_{ij} are S_{12} and S_{21}, both of which are equal to $\frac{1}{2}\partial U_1/\partial x_2$. The turbulence production then is $-\overline{u_1 u_2}\, S_{12} - \overline{u_2 u_1}\, S_{21} = -\overline{u_1 u_2}\, \partial U_1/\partial x_2$. If the Reynolds number is large, the energy budget (3.2.1) reads

$$0 = -\overline{u_1 u_2}\frac{\partial U_1}{\partial x_2} - \frac{\partial}{\partial x_2}\left(\frac{1}{\rho}\,\overline{u_2 p} + \tfrac{1}{2}\,\overline{u_i u_i u_2}\right) - \epsilon. \tag{3.2.38}$$

All of these terms are of order u^3/ℓ; the viscous-transport term, which is much smaller (3.2.26), has been neglected.

The main features of the energy budget have already been discussed. In this simple geometry, it is worthwhile to compare (3.2.38) with the equations for the kinetic energy of the three velocity components individually. These equations are obtained in the same way as the equation for $\frac{1}{2}\overline{u_i u_i}$; if viscous transport is neglected and if the Reynolds number is so large that the dissipative structure can be assumed to be isotropic, the equations for $\frac{1}{2}\overline{u_1^2}$, $\frac{1}{2}\overline{u_2^2}$, and $\frac{1}{2}\overline{u_3^2}$ are, respectively,

$$0 = -\overline{u_1 u_2}\frac{\partial U_1}{\partial x_2} + \frac{1}{\rho}\,\overline{p\frac{\partial u_1}{\partial x_1}} - \frac{\partial}{\partial x_2}\,\overline{(\tfrac{1}{2}u_1^2 u_2)} \qquad - \tfrac{1}{3}\epsilon, \tag{3.2.39}$$

$$0 = \qquad 0 \qquad + \frac{1}{\rho}\,\overline{p\frac{\partial u_2}{\partial x_2}} - \frac{\partial}{\partial x_2}\,\overline{(p/\rho + \tfrac{1}{2}u_2^2)u_2} - \tfrac{1}{3}\epsilon, \tag{3.2.40}$$

$$0 = \qquad 0 \qquad + \frac{1}{\rho}\,\overline{p\frac{\partial u_3}{\partial x_3}} - \frac{\partial}{\partial x_2}\,\overline{(\tfrac{1}{2}u_3^2 u_2)} \qquad - \tfrac{1}{3}\epsilon. \tag{3.2.41}$$

The sum of these three equations equals (3.2.38), as it should. Note that because of incompressibility

$$\overline{p\frac{\partial u_1}{\partial x_1}} + \overline{p\frac{\partial u_2}{\partial x_2}} + \overline{p\frac{\partial u_3}{\partial x_3}} = \overline{p\frac{\partial u_i}{\partial x_i}} = 0. \tag{3.2.42}$$

Comparing (3.2.38) with (3.2.39–3.2.41), we see that the entire production of kinetic energy occurs in the equation for $\frac{1}{2}\overline{u_1^2}$ (3.2.39) and that the equations for $\frac{1}{2}\overline{u_2^2}$ and $\frac{1}{2}\overline{u_3^2}$ have no production terms. The u_2 and u_3 components must thus receive their energy from the pressure interaction terms listed in (3.2.42). The transport terms in (3.2.39–3.2.41) could import energy from elsewhere, but that would not explain how the u_2 and u_3 components can have energy at all: $\frac{1}{2}\overline{u_2^2}$ and $\frac{1}{2}\overline{u_3^2}$ have to be generated somehow. Because the sum of the pressure terms is equal to zero, by (3.2.42), the pressure terms exchange energy between components, without changing the total amount of energy. Also, if $\frac{1}{2}\overline{u_2^2}$ and $\frac{1}{2}\overline{u_3^2}$ are to maintain themselves, notwithstanding dissipative losses, $\overline{p\, \partial u_2/\partial x_2}$ and $\overline{p\, \partial u_3/\partial x_3}$ must be positive, so that $\overline{p\, \partial u_1/\partial x_1}$ must be negative. This, of course, can occur only if the turbulence is not isotropic. Indeed, in most shear flows $\frac{1}{2}\overline{u_1^2}$ is roughly twice as large as $\frac{1}{2}\overline{u_2^2}$ and $\frac{1}{2}\overline{u_3^2}$. In summary: the u_1 component has more energy than the other components because it receives all of the production of kinetic energy; the transfer of energy to the other components is performed by nonlinear pressure-velocity interactions.

3.3 Vorticity dynamics

All turbulent flows are characterized by high levels of fluctuating vorticity. This is the feature that distinguishes turbulence from other random fluid motions like ocean waves and atmospheric gravity waves. Therefore, we have to make a careful study of the role of vorticity fluctuations in the dynamics of turbulence.

Recalling from Section 2.3 that Reynolds stresses may be associated with eddies whose vorticity is roughly aligned with the mean strain rate, we first show that the turbulence terms in the equations for the mean flow are associated with transport and stretching of vorticity. We then turn to a study of the vorticity equation. We shall find that vorticity can indeed be amplified by line stretching due to the strain rate. The equation for the mean vorticity in a turbulent shear flow also will be explored; the interactions between velocity and vorticity fluctuations again include both transport and stretching.

Because the scale of eddies that are stretched by a strain rate decreases, the energy transfer from large eddies to small eddies may be considered in terms of vortex stretching. We shall study the mean-square vorticity fluctuations

$\overline{\omega_i \omega_i}$ in detail. The ultimate energy transfer, the dissipation of kinetic energy into heat, will turn out to be approximately equal to $\nu\, \overline{\omega_i \omega_i}$ if the Reynolds number is large. In summary, this section attempts to explain what we mean when we say that turbulence is rotational and dissipative.

Vorticity vector and rotation tensor The vorticity is the curl of the velocity vector:

$$\tilde{\omega}_i = \epsilon_{ijk} \frac{\partial \tilde{u}_k}{\partial x_j}. \tag{3.3.1}$$

This relation shows that $\tilde{\omega}_i$ is related to the deformation rate $\partial \tilde{u}_i / \partial x_j$. The deformation rate can be split up into a symmetric and a skew-symmetric part:

$$\frac{\partial \tilde{u}_i}{\partial x_j} = \tilde{s}_{ij} + \tilde{r}_{ij}. \tag{3.3.2}$$

The strain rate $\tilde{s}_{ij}$ has been introduced before. The skew-symmetric tensor $\tilde{r}_{ij}$ is called the rotation tensor; it is defined by

$$\tilde{r}_{ij} \equiv \frac{1}{2}\left(\frac{\partial \tilde{u}_i}{\partial x_j} - \frac{\partial \tilde{u}_j}{\partial x_i}\right). \tag{3.3.3}$$

Since the alternating tensor ϵ_{ijk} in the definition of $\tilde{\omega}_i$ is a skew-symmetric tensor (it is +1 if i, j, k are in cyclic order, −1 if i, j, k are in anticyclic order, 0 if any two of i, j, k are equal), the vorticity vector is related only to the skew-symmetric part of $\partial \tilde{u}_i / \partial x_j$:

$$\tilde{\omega}_i = \epsilon_{ijk} \tilde{r}_{kj}. \tag{3.3.4}$$

Conversely, with some tensor algebra it is found that

$$\tilde{r}_{ij} = -\tfrac{1}{2}\, \epsilon_{ijk}\, \tilde{\omega}_k. \tag{3.3.5}$$

The one-to-one relation between the vorticity vector and the rotation tensor is due to the fact that $\tilde{r}_{ij}$ has only three independent components which, if so desired, may be represented as the components of the axial vector $\tilde{\omega}_i$.

Vortex terms in the equations of motion The vorticity equation is obtained by taking the curl of the Navier-Stokes equations. Before we perform this operation, we want to look at the way in which vorticity appears in the

Navier-Stokes equations themselves. If we treat the inertia term $\tilde{u}_j\, \partial\tilde{u}_i/\partial x_j$ as a gradient of a stress, we may write

$$\frac{\partial\tilde{u}_i}{\partial t} = -\frac{1}{\rho}\frac{\partial\tilde{p}}{\partial x_i} - \frac{\partial}{\partial x_j}(\tilde{u}_i\tilde{u}_j) + \nu\,\frac{\partial^2\tilde{u}_i}{\partial x_j\partial x_j}. \tag{3.3.6}$$

Here, the continuity equation $\partial\tilde{u}_j/\partial x_j = 0$ has been used. This particular way of writing the Navier-Stokes equations serves as a reminder that the Reynolds stress is the contribution of the velocity fluctuations to the convective terms in the equation of motion.

The convective stress term may be decomposed as follows:

$$\begin{aligned}\frac{\partial}{\partial x_j}(\tilde{u}_i\tilde{u}_j) &= \tilde{u}_j\frac{\partial\tilde{u}_i}{\partial x_j} = \tilde{u}_j\left(\frac{\partial\tilde{u}_i}{\partial x_j} - \frac{\partial\tilde{u}_j}{\partial x_i}\right) + \tilde{u}_j\frac{\partial\tilde{u}_j}{\partial x_i}\\ &= 2\tilde{u}_j\tilde{r}_{ij} + \frac{\partial}{\partial x_i}(\tfrac{1}{2}\,\tilde{u}_j\tilde{u}_j)\\ &= -\epsilon_{ijk}\,\tilde{u}_j\,\tilde{\omega}_k + \frac{\partial}{\partial x_i}(\tfrac{1}{2}\,\tilde{u}_j\tilde{u}_j).\end{aligned} \tag{3.3.7}$$

The viscous term may be expressed in terms of vorticity by putting

$$\begin{aligned}\nu\frac{\partial^2\tilde{u}_i}{\partial x_j\partial x_j} &= \nu\frac{\partial}{\partial x_j}\left(\frac{\partial\tilde{u}_i}{\partial x_j} - \frac{\partial\tilde{u}_j}{\partial x_i}\right) + \nu\frac{\partial}{\partial x_i}\left(\frac{\partial\tilde{u}_j}{\partial x_j}\right)\\ &= 2\nu\frac{\partial}{\partial x_j}\tilde{r}_{ij} + 0\\ &= -\nu\,\epsilon_{ijk}\frac{\partial\tilde{\omega}_k}{\partial x_j}.\end{aligned} \tag{3.3.8}$$

The continuity equation $\partial u_j/\partial x_j = 0$ was used to remove the second term.

If (3.3.7) and (3.3.8) are substituted into (3.3.6), there results

$$\frac{\partial\tilde{u}_i}{\partial t} = -\frac{\partial}{\partial x_i}\left(\frac{\tilde{p}}{\rho} + \tfrac{1}{2}\,\tilde{u}_j\tilde{u}_j\right) + \epsilon_{ijk}\,\tilde{u}_j\tilde{\omega}_k - \nu\epsilon_{ijk}\frac{\partial\tilde{\omega}_k}{\partial x_j}. \tag{3.3.9}$$

In irrotational flow, $\tilde{\omega}_k = 0$ by definition, so that the viscous term and the vorticity part of the inertia term vanish. The inertia term then reduces to the gradient of the dynamic pressure $\frac{1}{2}\rho\,\tilde{u}_j\tilde{u}_j$ and (3.3.9) reduces to the Bernoulli equation. In turbulent flow, of course, neither of these conditions is satisfied.

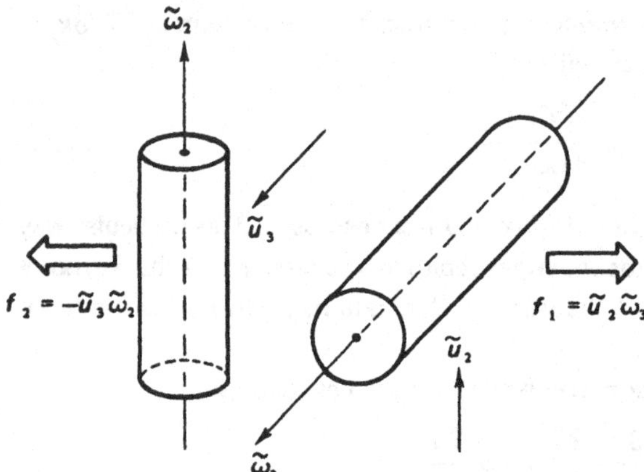

Figure 3.4. The vorticity-velocity cross product generates the body forces (per unit mass) f_1 and f_2.

The cross-product term $\epsilon_{ijk}\,\tilde{u}_j\,\tilde{\omega}_k$ is crucial to turbulence theory. It is analogous to the Coriolis force $2\,\epsilon_{ijk}\,\tilde{u}_j\,\tilde{\Omega}_k$ that would appear in the equation of motion if the coordinate system were rotating with an angular velocity $\tilde{\Omega}_k$ (the factor 2 is absent from the vorticity term because $\tilde{\omega}_k$ is twice the angular velocity of a small fluid element). The vortex term is also related to the lift force (Magnus effect) experienced by a vortex line exposed to a velocity $\tilde{u}_j$. A graphical interpretation of the "vortex force" may be helpful. In the equation for $\tilde{u}_1$, the term $\epsilon_{ijk}\,\tilde{u}_j\,\tilde{\omega}_k$ becomes $\tilde{u}_2\,\tilde{\omega}_3 - \tilde{u}_3\,\tilde{\omega}_2$. Figure 3.4 illustrates the geometry involved.

Reynolds stress and vorticity In turbulent flow, cross-product forces arise both from $U_j\,\partial U_i/\partial x_j$ and from $\partial(\overline{u_iu_j})/\partial x_j$. The instantaneous vorticity $\tilde{\omega}_i$ is decomposed into a mean vorticity Ω_i and vorticity fluctuations ω_i:

$$\tilde{\omega}_i = \Omega_i + \omega_i, \quad \overline{\omega_i} = 0. \tag{3.3.10}$$

If we assume that the flow is steady in the mean, so that we can use time averages, the equation for the mean velocity U_i may be written as

$$0 = -\frac{\partial}{\partial x_i}\left(\frac{P}{\rho} + \tfrac{1}{2}\,U_jU_j + \tfrac{1}{2}\,\overline{u_ju_j}\right) + \epsilon_{ijk}\,(U_j\Omega_k + \overline{u_j\omega_k}) + \nu\,\frac{\partial^2 U_i}{\partial x_j \partial x_j}. \tag{3.3.11}$$

Clearly, Reynolds-stress gradients contain both a dynamic-pressure gradient and an interaction term between the vorticity fluctuations and the velocity fluctuations. In many turbulent flows the contribution of the turbulence to the dynamic pressure is insignificant because $\frac{1}{2}\overline{u_j u_j} \ll \frac{1}{2}U_j U_j$. The dynamic significance of the Reynolds stress is then associated mainly with the interaction between velocity and vorticity. For a closer look at this interaction, let us consider a two-dimensional mean flow in which $U_1 \gg U_2$, $U_3 = 0$, and in which downstream derivatives of mean quantities are small compared to cross-stream derivatives ($\partial/\partial x_1 \ll \partial/\partial x_2$). This corresponds to most boundary-layer and wake flows (see Chapters 4 and 5). Under these conditions, the only nonzero component of Ω_i is $\Omega_3 = \partial U_2/\partial x_1 - \partial U_1/\partial x_2$. Because $U_2 \ll U_1$ and $\partial/\partial x_1 \ll \partial/\partial x_2$, the vorticity component Ω_3 is approximately equal to $-\partial U_1/\partial x_2$.

In the equation for U_1 the vorticity cross-product terms associated with the mean flow are $U_2\Omega_3$ and $-U_3\Omega_2$. The first of these is equal to $-U_2\,\partial U_1/\partial x_2 + U_2\,\partial U_2/\partial x_1$, the second is zero because $U_3 = 0$, $\Omega_2 = 0$. Also, $-\partial(\frac{1}{2}U_j U_j)/\partial x_1$ is equal to $-U_1\,\partial U_1/\partial x_1 - U_2\,\partial U_2/\partial x_1$ in this flow; the small term $U_2\,\partial U_2/\partial x_1$ cancels the same term generated by $U_2\Omega_3$. If we neglect the viscous term and the contribution of the turbulence to the dynamic pressure, the equation for U_1 may be written as

$$U_1\frac{\partial U_1}{\partial x_1} + U_2\frac{\partial U_1}{\partial x_2} = -\frac{1}{\rho}\frac{\partial P}{\partial x_1} + \overline{u_2\omega_3} - \overline{u_3\omega_2}. \tag{3.3.12}$$

Comparing (2.1.23) and (3.3.12) and observing that $\partial\overline{u_1^2}/\partial x_1 \ll \partial(\overline{u_1u_2})/\partial x_2$, we find that the vortex terms represent the cross-stream derivative of the Reynolds shear stress $-\overline{u_1u_2}$:

$$\frac{\partial}{\partial x_2}(-\overline{u_1u_2}) = \overline{u_2\omega_3} - \overline{u_3\omega_2}. \tag{3.3.13}$$

This result can be obtained also by substituting $\omega_3 = \partial u_2/\partial x_1 - \partial u_1/\partial x_2$ and $\omega_2 = \partial u_1/\partial x_3 - \partial u_3/\partial x_1$ into $\overline{u_2\omega_3} - \overline{u_3\omega_2}$ and neglecting all terms that can be written as gradients of dynamic pressures.

Some understanding of the turbulent vorticity terms in (3.3.13) may be obtained by employing the estimate

$$-\overline{u_1u_2} \sim \mathscr{u}\ell\,\partial U_1/\partial x_2. \tag{3.3.14}$$

If $\mathscr{u}$ is approximately independent of x_2 (this is true for many turbulent shear flows), the Reynolds-stress gradient becomes

$$\frac{\partial}{\partial x_2}(-\overline{u_1 u_2}) \sim u\ell \frac{\partial^2 U_1}{\partial x_2^2} + u \frac{\partial \ell}{\partial x_2} \frac{\partial U_1}{\partial x_2}. \tag{3.3.15}$$

Of course, (3.3.15) needs to be viewed with considerable reservation because (3.3.14) is a scaling law, not an equation. Because $\partial U_1/\partial x_2 = -\Omega_3$ approximately, (3.3.15) may be written as

$$\frac{\partial}{\partial x_2}(-\overline{u_1 u_2}) \sim -u\ell \frac{\partial \Omega_3}{\partial x_2} - u\, \Omega_3 \frac{\partial \ell}{\partial x_2}. \tag{3.3.16}$$

Let us now consider $\overline{u_2 \omega_3}$ and $\overline{u_3 \omega_2}$. In the flow treated here, the only nonzero component of Ω_i is Ω_3. If vorticity can be transported in the x_2 direction by u_2 in the same way as momentum is transported, we should be able to write

$$\overline{u_2 \omega_3} \sim -u\ell\, \partial \Omega_3/\partial x_2. \tag{3.3.17}$$

The adoption of this expression constitutes a mixing-length theory of vorticity transfer (Taylor, 1932). Of course, (3.3.17) does not need to be the same as the first term on the right-hand side of (3.3.16), because the numerical coefficients involved, which have been omitted from (3.3.16) and (3.3.17), are not necessarily equal. However, it is clear that the other term, $\overline{u_3 \omega_2}$, cannot be represented by an expression like (3.3.17) because $\Omega_2 = 0$. From a comparison of (3.3.13) and (3.3.16) we conclude that the nature of $\overline{u_3 \omega_2}$ is associated with a change-of-scale effect:

$$\overline{u_3 \omega_2} \sim u \Omega_3 \frac{\partial \ell}{\partial x_2}. \tag{3.3.18}$$

The term $\overline{u_3 \omega_2}$ may be called a vortex-stretching force, since it is associated with the change of size of eddies with vorticity of order Ω_3 (see also the discussion following (3.3.35)).

The relative contributions of $\overline{u_2 \omega_3}$ and $\overline{u_3 \omega_2}$ to $\partial(-\overline{u_1 u_2})/\partial x_2$ apparently depend on the kind of flow considered. If the length scale ℓ is approximately constant across the flow, the vortex-stretching force (3.3.18) should be negligible; the Reynolds-stress gradient may then be interpreted as vorticity transport, which should scale according to (3.3.17). This may explain why vorticity transport theory has had some success in the description of turbulent wakes and jets: in those flows, the length scale is roughly constant in the cross-stream direction.

If the length scale ℓ changes in the x_2 direction, vorticity transport theory is inadequate. A case in point is the surface layer with constant stress ($-\overline{u_1 u_2} = u_*{}^2$). In this flow,

$$-\overline{u_1 u_2} = u_*^2 = \kappa x_2 u_* \frac{\partial U_1}{\partial x_2}, \tag{3.3.19}$$

so that

$$\frac{\partial}{\partial x_2}(-\overline{u_1 u_2}) = 0 = \kappa x_2 u_* \frac{\partial^2 U_1}{\partial x_2^2} + \kappa u_* \frac{\partial U_1}{\partial x_2}. \tag{3.3.20}$$

According to (3.3.19), $\partial U_1/\partial x_2 = u_*/\kappa x_2$, so that $\partial^2 U_1/\partial x_2^2 < 0$. The vorticity-transport term $\kappa x_2 u_*\ \partial^2 U_1/\partial x_2^2 = -\kappa x_2 u_*\ \partial \Omega_3/\partial x_2$ thus is a deceleration. The deceleration of this flow is avoided because the vortex-stretching force $\kappa u_*\ \partial U_1/\partial x_2 = -\kappa u_* \Omega_3$ balances the vorticity-transport force.

One final observation needs to be made. If the local length scale of the mean-flow field is comparable to the eddy size ℓ, the order of magnitude of $\overline{u_2 \omega_3}$ and $\overline{u_3 \omega_2}$ is u^2/ℓ. Now, as we see later in this section, ω_i is of order u/λ, so that the correlation coefficient between ω_i and u_j is of order λ/ℓ. This is similar to the correlation between u_i and s_{ij} which was discussed earlier; the correlation is poor because most contributions to ω_i are made at high frequencies while most of u_j is associated with low frequencies.

The vorticity equation Let us return to the vorticity equation. This equation is obtained by applying the operator "curl" ($\epsilon_{pqi}\ \partial/\partial x_q$) to the Navier-Stokes equation (3.3.9):

$$\frac{\partial \tilde{\omega}_p}{\partial t} = -\epsilon_{pqi} \frac{\partial^2}{\partial x_i \partial x_q}\left(\frac{\tilde{p}}{\rho} + \frac{1}{2}\tilde{u}_j \tilde{u}_j\right)$$

$$+ (\delta_{pj}\delta_{qk} - \delta_{pk}\delta_{qj})\left(\frac{\partial}{\partial x_q}\tilde{u}_j \tilde{\omega}_k - \nu \frac{\partial^2 \tilde{\omega}_k}{\partial x_q \partial x_j}\right). \tag{3.3.21}$$

Here, the tensor identity $\epsilon_{pqi}\epsilon_{ijk} = \delta_{pj}\delta_{qk} - \delta_{pk}\delta_{qj}$ has been used. The pressure term in (3.3.21) is zero because it involves the product of the skew-symmetric tensor ϵ_{pqi} and the symmetric tensor operator $\partial^2/\partial x_i \partial x_q$. Accounting for all of the Kronecker deltas in (3.3.21), we obtain

$$\frac{\partial \tilde{\omega}_p}{\partial t} = \tilde{\omega}_k \frac{\partial \tilde{u}_p}{\partial x_k} - \tilde{u}_k \frac{\partial \tilde{\omega}_p}{\partial x_k} - \nu \frac{\partial}{\partial x_p}\left(\frac{\partial \tilde{\omega}_k}{\partial x_k}\right) + \nu \frac{\partial^2 \tilde{\omega}_p}{\partial x_k \partial x_k}. \tag{3.3.22}$$

The first of the viscous terms in (3.3.22) is zero because vorticity has zero divergence (the divergence of the curl of a vector is zero):

$$\frac{\partial \tilde{\omega}_k}{\partial x_k} = \epsilon_{ijk} \frac{\partial^2 \tilde{u}_j}{\partial x_i \partial x_k} = 0. \tag{3.3.23}$$

The final form of the vorticity equation is (changing p to i and the dummy index k to j for convenience)

$$\frac{\partial \tilde{\omega}_i}{\partial t} + \tilde{u}_j \frac{\partial \tilde{\omega}_i}{\partial x_j} = \tilde{\omega}_j \frac{\partial \tilde{u}_i}{\partial x_j} + \nu \frac{\partial^2 \tilde{\omega}_i}{\partial x_j \partial x_j}. \tag{3.3.24}$$

In keeping with the form of the Navier-Stokes equations introduced in Chapter 2, (3.3.24) is valid for an incompressible constant-property fluid. Before we interpret the first term on the right-hand side of (3.3.24), we want to show that the skew-symmetric part $\tilde{r}_{ij}$ of $\partial \tilde{u}_i/\partial x_j$ does not contribute to it. For this purpose, $\partial \tilde{u}_i/\partial x_j$ is split up into $\tilde{r}_{ij}$ and $\tilde{s}_{ij}$, such that

$$\tilde{\omega}_j \frac{\partial \tilde{u}_i}{\partial x_j} = \tilde{\omega}_j \tilde{s}_{ij} + \tilde{\omega}_j \tilde{r}_{ij}. \tag{3.3.25}$$

Because of the definition of $\tilde{r}_{ij}$, the second term in (3.3.25) becomes

$$\tilde{\omega}_j \tilde{r}_{ij} = -\tfrac{1}{2}\,\epsilon_{ijk}\,\tilde{\omega}_j \tilde{\omega}_k. \tag{3.3.26}$$

Since j and k are dummy indices they may be interchanged to yield

$$-\tfrac{1}{2}\,\epsilon_{ijk}\,\tilde{\omega}_j \tilde{\omega}_k = -\tfrac{1}{2}\,\epsilon_{ikj}\,\tilde{\omega}_j \tilde{\omega}_k. \tag{3.3.27}$$

Again interchanging the indices j and k in ϵ_{ikj}, we obtain a change in sign because ϵ_{ijk} is skew-symmetric. Hence, we find

$$-\tfrac{1}{2}\,\epsilon_{ijk}\,\tilde{\omega}_j \tilde{\omega}_k = \tfrac{1}{2}\,\epsilon_{ijk}\,\tilde{\omega}_j \tilde{\omega}_k. \tag{3.3.28}$$

This can be true only if this term is zero. Consequently, only the term in $\tilde{s}_{ij}$ survives in (3.3.25). The vorticity equation then may be written as

$$\frac{\partial \tilde{\omega}_i}{\partial t} + \tilde{u}_j \frac{\partial \tilde{\omega}_i}{\partial x_j} = \tilde{\omega}_j \tilde{s}_{ij} + \nu \frac{\partial^2 \tilde{\omega}_i}{\partial x_j \partial x_j}. \tag{3.3.29}$$

The term $\tilde{\omega}_j \tilde{s}_{ij}$ represents amplification and rotation of the vorticity vector by the strain rate. In the context of this section, the turning of vortex axes by the strain rate is of minor importance; we shall concentrate on the components of $\tilde{\omega}_j \tilde{s}_{ij}$ that represent vortex stretching.

Vorticity apparently can be amplified by stretching of present vorticity by the strain rate $\tilde{s}_{ij}$. On the other hand, vorticity is decreased in an environment where squeezing ($\tilde{s}_{ij} < 0$) occurs.

This "source" or "sink" for vorticity is the most interesting term of the vorticity equation. It is essential to recognize that the term does not occur in two-dimensional flow. Suppose a flow is entirely in the x_1, x_2 plane. Then $\tilde{\omega}_1$ and $\tilde{\omega}_2$ are zero, so that the only nonzero vorticity component is $\tilde{\omega}_3$. The vortex-stretching term then becomes $\tilde{\omega}_3 \tilde{s}_{i3}$. However, in a two-dimensional flow only $\tilde{s}_{12}$ $(=\tilde{s}_{21})$, $\tilde{s}_{11}$, and $\tilde{s}_{22}$ can be different from zero. A two-dimensional flow cannot turn or stretch the vorticity vector.

A simple illustration of vortex stretching is the accelerated flow in a wind-tunnel contraction. Here (Figure 3.5) $\tilde{s}_{11}$ is positive, so that $\tilde{s}_{22}$ and $\tilde{s}_{33}$ must be negative to satisfy the continuity equation ($\tilde{s}_{ii} = 0$). In this kind of flow, $\tilde{\omega}_1$ is increased by vortex stretching, while $\tilde{\omega}_2$ and $\tilde{\omega}_3$ are attenuated.

The change of vorticity by vortex stretching is a consequence of the conservation of angular momentum. The angular momentum of a material

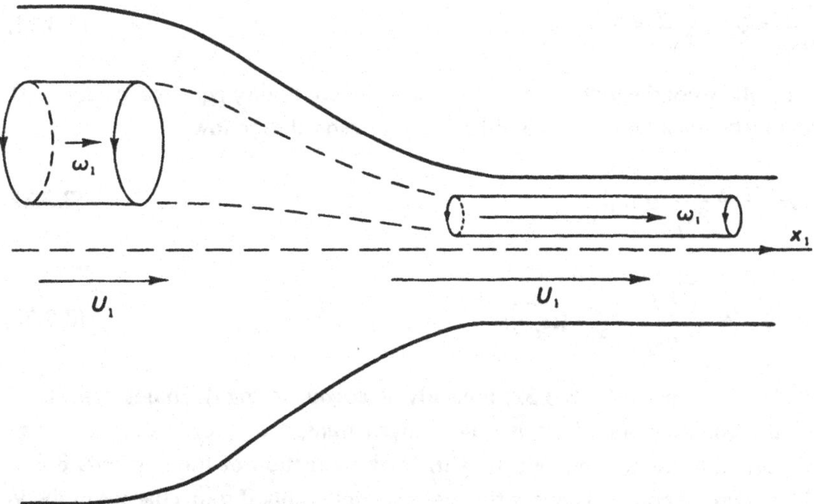

Figure 3.5. Vortex stretching in a wind-tunnel contraction. As the flow speeds up from left to right, the vorticity component ω_1 is amplified because angular momentum has to be conserved.

volume element would remain constant if viscous effects were absent; if the fluid element is stretched so that its cross-sectional area and moment of inertia become smaller, the component of the angular velocity in the direction of the stretching must increase in order to conserve angular momentum. Vortex stretching always involves a change of length scale, as Figure 3.5 illustrates. For a full account of vorticity kinematics, readers should consult general texts in fluid dynamics (for example, Batchelor, 1967).

Vorticity in turbulent flows In turbulent flow, the vorticity is decomposed into a mean vorticity Ω_i and vorticity fluctuations ω_i according to (3.3.10). After substituting (3.3.10) and the corresponding Reynolds decompositions for $\tilde{u}_i$ and $\tilde{s}_{ij}$ into (3.3.29) and taking the average of all terms in the equation, we obtain

$$U_j \frac{\partial \Omega_i}{\partial x_j} = -\overline{u_j \frac{\partial \omega_i}{\partial x_j}} + \overline{\omega_j s_{ij}} + \Omega_j S_{ij} + \nu \frac{\partial^2 \Omega_i}{\partial x_j \partial x_j} . \tag{3.3.30}$$

The mean flow has been assumed to be steady.

From (3.3.10) and (3.3.23) we conclude that both the mean vorticity and the fluctuating vorticity are solenoidal (that is, divergenceless):

$$\frac{\partial \Omega_i}{\partial x_i} = 0, \quad \frac{\partial \omega_i}{\partial x_i} = 0. \tag{3.3.31}$$

With the second equation in (3.3.31) and the continuity equation $\partial u_i/\partial x_i = 0$, the turbulence terms in (3.3.30) can be rearranged as follows:

$$\overline{u_j \frac{\partial \omega_i}{\partial x_j}} = \frac{\partial}{\partial x_j} (\overline{u_j \omega_i}), \tag{3.3.32}$$

$$\overline{\omega_j s_{ij}} = \overline{\omega_j \frac{\partial u_i}{\partial x_j}} = \frac{\partial}{\partial x_j} (\overline{\omega_j u_i}). \tag{3.3.33}$$

The term given in (3.3.32) is clearly analogous to the Reynolds-stress term in the equation for U_i; it is due to mean transport of ω_i through its interaction with fluctuating velocities u_j in the direction of the gradients $\partial/\partial x_j$. This term, of course, changes the mean vorticity only if $\overline{u_j \omega_i}$ changes in the x_j direction. Properly speaking, (3.3.32) is a transport "divergence."

The term given in (3.3.33) is the gain (or loss) of mean vorticity caused by the stretching and rotation of fluctuating vorticity components by fluctuating strain rates.

Two-dimensional mean flow In a flow with $U_3 = 0, \Omega_1 = \Omega_2 = 0, \partial/\partial x_3 = 0$, and $\partial/\partial x_1 << \partial/\partial x_2$ (whose equation of motion was discussed earlier), the major turbulence terms in the equation for Ω_3 are

$$\overline{u_j \frac{\partial \omega_3}{\partial x_j}} \cong \frac{\partial}{\partial x_2} (\overline{u_2 \omega_3}), \tag{3.3.34}$$

$$\overline{\omega_j s_{3j}} \cong \frac{\partial}{\partial x_2} (\overline{u_3 \omega_2}). \tag{3.3.35}$$

The products $\overline{u_2 \omega_3}$ and $\overline{u_3 \omega_2}$ are related to the Reynolds-stress gradient by (3.3.13); $\overline{u_2 \omega_3}$ was interpreted as a body force arising from transport of ω_3 by u_2 in a field with a mean gradient $\partial \Omega_3/\partial x_2$, whereas $\overline{u_3 \omega_2}$ was interpreted as a body force associated with the change of size of eddies in a flow field with a varying length scale. The vortex-stretching nature of $\overline{u_3 \omega_2}$ is confirmed by (3.3.35). The cross-stream gradients of these body forces are sources or sinks for mean vorticity. In a surface layer with constant stress, the mean vorticity Ω_3 is constant along streamlines; from (3.3.17, 3.3.34) and (3.3.18, 3.3.35) we may conclude that Ω_3 is maintained because the gain of mean vorticity due to a net transport surplus is balanced by the loss of mean vorticity due to the transfer of vorticity to the turbulence by vortex stretching. A more comprehensive interpretation of (3.3.34) and (3.3.35) becomes extremely involved. Even if (3.3.17) and (3.3.18) are adopted as crude models of $\overline{u_2 \omega_3}$ and $\overline{u_3 \omega_2}$, respectively, it would be presumptuous to differentiate these equations in order to obtain models for (3.3.34, 3.3.35), because that would amount to differentiating the Reynolds-stress scaling law (3.3.14) twice. In vorticity-transfer theory, of course, the term $\overline{\omega_j s_{ij}}$ is ignored and the transport term (3.3.34) is scaled on basis of (3.3.17).

In the discussion following (3.3.20) we found that $\overline{u_2 \omega_3}$ and $\overline{u_3 \omega_2}$ both are of order u^2/ℓ. The cross-stream gradients $(\partial/\partial x_2)$ should scale with the local length scale of the mean flow, which is comparable to ℓ, in flows without multiple scales. Therefore, (3.3.34) and (3.3.35) are of order u^2/ℓ^2.

The dynamics of $\Omega_i\Omega_i$ An equation for the square of the mean vorticity is needed because the interaction between mean and fluctuating vorticities can be studied only in terms of $\Omega_i\Omega_i$ and $\overline{\omega_i\omega_i}$. Multiplying (3.3.30) by Ω_i and rearranging terms, we find

$$U_j \frac{\partial}{\partial x_j}\left(\tfrac{1}{2}\,\Omega_i\Omega_i\right) = -\frac{\partial}{\partial x_j}\left(\Omega_i\,\overline{\omega_i u_j}\right) + \overline{u_j\omega_i}\,\frac{\partial \Omega_i}{\partial x_j} + \Omega_i\Omega_j S_{ij}$$
$$+\,\Omega_i\,\overline{\omega_j s_{ij}} + \nu\,\frac{\partial^2}{\partial x_j \partial x_j}\left(\tfrac{1}{2}\,\Omega_i\Omega_i\right) - \nu\,\frac{\partial \Omega_i}{\partial x_j}\frac{\partial \Omega_i}{\partial x_j}. \tag{3.3.36}$$

The first term on the right-hand side of (3.3.36) is the transport of $\Omega_i\Omega_i$ by turbulent vorticity-velocity interactions. This term is equivalent to the turbulent transport term of U_iU_i. The second term on the right-hand side of (3.3.36) is like the turbulence-production term in the energy equation. We may call it gradient production of $\overline{\omega_i\omega_i}$, in anticipation of the occurrence of the same term (with opposite sign) in the equation for $\overline{\omega_i\omega_i}$. The third term is stretching or shrinking of mean vorticity by the mean strain rate. The fourth term is amplification or attenuation of $\Omega_i\Omega_i$ caused by the stretching of fluctuating vorticity components by fluctuating strain rates. The fifth term is viscous transport of $\Omega_i\Omega_i$, and the sixth is viscous dissipation of $\Omega_i\Omega_i$.

The mean vorticity Ω_i is of order u/ℓ. Because $\overline{\omega_i u_j} \sim u^2/\ell$ and $\overline{\omega_j s_{ij}} \sim u^2/\ell^2$, the viscous terms in (3.3.36) are of order $(u^3/\ell^3)\,(\nu/u\ell)$, and all the other terms are of order u^3/ℓ^3. Generally speaking, therefore, only the viscous terms can be neglected. In a two-dimensional flow in the x_1, x_2 plane the only nonzero component of Ω_i is Ω_3. At large Reynolds numbers, (3.3.36) may then be approximated by

$$U_j \frac{\partial}{\partial x_j}\left(\tfrac{1}{2}\,\Omega_3\Omega_3\right) = -\frac{\partial}{\partial x_j}\left(\Omega_3\,\overline{\omega_3 u_j}\right) + \overline{u_j\omega_3}\,\frac{\partial \Omega_3}{\partial x_j} + \Omega_3\,\overline{\omega_j s_{3j}}. \tag{3.3.37}$$

The stretching term $\Omega_i\Omega_j S_{ij}$ is zero in two-dimensional flow. If the flow involves no change of length scale, the last term of (3.3.37) may be neglected (see the discussion following (3.3.35)).

The equation for $\overline{\omega_i\omega_i}$ The equation of the mean-square vorticity fluctuations is obtained by a procedure exactly similar to the one followed for the equation of the turbulent kinetic energy. We leave the algebra as an exercise for the reader; the final result is, if the flow is steady in the mean,

$$U_j \frac{\partial}{\partial x_j}(\tfrac{1}{2}\,\overline{\omega_i\omega_i}) = -\overline{u_j\omega_i}\,\frac{\partial \Omega_i}{\partial x_j} - \frac{1}{2}\frac{\partial}{\partial x_j}(\overline{u_j\omega_i\omega_i}) + \overline{\omega_i\omega_j s_{ij}} + \overline{\omega_i\omega_j}\,S_{ij}$$

$$+ \Omega_j\,\overline{\omega_i s_{ij}} + \nu\frac{\partial^2}{\partial x_j \partial x_j}(\tfrac{1}{2}\,\overline{\omega_i\omega_i}) - \nu\,\overline{\frac{\partial \omega_i}{\partial x_j}\frac{\partial \omega_i}{\partial x_j}}. \tag{3.3.38}$$

The first term on the right-hand side of (3.3.38) is the gradient production of $\overline{\omega_i\omega_i}$. This term exchanges vorticity between $\overline{\omega_i\omega_i}$ and $\Omega_i\Omega_i$ in the same way as turbulent energy production ($-\overline{u_iu_j}\,S_{ij}$) exchanges energy between U_iU_i and $\overline{u_iu_i}$.

The second term is the transport of mean-square turbulent vorticity by turbulent velocity fluctuations. This term is analogous to the transport term $\partial(\overline{u_iu_iu_j})/\partial x_j$ in the equation for $\overline{u_iu_i}$.

The third term is the production of mean-square turbulent vorticity by turbulent stretching of turbulent vorticity. We shall soon see that this is one of the dominant terms in the equation for $\overline{\omega_i\omega_i}$.

The fourth term is the production (or removal, as the case may be) of turbulent vorticity caused by the stretching (or squeezing) of vorticity fluctuations by the mean rate of strain S_{ij}.

The fifth term is a mixed production term. It occurs in the equation for $\Omega_i\Omega_i$ with the same sign. Evidently, the stretching of fluctuating vorticity by strain-rate fluctuations produces $\Omega_i\Omega_i$ and $\overline{\omega_i\omega_i}$ at the same rate.

The sixth and seventh terms on the right-hand side of (3.3.38) are viscous transport and dissipation of $\overline{\omega_i\omega_i}$, respectively.

Turbulence is rotational The equation for $\overline{\omega_i\omega_i}$ looks nearly intractable. However, if the Reynolds number is large, a very simple approximate form of (3.3.38) can be obtained, because strain-rate fluctuations are much larger than the mean strain rate and vorticity fluctuations are much larger than the mean vorticity:

$$\overline{s_{ij}s_{ij}} = \mathcal{O}\,(u/\lambda)^2, \quad S_{ij}S_{ij} = \mathcal{O}\,(u/\ell)^2, \tag{3.3.39}$$

$$\overline{\omega_i\omega_i} = \mathcal{O}\,(u/\lambda)^2, \quad \Omega_i\Omega_i = \mathcal{O}\,(u/\ell)^2. \tag{3.3.40}$$

As before, $\mathcal{O}$ stands for "order of magnitude." The estimates for s_{ij}, S_{ij}, and Ω_i were obtained earlier; we have to prove that the first of (3.3.40) is a valid statement before we can proceed. Some tensor algebra applied to the definitions of s_{ij}, r_{ij}, and ω_i yields

$$\overline{\omega_i\omega_i} = 2\,\overline{r_{ij}r_{ij}}, \tag{3.3.41}$$

$$\overline{s_{ij}s_{ij}} - \overline{r_{ij}r_{ij}} = \partial^2(\overline{u_iu_j})/\partial x_i\partial x_j. \tag{3.3.42}$$

Now, $\overline{s_{ij}s_{ij}}$ is of order u^2/λ^2, but the right-hand side of (3.3.42) is of order u^2/ℓ^2. Consequently, at large Reynolds numbers (3.3.42) is approximated by

$$\overline{s_{ij}s_{ij}} \cong \overline{r_{ij}r_{ij}}. \tag{3.3.43}$$

Substituting this into (3.3.41), we find

$$\overline{\omega_i\omega_i} \cong 2\,\overline{s_{ij}s_{ij}}. \tag{3.3.44}$$

From this we conclude that ω_i is of order u/λ, just like s_{ij}. This proves that the first of (3.3.40) is a valid statement if the Reynolds number is large enough. Turbulence indeed is rotational, with large vorticity fluctuations.

The strain-rate fluctuations are associated with viscous dissipation of turbulent energy. We recall that the dissipation rate ϵ is defined by

$$\epsilon \equiv 2\nu\,\overline{s_{ij}s_{ij}}. \tag{3.3.45}$$

Because of (3.3.44), this may be rewritten as

$$\epsilon \cong \nu\,\overline{\omega_i\omega_i}. \tag{3.3.46}$$

This relation shows that dissipation of energy is also associated with vorticity fluctuations. This is a useful result, but it should be kept in mind that a causal relation exists only between the strain-rate fluctuations and the dissipation rate. Indeed, (3.3.44) states merely that in flows with high Reynolds numbers the symmetric and skew-symmetric parts of the deformation-rate tensor have about the same mean-square value.

An approximate vorticity budget The estimates (3.3.39) and (3.3.40) should enable us to simplify the vorticity budget (3.3.38) appreciably. However, many of the terms in (3.3.38) contain mixed products like $\overline{\omega_iu_j}$ and $\overline{\omega_is_{ij}}$, which have to be estimated with care because they are nonzero due to the distorting effect of the mean strain rate S_{ij}. From (3.3.13) we concluded before that

$$\overline{u_i\omega_j} = \mathcal{O}(u^2/\ell); \tag{3.3.47}$$

from (3.3.13) and (3.3.33) we concluded that

$$\overline{\omega_j s_{ij}} = \mathcal{O}(u^2/\ell^2). \tag{3.3.48}$$

We also need the orders of magnitude of $\overline{\omega_i\omega_j}$ and of $\overline{u_j\omega_i\omega_i}$. The diagonal components of $\overline{\omega_i\omega_j}$ are of order u^2/λ^2, but the off-diagonal components are different from zero only in response to a mean strain rate. The mean strain rate S_{ij} is of order u/ℓ so that it can only weakly affect the vorticity structure whose characteristic frequency is u/λ. Therefore, we expect that the effect of S_{ij} should be proportional to the time-scale ratio $(\lambda/u)/(\ell/u) = \lambda/\ell$:

$$\overline{\omega_i\omega_j} = \frac{u^2}{\lambda^2}\left(a\delta_{ij} + b_{ij}\frac{\lambda}{\ell} + \ldots\right). \tag{3.3.49}$$

The coefficients a and b_{ij} should be of order one. The discount for the time-scale ratio λ/ℓ applied here is analogous to the discount needed in $\overline{u_i\omega_j}$. The term $\overline{\omega_i\omega_j}\,S_{ij}$ in (3.3.38) becomes

$$\overline{\omega_i\omega_j}\;S_{ij} = \frac{u^2}{\lambda^2}\left(aS_{ii} + b_{ij}\frac{\lambda}{\ell}S_{ij} + \ldots\right). \tag{3.3.50}$$

Because $S_{ii} = 0$ as a result of incompressibility, and $b_{ij}\,S_{ij} \sim u/\ell$, we find that

$$\overline{\omega_i\omega_j}\;S_{ij} = \mathcal{O}(u^3/\lambda\ell^2). \tag{3.3.51}$$

The transport term $\partial(\overline{u_j\omega_i\omega_i})/\partial x_j$ may be written as

$$\frac{\partial}{\partial x_j}\overline{u_j\omega_i\omega_i} = \overline{u_j\frac{\partial}{\partial x_j}(\omega_i\omega_i)}. \tag{3.3.52}$$

This term does not depend on the mean strain rate but on inhomogeneity in the distribution of mean square vorticity. If we assume that turbulent motion is an effective "mixer" of vorticity, u_j should be well correlated with the gradients of $\overline{\omega_i\omega_i}$, so that

$$\frac{\partial}{\partial x_j}(\overline{u_j\omega_i\omega_i}) = \mathcal{O}\left(\frac{u}{\ell}\cdot\frac{u^2}{\lambda^2}\right) = \mathcal{O}\left(\frac{u^3}{\lambda^2\ell}\right). \tag{3.3.53}$$

With the results obtained above, most of the terms of (3.3.38) can be estimated. We obtain

$$-\overline{u_j\omega_i}\frac{\partial\Omega_i}{\partial x_j} = \mathcal{O}\left(\frac{u^2}{\ell}\cdot\frac{u}{\ell^2}\right) = \mathcal{O}\left(\frac{u^3}{\lambda^3}\cdot\frac{\lambda^3}{\ell^3}\right), \tag{3.3.54}$$

$$\Omega_j\,\overline{\omega_i s_{ij}} = \mathcal{O}\left(\frac{u}{\ell}\cdot\frac{u^2}{\ell^2}\right) = \mathcal{O}\left(\frac{u^3}{\lambda^3}\cdot\frac{\lambda^3}{\ell^3}\right), \tag{3.3.55}$$

$$\nu\frac{\partial^2}{\partial x_j\,\partial x_j}\left(\tfrac{1}{2}\,\overline{\omega_i\omega_i}\right) = \mathcal{O}\left(\frac{\nu}{\ell^2}\cdot\frac{u^2}{\lambda^2}\right) = \mathcal{O}\left(\frac{u^3}{\lambda^3}\cdot\frac{\lambda^3}{\ell^3}\right), \tag{3.3.56}$$

$$\overline{\omega_i\omega_j}\,S_{ij} = \mathcal{O}\left(\frac{u^2}{\lambda\ell}\cdot\frac{u}{\ell}\right) = \mathcal{O}\left(\frac{u^3}{\lambda^3}\cdot\frac{\lambda^2}{\ell^2}\right), \tag{3.3.57}$$

$$\frac{1}{2}\frac{\partial}{\partial x_j}\overline{(u_j\omega_i\omega_i)} = \mathcal{O}\left(\frac{u}{\ell}\cdot\frac{u^2}{\lambda^2}\right) = \mathcal{O}\left(\frac{u^3}{\lambda^3}\cdot\frac{\lambda}{\ell}\right), \tag{3.3.58}$$

$$U_j\frac{\partial}{\partial x_j}\left(\tfrac{1}{2}\,\overline{\omega_i\omega_i}\right) = \mathcal{O}\left(\frac{u}{\ell}\cdot\frac{u^2}{\lambda^2}\right) = \mathcal{O}\left(\frac{u^3}{\lambda^3}\cdot\frac{\lambda}{\ell}\right), \tag{3.3.59}$$

$$\overline{\omega_i\omega_j s_{ij}} = \mathcal{O}\left(\frac{u^2}{\lambda^2}\cdot\frac{u}{\lambda}\right) = \mathcal{O}\left(\frac{u^3}{\lambda^3}\cdot 1\right), \tag{3.3.60}$$

$$\nu\overline{\frac{\partial\omega_i}{\partial x_j}\frac{\partial\omega_i}{\partial x_j}} = ? \tag{3.3.61}$$

In the stretching term (3.3.60), no prorating with λ/ℓ is necessary, because ω_i operates on the same time scale as s_{ij}. The viscous dissipation term (3.3.61) has been left undecided, since we expect dissipation of vorticity to occur mainly at length scales smaller than λ. In the viscous diffusion term (3.3.56), the relation $\ell^2/\lambda^2 \sim u\ell/\nu$ has been used. In the transport term (3.3.59), the operator $U_j\partial/\partial x_j$ has been estimated as u/ℓ; that choice is consistent with the estimates used in the equations for the mean flow and the turbulent kinetic energy (see 3.2.28, 3.2.31, 3.2.32).

The expressions (3.3.54) through (3.3.60) have been arranged in increasing order of magnitude. If the Reynolds number is large, all of the terms (3.3.54) through (3.3.59) are smaller than the turbulent stretching term (3.3.60) by at least a factor of λ/ℓ, which is of order $R_\ell^{-1/2}$. Therefore, at sufficiently high

Reynolds numbers the turbulent vorticity budget (3.3.38) may be approximated as (Taylor, 1938)

$$\overline{\omega_i \omega_j s_{ij}} = \nu \overline{\frac{\partial \omega_i}{\partial x_j} \frac{\partial \omega_i}{\partial x_j}} \,. \tag{3.3.62}$$

The budget of mean-square vorticity fluctuations is thus approximately independent of the structure of the mean flow. Turbulent vorticity fluctuations, unlike turbulent velocity fluctuations, do not need the continued presence of a source term associated with the mean flow field. Of course, in the absence of a source of energy, turbulent vorticity fluctuations will decay, too. Also, the rate of change of $\overline{\omega_i \omega_i}$, as represented by (3.3.59), is small compared to the rate at which turbulent vortex stretching occurs. In Chapter 8 it will be shown that these conclusions lead to the concept of an equilibrium spectrum of turbulence at small scales.

The right-hand side of (3.3.62) is quadratic in $\partial \omega_i / \partial x_j$, so that it is always positive. Hence, the left-hand side is positive, too. This implies that, on the average, there is more turbulent vortex stretching than vortex squeezing: vortex stretching transfers turbulent vorticity (and the energy associated with it) from large-scale fluctuations to small-scale fluctuations. In this way turbulence obtains the broad energy spectrum that is observed experimentally, and in this way the very smallest eddies (which suffer rapid viscous decay) are continually being supplied with new energy. The approximate vorticity budget (3.3.62) is just as essential to understanding turbulence dynamics as the approximate energy budget (3.2.6). The relationship between these two budgets, incidentally, is a close one: viscous dissipation of vorticity prevents vorticity production ($\overline{\omega_i \omega_j s_{ij}}$) from increasing $\overline{\omega_i \omega_i}$ without limit, while viscous dissipation of energy (which is proportional to $\overline{\omega_i \omega_i}$) prevents the energy production ($-\overline{u_i u_j} S_{ij}$) from increasing $\overline{u_i u_i}$ without limit. Vortex stretching makes $\overline{\omega_i \omega_i}$ as large as viscosity will permit; at large Reynolds numbers the mean-square strain-rate fluctuations keep pace, so that the turbulent energy is subject to rapid dissipation.

Two points need to be emphasized. First, in two-dimensional "turbulence" there is no vortex stretching, so that the vorticity budget (3.3.62) is irrelevant in that case. This implies that the spectral energy-transfer concepts developed here do not apply to two-dimensional stochastic flow fields.

Second, vorticity amplification is a result of the kinematics of turbulence. As an example, take a situation in which the principal axes of the instantaneous strain rate are aligned with the coordinate system, so that s_{ij} has only diagonal components (s_{11}, s_{22}, and s_{33}). Let us assume for simplicity that $s_{22} = s_{33}$, so that, by virtue of continuity, $s_{11} = -2s_{22}$. The term $\omega_i\omega_j s_{ij}$ becomes, if we also assume that $\omega_2^2 - \omega_3^2$,

$$\omega_1^2 s_{11} + \omega_2^2 s_{22} + \omega_3^2 s_{33} = s_{11}(\omega_1^2 - \omega_2^2). \tag{3.3.63}$$

If $s_{11} > 0$, ω_1^2 is amplified (see Figure 3.4), but ω_2^2 and ω_3^2 are attenuated because s_{22} and s_{33} are negative. Thus, $\omega_1^2 - \omega_2^2$ tends to become positive if s_{11} is positive. Again, if $s_{11} < 0$, ω_1^2 decreases, but ω_2^2 and ω_3^2 increase, so that $\omega_1^2 - \omega_2^2 < 0$, making the stretching term positive again.

Multiple length scales If the vorticity gradients $\partial\omega_i/\partial x_j$ in (3.3.62) were estimated as u/λ^2, the dissipation term would be smaller than the stretching term. However, λ is not the proper length scale for estimates of ω_i and u is not the proper velocity scale; all we know is that the ratio u/λ is the order of magnitude of ω_i. Clearly, we need a new length scale. Calling it δ, using (3.3.60), and requiring that the two sides of (3.3.62) have the same order of magnitude, we obtain

$$\nu \frac{u^2}{\lambda^2\delta^2} = \mathcal{O}\left(\frac{u^3}{\lambda^3}\right). \tag{3.3.64}$$

The ratio δ/λ becomes

$$\delta/\lambda = \mathcal{O}(\nu/u\lambda)^{1/2} = \mathcal{O}(R_\lambda^{-1/2}). \tag{3.3.65}$$

Comparing this with (3.2.18), we see that δ is proportional to the Kolmogorov microscale η. The Kolmogorov microscale thus has a role in the turbulent vorticity budget which is comparable to the role of the Taylor microscale in the turbulent energy budget. Since vortex stretching is the only known spectral energy-transfer mechanism, η is the smallest length scale possible: the dynamics of $\overline{(\partial\omega_i/\partial x_j)^2}$ would not lead to a length scale smaller than η.

Since the vorticity budget is approximately independent of the structure of the mean flow, vorticity dynamics can be studied more easily in the wave-number (spectral) domain than in the spatial domain. This subject, therefore, is taken up again in Chapter 8.

Stretching of magnetic field lines The dynamics of the fluctuating vorticity is representative of the dynamics of other axial vector fields in turbulent flow. For example, magnetic field lines in a conducting fluid are stretched by fluctuating strain rates much like vortex lines. In incompressible fluids with constant properties, charge equilibrium, negligible displacement currents and radiation, the equation for the magnetic field is the same as the equation for vorticity. If the magnetic energy is small compared to the kinetic energy, the magnetic field is a passive contaminant which does not change the velocity field appreciably. In that case magnetic-field fluctuations are intensified only by fluctuating strain rates, and an approximate equation for the fluctuations h_i of the magnetic field reads, in analogy with (3.3.62) (Saffman, 1963),

$$\overline{h_i h_j s_{ij}} = \gamma_m \overline{\frac{\partial h_i}{\partial x_j} \frac{\partial h_i}{\partial x_j}} . \tag{3.3.66}$$

This equation states that the amplification of $\overline{h_i h_i}$ by strain-rate fluctuations is kept in balance by ohmic dissipation of $\overline{h_i h_i}$ (the right-hand side of (3.3.66) is proportional to j^2/σ, where j is the current density and σ is the electrical conductivity).

If the magnetic diffusivity γ_m differs from ν, the dissipative length scale of the magnetic-field fluctuations is different from the Kolmogorov microscale η. If the dissipative length scale for h_i is called η_m and if the rms value of h_i is called $\mathscr{h}$, we may estimate (3.3.67) by

$$\mathscr{h}^2 \cdot u/\lambda \sim \gamma_m \mathscr{h}^2/\eta_m^2 . \tag{3.3.67}$$

Because the magnetic-field fluctuations are generated by fluctuating strain rates, the correlation coefficient between $h_i h_j$ and s_{ij} should be of order one. Because we are interested only in estimates for scales, we ignore all numerical factors that are of order one. Using the scale relation $u/\lambda \sim (\epsilon/\nu)^{1/2}$ and the definition of η $(\eta = (\nu^3/\epsilon)^{1/4})$, and absorbing numerical coefficients in the definition of η_m, we obtain

$$\eta_m/\eta = (\gamma_m/\nu)^{1/2} . \tag{3.3.68}$$

If the fluid is a very good conductor of electricity so that $\gamma_m/\nu \ll 1$, this implies that the spectrum of $\overline{h_i h_i}$ extends to scales much smaller than η. The possibility of achieving scales smaller than η, even though h_i is a passive contaminant, arises because the strain rate stretches the magnetic field into thin filaments if the magnetic diffusivity is small. The scale-reducing effect of the strain rate proceeds until it is checked by the magnetic diffusivity (see Figure 3.6). This effect is similar to that observed in mixing paint of different colors. The diffusivity of pigment is quite small relative to the kinematic viscosity of paint; it takes long, patient stirring before the filaments of different color have become so thin and so close together that the molecular diffusivity of pigment can homogenize the mixture.

In interstellar gas clouds consisting mainly of ionized hydrogen, γ_m/ν may be as small as 10^{-8}, so that the smallest magnetic eddies are quite small compared to η. In liquid metals and electrolytes, on the other hand, $\gamma_m/\nu \gg 1$, so that the smallest magnetic eddies are large compared to η. If this is the case, the estimate $s_{ij} \sim (\epsilon/\nu)^{1/2}$ has to be revised, because the strain rate at scales comparable to the magnetic microscale η_m is smaller than $(\epsilon/\nu)^{1/2}$ if $\eta_m \gg \eta$. In other words, the viscosity cannot be used as a scaling parameter at scales large compared to η. The only alternative is to construct a strain rate from ϵ and η_m; this yields $s_{ij} \sim \epsilon^{1/3}\, \eta_m^{-2/3}$ (see also Section 8.6). If we use this instead of u/λ in (3.3.67), we obtain

$$\eta_m = (\gamma_m^3/\epsilon)^{1/4}, \quad \eta_m/\eta = (\gamma_m/\nu)^{3/4}. \tag{3.3.69}$$

A note of warning is in order, because there may be no magnetic eddies at all if γ_m/ν is large enough. In mercury, $\gamma_m/\nu = 7 \times 10^6$, so that the *magnetic Reynolds number* $R_m = u\ell/\gamma_m$ is less than one if $R = u\ell/\nu < 7 \times 10^6$. If $R_m < 1$, the generation of magnetic-field fluctuations is prevented by the magnetic diffusivity, much as turbulent motion cannot exist if $R < 1$. In that case,

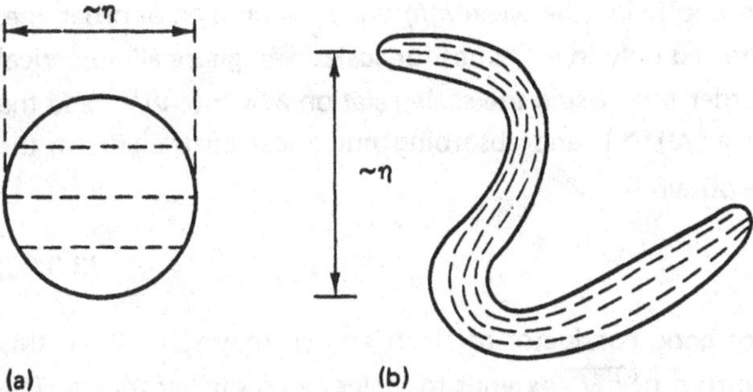

Figure 3.6. A magnetic eddy (a) of scale η is stretched by the strain rate into a thin filament (b). If $\gamma_m \ll \nu$, the gradients in magnetic field intensity can become quite steep (the dashed lines represent surfaces of constant h).

there can be only a mean magnetic field, which affects the velocity turbulence if it is strong enough.

3.4
The dynamics of temperature fluctuations

The equations governing turbulent fluctuations of vectors (such as vorticity) are complicated because vectors interact with a flow field in a variety of ways. However, scalar contaminants (such as temperature) are governed by fairly simple equations, as we have seen in Chapter 2. We shall discuss the dynamics of temperature fluctuations in an incompressible turbulent flow as an example of the dynamics of all other passive scalar contaminants.

The equation governing the dynamics of $\overline{\theta^2}$ in a steady flow is obtained in exactly the same way as the equations for $\overline{u_i u_i}$ and $\overline{\omega_i \omega_i}$. The result is

$$U_j \frac{\partial}{\partial x_j}(\tfrac{1}{2}\overline{\theta^2}) = -\frac{\partial}{\partial x_j}\left[\tfrac{1}{2}\overline{\theta^2 u_j} - \gamma \frac{\partial}{\partial x_j}(\tfrac{1}{2}\overline{\theta^2})\right] - \overline{\theta u_j}\frac{\partial \Theta}{\partial x_j} - \gamma \overline{\frac{\partial \theta}{\partial x_j}\frac{\partial \theta}{\partial x_j}}. \tag{3.4.1}$$

The rate of change of $\overline{\theta^2}$ is thus controlled by turbulent and molecular transport of $\overline{\theta^2}$ (the first two terms on the right-hand side of the equation), by gradient production (which is like the production term of turbulent kinetic energy), and by molecular dissipation (γ is the thermal diffusivity). In a steady homogeneous shear flow, (3.4.1) reduces to

$$-\overline{\theta u_j}\frac{\partial \Theta}{\partial x_j} = \gamma \overline{\frac{\partial \theta}{\partial x_j}\frac{\partial \theta}{\partial x_j}}, \tag{3.4.2}$$

which states that gradient production of $\overline{\theta^2}$ is balanced by the molecular "smearing" of temperature fluctuations.

If there is only one temperature scale and one length scale, $\overline{\theta u_j}$ is of order $\theta' u$ and $\partial\Theta/\partial x_j$ is of order θ'/ℓ (θ' is the rms temperature fluctuation). The left-hand side of (3.4.2) is then of order $\overline{\theta^2} u/\ell$, which is consistent with the idea that spectral transfer of temperature fluctuations toward the dissipative range of eddy sizes should proceed at a rate dictated by the characteristic time of large eddies (ℓ/u) and the amount of $\overline{\theta^2}$ that is involved.

Microscales in the temperature field The right-hand side of (3.4.2) requires the introduction of a Taylor microscale for the temperature fluctuations. Let us define

$$\overline{(\partial\theta/\partial x_1)^2} \equiv 2\overline{\theta^2}/\lambda_\theta^2 . \tag{3.4.3}$$

The coefficient 2 in (3.4.3) is a normalization factor, which brings (3.4.3) into agreement with the expressions used in the literature (see also Chapter 6). If the small-scale structure of the temperature field is isotropic, $\overline{(\partial\theta/\partial x_1)^2} = \overline{(\partial\theta/\partial x_2)^2} = \overline{(\partial\theta/\partial x_3)^2}$, so that the right-hand side of (3.4.2) becomes

$$\gamma \overline{\frac{\partial\theta}{\partial x_j}\frac{\partial\theta}{\partial x_j}} = 6\gamma \frac{\overline{\theta^2}}{\lambda_\theta^2}. \tag{3.4.4}$$

An estimate for λ_θ is obtained by requiring that both sides of (3.4.2) have the same order of magnitude. Recalling that $-\overline{\theta u_j}\,\partial\Theta/\partial x_j \sim \theta^2 u/\ell$ (as discussed previously) and that $(\ell/\lambda)^2 \sim u\ell/\nu$, we find

$$\lambda_\theta/\lambda = C(\gamma/\nu)^{1/2}. \tag{3.4.5}$$

The constant C is of order one (Corrsin, 1951).

The Taylor microscale for temperature, λ_θ, is an artificial length scale, just like λ. If we want to determine the dissipative eddy size of the temperature field, we have to consult the equation governing temperature gradients. In analogy with (3.3.62) and (3.3.66), the equation for $\overline{(\partial\theta/\partial x_i)(\partial\theta/\partial x_j)}$ may be approximated by (Corrsin, 1953)

$$-\overline{\frac{\partial\theta}{\partial x_i}\frac{\partial\theta}{\partial x_j}}s_{ij} = \gamma \overline{\frac{\partial^2\theta}{\partial x_i \partial x_j}\frac{\partial^2\theta}{\partial x_i \partial x_j}}. \tag{3.4.6}$$

If $\gamma < \nu$, most of the dissipation of temperature-gradient fluctuations occurs at scales smaller than η, so that the temperature field is exposed to the entire spectrum of strain-rate fluctuations. Consequently, the proper estimate for s_{ij} is $(\epsilon/\nu)^{1/2}$ in this case. In analogy with (3.3.68), the temperature microscale η_θ is then given by (Batchelor, 1959; see also Section 8.6)

$$\eta_\theta/\eta = (\gamma/\nu)^{1/2}. \tag{3.4.7}$$

If the thermal diffusivity γ and the kinematic viscosity ν are approximately equal (as in gases), temperature fluctuations extend to scales as small as η. In liquids, the microscales may be different. For water, the Prandtl number ν/γ is about 7, so that temperature fluctuations extend to scales almost 3 times as small as η. The creation of very small temperature eddies in a fluid with a large Prandtl number is due to the straining effect illustrated in Figure 3.6.

If $\gamma > \nu$, so that the Prandtl number is smaller than one, η_θ is larger than η. In this case, even the very smallest temperature eddies are not exposed to the entire spectrum of strain-rate fluctuations. If $\gamma >> \nu$, the effective value of the strain rate must be independent of ν. This leads to $s_{ij} \sim \epsilon^{1/3}\, \eta_\theta^{-2/3}$; in analogy with (3.3.69), the temperature microscale becomes (Oboukhov, 1949; Corrsin, 1951; see also Section 8.6)

$$\eta_\theta = (\gamma^3/\epsilon)^{1/4}, \quad \eta_\theta/\eta = (\gamma/\nu)^{3/4}. \tag{3.4.8}$$

This estimate applies to liquid metals and electrolytes, in which the Prandtl number is small (for mercury, $\nu/\gamma = 0.028$).

Buoyant convection One interesting group of problems arises when temperature is not a passive but an active contaminant which can contribute to the generation of velocity fluctuations. The case we have in mind is thermal convection in gases exposed to a gravity field. Temperature fluctuations cause density fluctuations in a gas at essentially constant pressure (that is, very low Mach number). The density fluctuations cause a fluctuating body force $g_i\rho'/\bar{\rho}$ (g_i is the vector acceleration of gravity, ρ' is the density fluctuation, and ρ is the mean density). In the *Boussinesq approximation*, the fluctuating body force is written as $-g_i\vartheta/\Theta_0$, where Θ_0 is the mean temperature of an adiabatic atmosphere and ϑ is the difference between the actual temperature and Θ_0. The adiabatic temperature Θ_0 changes in the direction of the gravity vector in response to the gravity-induced pressure gradient, but the length scale involved is large, so that Θ_0 may be treated as a constant in many problems (Lumley and Panofsky, 1964).

The temperature difference ϑ is decomposed into a mean value $\bar{\vartheta}$ and fluctuations $\theta\,(\bar{\theta} = 0)$. If $U_j = 0$, the fluctuating body force performs work at a mean rate $-g_j\,\overline{\theta u_j}/\Theta_0$. This work, called *buoyant production*, must be added as a source term in the budget of turbulent kinetic energy. The heat flux $c_p\rho\overline{\theta u_j}$ then assumes a dual role, because it occurs in production terms for both $\frac{1}{2}\overline{u_iu_i}$ and $\overline{\theta^2}$.

In a flow that is steady and homogeneous in the x_1, x_2 plane and in which the only nonzero components of U_j and g_j are $U_1 = U_1\,(x_3)$ and $g_3 = -g$ (it is consistent with geophysical practice to take the x_3 direction vertically upwards), the heat and momentum fluxes $\rho c_p\,\overline{\theta u_3}$ and $\rho\overline{u_1u_3}$ are constant if molecular transport of $\bar{\vartheta}$ and U_1 in the x_3 direction can be neglected. The equations for $\frac{1}{2}\overline{u_iu_i}$ and $\overline{\theta^2}$ reduce to

$$0 = -\overline{u_1u_3}\frac{\partial U_1}{\partial x_3} + \frac{g}{\Theta_0}\overline{u_3\theta} - \frac{\partial}{\partial x_3}\left(\frac{1}{2}\overline{u_iu_iu_3} + \frac{1}{\rho}\overline{pu_3}\right) - \nu\overline{\frac{\partial u_i}{\partial x_j}\frac{\partial u_i}{\partial x_j}}, \tag{3.4.9}$$

$$0 = -\overline{\theta u_3}\frac{\partial \overline{\vartheta}}{\partial x_3} - \frac{\partial}{\partial x_3}(\tfrac{1}{2}\overline{\theta^2 u_3}) - \gamma \overline{\frac{\partial \theta}{\partial x_j}\frac{\partial \theta}{\partial x_j}}. \tag{3.4.10}$$

In these equations the terms representing transport of kinetic energy and temperature variance by molecular motion have been neglected because they are ordinarily very small. The mean temperature gradient $\partial\overline{\vartheta}/\partial x_3$ is equal to the actual temperature gradient minus the gravity-induced temperature gradient $\partial\Theta_0/\partial x_3 = -g/c_p$ which would exist in a flow without heat transfer (the set $\partial\Theta_0/\partial x_3 = -g/c_p$, $(1/\rho_0)\ \partial P_0/\partial x_3 = -g$, $P_0/\rho_0 = R\Theta_0$ defines a perfect-gas atmosphere in which the entropy is constant).

The two equations (3.4.9, 3.4.10) are used in the study of atmospheric turbulence. The outstanding feature of these equations, of course, is the buoyant production of kinetic energy. Apparently, there exist situations in which turbulence need not be maintained by shear stresses because it can be maintained by fluctuating buoyancy forces. Turbulence driven by body forces is not nearly as well understood as turbulence driven by shear stresses; for example, no satisfactory theory of atmospheric turbulence in unstable conditions ($\partial\overline{\vartheta}/\partial x_3 < 0$) exists.

Richardson numbers Some of the parameters governing (3.4.9, 3.4.10) need to be introduced. The most obvious one is the ratio of buoyant production to stress production of turbulent kinetic energy. This parameter is called the *flux Richardson number*; it is defined as

$$R_f \equiv \frac{g}{\Theta_0} \frac{\overline{u_3\theta}}{\overline{u_1u_3}\ \partial U_1/\partial x_3}. \tag{3.4.11}$$

If the heat transfer is upward ($\overline{u_3\theta} > 0$), the value of R_f is negative because $\overline{u_1u_3} < 0$ if $\partial U_1/\partial x_3 > 0$. As (3.4.9) indicates, the production of turbulent kinetic energy is increased in this case. Upward heat flux generally corresponds to $\partial\overline{\vartheta}/\partial x_3 < 0$; this is called an *unstable* atmosphere. If the heat transfer is downward ($\overline{\theta u_3} < 0$), $R_f > 0$, and the buoyant-production term becomes negative, indicating that kinetic energy is lost. Negative values of $\overline{\theta u_3}$ generally correspond to positive values of $\partial\overline{\vartheta}/\partial x_3$; this is called *stable* stratification. If a positive R_f becomes large enough, it leads to complete suppression of all turbulence.

If we define an eddy viscosity and an eddy conductivity by

$$-\overline{u_1 u_3} \equiv \nu_T \partial U_1/\partial x_3, \tag{3.4.12}$$

$$-\overline{\theta u_3} \equiv \gamma_T \partial\overline{\vartheta}/\partial x_3, \tag{3.4.13}$$

the flux Richardson number may be written as

$$R_f = \frac{\gamma_T g}{\nu_T \Theta_0} \frac{\partial\overline{\vartheta}/\partial x_3}{(\partial U_1/\partial x_3)^2}. \tag{3.4.14}$$

Apart from the "exchange" coefficients ν_T and γ_T, this expression contains variables that can be measured with relative ease. This suggests that a different parameter, the *gradient Richardson number*, should be useful:

$$R_g \equiv \frac{g}{\Theta_0} \frac{\partial\overline{\vartheta}/\partial x_3}{(\partial U_1/\partial x_3)^2}. \tag{3.4.15}$$

If ν_T and γ_T are approximately the same (which may be a very unreliable assumption if the absolute value of R_f is not small), the parameters R_f and R_g are approximately the same, too. Observations have shown that turbulence cannot be maintained if $R_f > 0.2$ approximately.

Buoyancy time scale The group $(g/\Theta_0)\partial\overline{\vartheta}/\partial x_3$ in (3.4.15) has dimensions $\sec^{-2}$. If $\partial\overline{\vartheta}/\partial x_3 > 0$ (stable conditions), we define

$$(g/\Theta_0)\partial\overline{\vartheta}/\partial x_3 \equiv N_b^2; \tag{3.4.16}$$

if $\partial\overline{\vartheta}/\partial x_3 < 0$ (unstable conditions), we define

$$-(g/\Theta_0)\partial\overline{\vartheta}/\partial x_3 \equiv T_b^{-2}. \tag{3.4.17}$$

The parameter N_b is called the Brunt-Väisälä frequency; it is the frequency of gravity waves in a stable atmosphere. In an unstable atmosphere, gravity waves are unstable and break up into turbulence. Therefore, if $\partial\overline{\vartheta}/\partial x_3 < 0$ we use the buoyancy time scale T_b. In sunny weather, T_b is typically of the order of a few minutes; more strongly unstable conditions correspond to smaller values of T_b. In a neutral atmosphere ($\partial\overline{\vartheta}/\partial x_3 = 0$), the time scale $T_b \to \infty$, and the frequency $N_b = 0$.

The mean wind gradient $\partial U_1/\partial x_3$ has the dimensions $\sec^{-1}$. If we define

$$\partial U_1/\partial x_3 \equiv T_s^{-1}, \tag{3.4.18}$$

we obtain

$$R_g = (N_b T_s)^2, \quad (\partial\bar{\vartheta}/\partial x_3 > 0), \tag{3.4.19}$$

$$R_g = -(T_s/T_b)^2, \quad (\partial\bar{\vartheta}/\partial x_3 < 0). \tag{3.4.20}$$

We conclude that the gradient Richardson number is the square of a ratio of time scales.

Monin-Oboukhov length In the surface layer of the atmosphere (which may extend up to several tens of meters above the surface), different parameters are important, so that the Richardson number is arranged in a different way. We assume that the wind profile is logarithmic: $\partial U_1/\partial x_3 = u_*/\kappa x_3$ (see Section 2.3). The Reynolds stress $-\rho\,\overline{u_1 u_3}$ is constant; it is put equal to ρu_*^2 (u_* is the friction velocity). The flux Richardson number then reads

$$R_f = -\frac{\kappa g x_3 \overline{\theta u_3}}{\Theta_0 u_*^{\,3}}. \tag{3.4.21}$$

The heat flux $H = \rho c_p \overline{\theta u_3}$; if we define a length L by

$$L \equiv -\frac{\Theta_0 u_*^{\,3}}{\kappa g \overline{\theta u_3}} = -\frac{c_p \rho \Theta_0 u_*^{\,3}}{\kappa g H}, \tag{3.4.22}$$

we obtain

$$R_f = x_3/L. \tag{3.4.23}$$

The length L is known as the *Monin-Oboukhov length scale.* Monin and Oboukhov have successfully used x_3/L as the basic independent variable for the description of the surface layer, both in stable and unstable conditions. The absolute value of L is seldom less than 10 m, so that the conditions in the lowest meter of the atmosphere are approximately neutral, except when the wind speed is very low.

Convection in the atmospheric boundary layer As an illustration of the complexity of the problems caused by buoyant production of turbulence, let us consider atmospheric boundary layers in unstable conditions ($\partial\bar{\vartheta}/\partial x_3 < 0$). In the surface layer of these boundary layers the absolute value of R_f is small, but at heights above 50 m, say, we may expect production by Reynolds stresses to be very small compared to buoyant production if the upward heat flux is appreciable (sunny afternoon weather). Also, the turbulence outside the surface layer is thoroughly mixed by the thermal convection, so that transport terms in the energy budget (3.4.9) should be small. An approximate energy budget for the turbulence above the surface layer then reads

$$\frac{g}{\Theta_0}\overline{\theta u_3} \cong \nu \overline{\frac{\partial u_i}{\partial x_j}\frac{\partial u_i}{\partial x_j}}. \tag{3.4.24}$$

Let us assume that θ and u_3 are well correlated, so that $\overline{\theta u_3} \sim tw$ if the rms values of θ and u_3 are represented by t and w. In turbulence with velocity scale w and length scale h, the dissipation rate is of order w^3/h (h scales with the height of the atmospheric boundary layer). Substituting the estimates $\overline{\theta u_3} \sim tw$ and $\epsilon \sim w^3/h$ into (3.4.24), we obtain

$$w^2 \sim g\,t\,h/\Theta_0. \tag{3.4.25}$$

This estimate states that a buoyant acceleration of order gt/Θ_0, acting over a distance h, produces kinetic energy of order gth/Θ_0.

If the heat flux $\rho c_p \overline{\theta u_3}$ throughout the boundary layer is of the same order as the heat flux in the surface layer, $\overline{\theta u_3}$ can be written in terms of the Monin-Oboukhov length L defined in (3.4.22) (note that the Monin-Oboukhov length is defined on basis of the surface heat flux). This yields

$$\overline{\theta u_3} \sim wt \sim -\Theta_0 u_*{}^3/gL. \tag{3.4.26}$$

Substituting for t with (3.4.26) in (3.4.25), we obtain

$$(w/u_*)^2 \sim (-h/L)^{2/3}. \tag{3.4.27}$$

As the heat flux increases, the value of $-L$ ($L<0$ if $\overline{\theta u_3}>0$) decreases. A value of $-L$ representative of strong convection is $-L = 10$ m; the height h is of the order of 1,000 m. We conclude from (3.4.27) that the kinetic energy $\frac{1}{2}w^2$ of the turbulence above the surface layer becomes large compared to u_*^2 if the upward heat flux is large (in the absence of heat transfer, $w \sim u_*$). This implies that the correlation between u_1 and u_3 is small under these conditions, because $u_1 \sim w$, $u_3 \sim w$, but $\overline{u_1 u_3} \sim u_*^2$. Turbulent eddies created by buoyancy forces apparently cause relatively little momentum transfer. This undermines the foundation on which eddy-viscosity and mixing-length expressions are based, so that they cannot be used in a complicated problem like this.

In a flow with temperature fluctuations of order t and with a length scale

h, the mean temperature gradient $\partial\bar{\vartheta}/\partial x_2$ is at most of order t/h if the thermal convection keeps the temperature field mixed. Thus, the buoyancy time scale T_b defined in (3.4.17) may be estimated as

$$T_b \sim (gt/\Theta_0 h)^{-1/2}. \tag{3.4.28}$$

Substituting for t with (3.4.26, 3.4.27), we obtain

$$T_b \sim (h/u_*)(-L/h)^{1/3}. \tag{3.4.29}$$

The height h of the boundary layer often is of order u_*/f, where f is the Coriolis parameter (Blackadar and Tennekes, 1968). If this is the case, (3.4.29) becomes

$$T_b f \sim (-L/h)^{1/3}. \tag{3.4.30}$$

Clearly, the problem of buoyant convection is one with two time scales, that is, T_b and f^{-1}, which may differ by an order of magnitude if $-L$ and h differ by a few orders of magnitude. As we have seen before, most problems in turbulence theory that involve more than one dynamically significant time or length scale are so complicated that no comprehensive solution is possible at the present state of the art.

Buoyancy-generated eddies cause relatively little momentum transport, but they are quite effective in transporting heat. In other words, the ratio of the turbulent diffusivities for heat and momentum is much larger than one, so that Reynolds' analogy (Section 2.4) does not apply.

Problems

3.1 Estimate the characteristic velocity of eddies whose size is equal to the Taylor microscale λ (see Problem 1.3). Use this estimate to show that eddies of this size contribute very little to the total dissipation rate.

3.2 Experimental evidence suggests that the dissipation rate is not evenly distributed over the volume occupied by a turbulent flow. The distribution of the dissipation rate appears to be intermittent, with large dissipation rates occupying a small volume fraction. Make a model of this phenomenon by assuming that all of the dissipation occurs in thin vortex tubes (diameter η, characteristic velocity $u = [\frac{1}{3}\overline{u_i u_i}]^{1/2}$). What is the volume fraction occupied by these tubes? Verify if the approximate vorticity budget (3.3.62) indeed holds for these vortex tubes.

3.3 A qualitative estimate of the effect of a wind-tunnel contraction (Figure 3.5) on turbulent motion can be obtained by assuming that the angular momentum of eddies does not change through the contraction. Let the contraction ratio, which is equal to the ratio of the mean velocity behind the contraction to that in front of the contraction, be equal to c. Show that the velocity fluctuations associated with an "eddy" aligned with the mean flow (as in Figure 3.5) increase by a factor $c^{1/2}$ and that those associated with an eddy perpendicular to the mean flow decrease by a factor c. Compute the effect of the contraction on the relative turbulence intensity u/U. Estimate the effect of the contraction on the rate of decay of velocity fluctuations. Is it feasible to design a contraction such that the evolution of turbulent velocity fluctuations during the contraction can be ignored?

3.4 A fully developed turbulent pipe flow of fluid with a Prandtl number equal to one is being cooled by the addition of a small volume of slightly cooler fluid over a cross section. Estimate the initial temperature fluctuation level. How many pipe diameters downstream are required before the temperature fluctuations have decayed to 1% of the initial level? For the purpose of this calculation, it may be assumed that the mean velocity in the pipe is approximately independent of position. Also, an estimate for the dissipation rate ϵ is needed; it can be obtained from momentum and energy integrals for pipe flow. For a prescribed decrease in mean temperature in the pipe, should one increase the volume flow of coolant and reduce the temperature difference or vice versa in order to reduce the temperature fluctuations?

4

BOUNDARY-FREE SHEAR FLOWS

Turbulent shear flows that occur in nature and in engineering are usually evolving — that is, in the flow direction the structure of the flow is changing. This change is sometimes due to external influences, such as pressure or temperature gradients, and sometimes due only to evolutionary influences inherent within the turbulence. At the present time, very few evolving flows are well understood; those evolving because of external influences are particularly difficult to understand, unless the variation of the external influence happens to match in some way the flow's own evolutionary tendencies. In Section 5.5 we encounter an example of such a flow. Here we shall limit ourselves to flows evolving under the influence of their own evolutionary tendencies. Even this class of flows is not generally understood; we shall further restrict the discussion to two-dimensional flows whose evolution is slow and whose dynamics is not affected by the presence of a solid surface.

4.1
Almost parallel, two-dimensional flows

There are two types of two-dimensional flows, the so-called plane flows and the axisymmetric flows. In both, the mean velocity field is entirely confined to planes. In the plane flows, mean flow in planes parallel to a given plane is identical; in the axisymmetric flows, mean flow in planes through the axis of symmetry is identical. We analyze in detail the plane flows (for algebraic simplicity) and give the results for the axisymmetric flows.

Plane flows Let us consider flows whose principal mean-velocity component is in the x direction, which are confined to the x,y plane, and which evolve slowly in the x direction. Thus,

$$U_i = \{U, V, 0\}, \quad \partial/\partial x \ll \partial/\partial y \text{ nearly everywhere.} \tag{4.1.1}$$

The classical flows falling within this class are wakes, jets, and shear layers (Figure 4.1). For these flows it is possible to simplify the equations of motion by discarding many terms that are small. To identify these terms, we must determine in what order the terms vanish as these flows become more and more nearly parallel. Slightly more complicated flows, such as jets flowing into a moving medium, are not treated here; they can be analyzed in the same way as the flows in Figure 4.1.

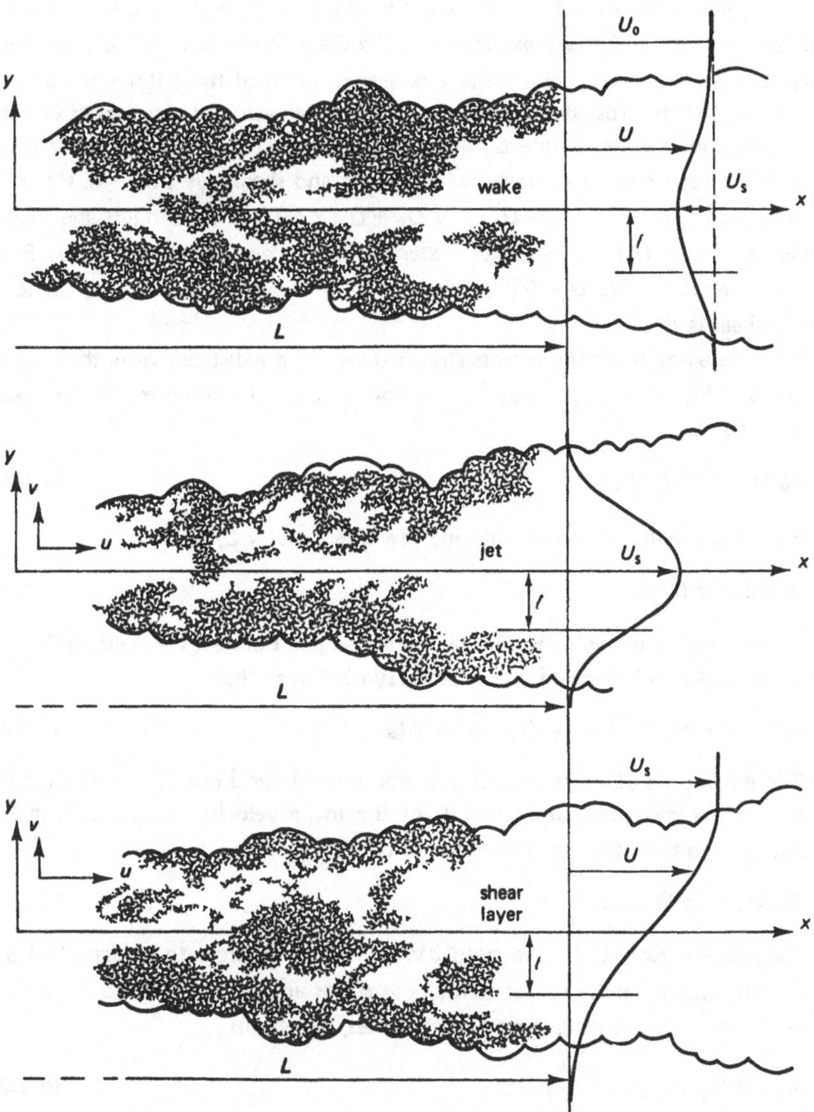

Figure 4.1. Plane turbulent wakes, jets, and shear layers (mixing layers).

Examining Figure 4.1, we can identify two velocity scales in the wake and one in the jet and the shear layer. In the wake, there is a scale U_0 for the velocity of the mean flow in the x direction; in all of these flows there is a scale U_s for the cross-stream variation of the mean velocity component in the x direction. Let us define U_s as the maximum value of $|U_0 - U|$. In wakes $U_s \ll U_0$ far from the obstacle, while in jets and shear layers $U_0 = 0$. Hence, in far wakes $U = U_0 + (U - U_0) = \mathcal{O}(U_0 + U_s) = \mathcal{O}(U_0)$, while in jets and shear layers $U = \mathcal{O}(U_s)$ (as before, $\mathcal{O}$ stands for "order of magnitude"). For convenience we use $U = \mathcal{O}(\tilde{U})$, where $\tilde{U} = U_0$ for wakes and $\tilde{U} = U_s$ for jets and shear layers.

If we agree to define a cross-stream scale ℓ as the distance from the center line at which $U - U_0$ is about $\frac{1}{2}U_s$ (a more precise selection is made later), we can write

$$\partial U/\partial y = \mathcal{O}(U_s/\ell). \tag{4.1.2}$$

We designate the scale of change in the x direction by L, so that

$$\partial U/\partial x = \mathcal{O}(U_s/L). \tag{4.1.3}$$

In addition to the velocity and length scales just defined, we need a velocity scale for the turbulence. Let us use the symbol u, so that

$$-\overline{uv} = \mathcal{O}(u^2), \quad \overline{u^2} = \mathcal{O}(u^2), \quad \overline{v^2} = \mathcal{O}(u^2). \tag{4.1.4}$$

The magnitude of u relative to U_s is determined later. Finally, we need a scale for the cross-stream component V of the mean velocity. This scale can be determined from the mean equation of continuity:

$$\partial U/\partial x + \partial V/\partial y = 0. \tag{4.1.5}$$

Because $\partial U/\partial x \sim U_s/L$, we need $\partial V/\partial y \sim U_s/L$ in order to balance (4.1.5). On the other hand, cross-stream length scales are proportional to ℓ, so that $\partial V/\partial y \sim V/\ell$. Equating these two estimates, we obtain

$$V = \mathcal{O}(U_s \ell/L). \tag{4.1.6}$$

The cross-stream momentum equation We are now in a position to examine the equations of motion in the limit as $\ell/L \to 0$, that is, as the flow becomes parallel. Let us first look at the equation for V, which governs the mean momentum in the cross-stream direction. This equation is

$$U\frac{\partial V}{\partial x}+V\frac{\partial V}{\partial y}+\frac{\partial}{\partial x}(\overline{uv})+\frac{\partial}{\partial y}(\overline{v^2})=-\frac{1}{\rho}\frac{\partial P}{\partial y}+\nu\left(\frac{\partial^2 V}{\partial x^2}+\frac{\partial^2 V}{\partial y^2}\right). \tag{4.1.7}$$

Expressing each term of (4.1.7) in the scales introduced earlier, we may identify their orders of magnitude as follows:

$$\begin{aligned}
U\frac{\partial V}{\partial x}&: \frac{\tilde{U}U_s\ell}{L\,L}=\left[\frac{\tilde{U}}{u}\frac{U_s}{u}\left(\frac{\ell}{L}\right)^2\right]\frac{u^2}{\ell},\\
V\frac{\partial V}{\partial x}&: \left(\frac{U_s\ell}{L}\right)^2\frac{1}{\ell}=\left[\left(\frac{U_s}{u}\right)^2\left(\frac{\ell}{L}\right)^2\right]\frac{u^2}{\ell},\\
\frac{\partial}{\partial x}(\overline{uv})&: \frac{u^2}{L}=\frac{\ell}{L}\cdot\frac{u^2}{\ell},\\
\frac{\partial}{\partial y}(\overline{v^2})&: \frac{u^2}{\ell}=1\cdot\frac{u^2}{\ell},\\
\frac{1}{\rho}\frac{\partial P}{\partial y}&: ?,\\
\nu\frac{\partial^2 V}{\partial x^2}&: \frac{\nu U_s\ell}{L}\frac{1}{L^2}=\left[\frac{U_s}{u}\frac{1}{R_\ell}\left(\frac{\ell}{L}\right)^3\right]\frac{u^2}{\ell},\\
\nu\frac{\partial^2 V}{\partial y^2}&: \frac{\nu U_s\ell}{L}\frac{1}{\ell^2}=\left[\frac{U_s}{u}\frac{1}{R_\ell}\left(\frac{\ell}{L}\right)\right]\frac{u^2}{\ell}.
\end{aligned} \tag{4.1.8}$$

Unless $u^2/(\tilde{U}U_s)\to 0$ as fast as $(\ell/L)^2$, the first, second, and third terms of (4.1.7) are negligible relative to $\partial\overline{v^2}/\partial y$. If the Reynolds number $R_\ell = u\ell/\nu$ is large enough, the viscous terms are also negligible compared to $\partial\overline{v^2}/\partial y$. There must be at least one term of the same order as $\partial\overline{v^2}/\partial y$ in order to balance the equation; inspection of (4.1.8) shows that only the pressure term can do this. Thus we obtain the following approximate form of (4.1.7):

$$\partial\overline{v^2}/\partial y=-(1/\rho)\,\partial P/\partial y. \tag{4.1.9}$$

This approximation is valid only if

$$\frac{\tilde{U}}{u}\frac{U_s}{u}\left(\frac{\ell}{L}\right)^2\to 0,\quad \frac{U_s}{u}\frac{1}{R_\ell}\left(\frac{\ell}{L}\right)\to 0 \tag{4.1.10}$$

in the limit as $\ell/L\to 0$; the conditions (4.1.10) need to be imposed to assure the negligibility of the first two and the last two terms of (4.1.7). We shall later show that (4.1.10) is always satisfied, provided that R_ℓ is sufficiently large.

Integration of (4.1.9) is straightforward; it yields

$$P/\rho + \overline{v^2} = P_0/\rho. \tag{4.1.11}$$

Here, P_0 is the pressure outside the turbulent part of the flow field ($y \to \pm\infty$). Equation (4.1.11) holds for all narrow, slowly evolving flows. We will assume that the imposed downstream pressure gradient $\partial P_0/\partial x = 0$. If P_0 were to vary in the x direction we could not state without hesitation that all derivatives in the downstream direction scale with L, since the variation of P_0 might introduce another scale.

We need the derivative of (4.1.11) with respect to x. Because $\partial P_0/\partial x = 0$, we obtain

$$(1/\rho)\, \partial P/\partial x + \partial \overline{v^2}/\partial x = 0. \tag{4.1.12}$$

The streamwise momentum equation The equation for U, which governs the downstream component of mean momentum, reads

$$U\frac{\partial U}{\partial x} + V\frac{\partial U}{\partial y} + \frac{\partial}{\partial x}(\overline{u^2} - \overline{v^2}) + \frac{\partial}{\partial y}(\overline{uv}) = \nu\left(\frac{\partial^2 U}{\partial x^2} + \frac{\partial^2 U}{\partial y^2}\right). \tag{4.1.13}$$

Here, (4.1.12) has been used to substitute for $\partial P/\partial x$. Using the scales already introduced, we estimate the orders of magnitude of the terms of (4.1.13) as

$$\begin{aligned}
U\frac{\partial U}{\partial x} &: \quad \tilde{U}\frac{U_s}{L} = \left[\frac{\tilde{U}}{u}\frac{U_s}{u}\frac{\ell}{L}\right]\frac{u^2}{\ell},\\
V\frac{\partial U}{\partial y} &: \quad \frac{U_s \ell}{L}\frac{U_s}{\ell} = \left[\left(\frac{U_s}{u}\right)^2\frac{\ell}{L}\right]\frac{u^2}{\ell},\\
\frac{\partial}{\partial x}(\overline{u^2} - \overline{v^2}) &: \quad \frac{u^2}{L} = \frac{\ell}{L}\cdot\frac{u^2}{\ell},\\
\frac{\partial}{\partial y}(\overline{uv}) &: \quad \frac{u^2}{\ell} = 1\cdot\frac{u^2}{\ell},\\
\nu\frac{\partial^2 U}{\partial x^2} &: \quad \frac{\nu U_s}{L^2} = \left[\frac{U_s}{u}\frac{1}{R_\ell}\left(\frac{\ell}{L}\right)^2\right]\frac{u^2}{\ell},\\
\nu\frac{\partial^2 U}{\partial y^2} &: \quad \frac{\nu U_s}{\ell^2} = \left[\frac{U_s}{u}\frac{1}{R_\ell}\right]\frac{u^2}{\ell}.
\end{aligned} \tag{4.1.14}$$

If we assume that R_ℓ is sufficiently large, we can make the viscous terms as small as desired. In the limit as $\ell/L \to 0$, the third term of (4.1.13) is also negligible. In order to balance the equation, at least one other term of the same order as $\partial(\overline{uv})/\partial y$ is needed. Of the remaining terms, the first is the largest because $\tilde{U} \geqslant U_s$. Thus, we must require that

$$\frac{\tilde{U}}{u}\frac{U_s}{u}\frac{\ell}{L} = \mathcal{O}(1); \tag{4.1.15}$$

that is, this nondimensional group must remain bounded as $\ell/L \to 0$.

Turbulent wakes There are two ways in which (4.1.15) can be satisfied. If, as one possible choice, we take $u/U_s = \mathcal{O}(1)$, (4.1.15) requires that

$$u/\tilde{U} = \mathcal{O}(\ell/L). \tag{4.1.16}$$

This situation occurs in far wakes. Far wakes have turbulence intensities of the order of the velocity defect; both of these are small relative to the mean velocity. As a wake evolves downstream and as ℓ/L becomes smaller, u/U keeps pace with it.

With $u = \mathcal{O}(U_s)$ and (4.1.16), the second term in (4.1.13, 4.1.14) is negligible relative to the first, so that the momentum equation for turbulent wakes far from an obstacle reduces to

$$U\,\partial U/\partial x + \partial(\overline{uv})/\partial y = 0. \tag{4.1.17}$$

We can make one further simplification. For wakes, $\tilde{U} = U_0$ and $u \sim U_s$, so that we may write, by virtue of (4.1.16),

$$(U - U_0)/U_0 = \mathcal{O}(U_s/U_0) = \mathcal{O}(\ell/L). \tag{4.1.18}$$

This implies that the undifferentiated U occurring in (4.1.17) may be replaced by U_0. Thus, (4.1.17) may be approximated by

$$U_0\,\partial U/\partial x + \partial(\overline{uv})/\partial y = 0. \tag{4.1.19}$$

This equation states that the net momentum flux due to the cross-stream velocity fluctuations v is replaced by x momentum carried by the mean flow in the streamwise direction.

Returning to the provisions expressed in (4.1.10), we see that the first is satisfied if $\ell/L \to 0$ and if (4.1.16) holds. The second provision is satisfied as long as $\ell/(R_\ell L) \to 0$. This condition can be met easily. If we examine (4.1.14), we see that the condition for neglecting the viscous terms in (4.1.13) is that $1/R_\ell \to 0$, which is more stringent than the second provision in (4.1.10). If the viscous terms are to be of the same order as the other terms that have been neglected, we must require the even stronger condition $1/R_\ell = \mathcal{O}(\ell/L)$. Hence, roughly speaking, (4.1.19) is a valid approximate equation of motion for far wakes provided $1/R_\ell \sim \ell/L \ll 1$.

Turbulent jets and mixing layers The second way in which (4.1.15) may be satisfied is by putting $\tilde{U} = U_s$, so that (4.1.15) becomes

$$u/U_s = \mathcal{O}(\ell/L)^{1/2}. \tag{4.1.20}$$

The choice describes jets and mixing layers, in which turbulence intensities are about half an order of magnitude (measured in terms of ℓ/L) smaller than the jet velocity or the velocity difference in the mixing layer (shear layer). With the choice (4.1.20) the first and second terms in (4.1.14) are of the same order, so that the appropriate momentum equation is

$$U\,\partial U/\partial x + V\,\partial U/\partial y + \partial(\overline{uv})/\partial y = 0. \tag{4.1.21}$$

Here, the x momentum removed by the cross-stream velocity fluctuations v is replaced by mean-flow convection carried by both the downstream and the cross-stream components of the mean velocity.

The provisions (4.1.10) need to be examined. We find that the first of these is satisfied if $\ell/L \to 0$ and if (4.1.20) holds. The second provision amounts to $(\ell/L)^{1/2}\,R_\ell^{-1} \to 0$. This appears to be an easy condition. From (4.1.14) we conclude that the condition for the negligibility of the major viscous term is that $(L/\ell)^{1/2}\,R_\ell^{-1} \to 0$, which is a fairly strong requirement. To assure that the viscous term is of the same order as the other terms which have been neglected, we need the even stronger condition $R_\ell = \mathcal{O}(L/\ell)^{3/2}$. We conclude that (4.1.21) is a correct approximation if $\ell/L \to 0$ and if $(L/\ell)^{1/2}\,R_\ell^{-1} \to 0$.

We shall find later that in wakes ℓ/L continually decreases downstream, so that (4.1.19) becomes a better approximation the farther downstream one goes. For mixing layers and jets, on the other hand, we shall find that ℓ/L is constant. The observed values of ℓ/L in jets and mixing layers are of the order 6×10^{-2}, so that the neglected terms in (4.1.21) amount to about 6% of the terms retained. In the various plane and axisymmetric wakes, jets, and shear layers we shall study the Reynolds number R_ℓ changes downstream in different ways. Hence, in each flow there are distinct regions in which the conditions on R_ℓ are satisfied.

The momentum integral Because (4.1.19) is a special case of (4.1.21), all relations based on (4.1.21) also hold for (4.1.19), so that we can confine further analysis to (4.1.21). If we subtract U_0 from U when the latter appears within the streamwise derivatives of (4.1.21), we obtain

$$U\frac{\partial}{\partial x}(U-U_0)+V\frac{\partial}{\partial y}(U-U_0)+\frac{\partial}{\partial y}(\overline{uv})=0. \tag{4.1.22}$$

This is legitimate because U_0 is not a function of position (the imposed pressure gradient is zero). The continuity equation $\partial U/\partial x+\partial V/\partial y=0$ may be used to rewrite the first two terms of (4.1.22) as

$$U_j\frac{\partial}{\partial x_j}(U-U_0)=\frac{\partial}{\partial x_j}[U_j(U-U_0)]. \tag{4.1.23}$$

Thus, (4.1.22) becomes

$$\frac{\partial}{\partial x}[U(U-U_0)]+\frac{\partial}{\partial y}[V(U-U_0)]+\frac{\partial}{\partial y}\overline{uv}=0. \tag{4.1.24}$$

In jets and wakes, $U-U_0$ vanishes at sufficiently large values of y and so does $\overline{uv}$. For those flows, we may integrate (4.1.24) with respect to y over the entire flow. The result is

$$\frac{d}{dx}\int_{-\infty}^{\infty}U(U-U_0)\,dy=0. \tag{4.1.25}$$

Consequently,

$$\rho\int_{-\infty}^{\infty}U(U-U_0)\,dy=M, \tag{4.1.26}$$

where M is a constant. This integral relation is clearly inapplicable to shear layers because their velocity defect is not integrable. For shear layers, the left-hand side of (4.1.25) is equal to V_0U_s, which is unknown because V_0, the value of V at $y\to+\infty$, is unknown.

The integral (4.1.26) may be identified with the mean momentum flux across planes normal to the x axis. For wakes, $\rho(U_0-U)$ is the net *momentum defect* per unit volume, while $U\,dy$ is the volume flux per unit depth. The integral (4.1.26) then is the net flux of momentum defect per unit depth. When we use the term *momentum defect*, we mean it in the following sense: if the wake were not present, the momentum per unit volume would be ρU_0. The difference $\rho(U_0 - U)$ is the momentum defect (or deficit). The constant M in (4.1.26) is the total momentum removed per unit time from the flow by the obstacle that produces the wake.

For jets, $U_0 = 0$, so that (4.1.26) simplifies to

$$\rho \int_{-\infty}^{\infty} U^2 \, dy = M. \tag{4.1.27}$$

Here, ρU is the mean momentum per unit volume and $U\,dy$ is the volume flux per unit depth (depth is the distance normal to the plane of the flow). Therefore, M is the total amount of momentum put into the jet at the origin per unit time.

Momentum thickness The momentum integral (4.1.26) can be used to define a length scale for turbulent wakes. Imagine that the flow past an obstacle produces a completely separated, stagnant region of width θ. The net momentum defect per unit volume is then ρU_0, because the wake contains no momentum. The total volume per unit time and depth is $U_0 \theta$, so that $\rho U_0^2 \theta$ represents the net momentum defect per unit time and depth. Thus,

$$-\rho U_0^2 \theta = M. \tag{4.1.28}$$

Equating (4.1.26) and (4.1.28), we obtain

$$\theta = \int_{-\infty}^{\infty} \frac{U}{U_0} \left(1 - \frac{U}{U_0}\right) dy. \tag{4.1.29}$$

The length θ defined this way is independent of x in a plane wake; it is called the *momentum thickness* of the wake.

The momentum thickness is related to the drag coefficient of the obstacle that produces the wake. The drag coefficient c_d is defined by

$$D \equiv c_d \tfrac{1}{2} \rho U_0^2 d, \tag{4.1.30}$$

where D is the drag per unit depth and d is the frontal height of the obstacle. Clearly, $D = -M$ because the drag D produces the momentum flux M. If we equate (4.1.28) and (4.1.30), we find

$$c_d = 2\theta/d. \tag{4.1.31}$$

If the obstacle is a circular cylinder, $c_d \sim 1$ for Reynolds numbers $(U_0 d/\nu)$ between 10^3 and 3×10^5, so that θ is about $\frac{1}{2}d$ in that range.

4.2
Turbulent wakes

Here we study self-preservation (invariance), the mean momentum budget, and the kinetic energy budget of turbulence in plane wakes.

Self-preservation In the preceding analysis, we assumed that the evolution of jets, wakes, and mixing layers is determined solely by the local scales of length and velocity. Let us evaluate this assumption. In general, we may expect that in wakes

$$(U_0 - U)/U_s = f(y/\ell, \ell/L, \ell U_s/\nu, U_s/U_0). \tag{4.2.1}$$

However, we have developed approximate equations that are valid for $\ell/L \to 0$, $\ell U_s/\nu \to \infty$, $U_s/U_0 \to 0$. Under these limit processes, the (presumably monotone) dependence of the function f on ℓ/L, $\ell U_s/\nu$, and U_s/U_0 is eliminated, because no monotone function can remain finite if it does not become asymptotically independent of very large or very small parameters. Therefore, we expect that only the length scale ℓ is relevant and that all properly nondimensionalized quantities are functions of y/ℓ only. In particular,

$$(U_0 - U)/U_s = f(y/\ell), \tag{4.2.2}$$

where, of course, ℓ may change downstream ($\ell = \ell(x)$). We expect that (4.2.2) is valid because it makes a statement about velocity differences, which are related to velocity gradients. Relations like (4.2.2) do not hold for the absolute velocity U, because the value of U_0 clearly could be changed without changing the form of $U_0 - U$.

In wakes, the turbulence intensity u is of order U_s, so that we expect that the Reynolds stress may be described by

$$-\overline{uv} = U_s^2 g(y/\ell). \tag{4.2.3}$$

The set (4.2.2, 4.2.3) constitutes the *self-preservation hypothesis*: the velocity defect and the Reynolds stress become invariant with respect to x if they are expressed in terms of the local length and velocity scales ℓ and U_s.

In order to test the feasibility of (4.2.2, 4.2.3), we must substitute these expressions into the equation of motion (4.1.19). Let us define $\xi = y/\ell$, so that we may write

$$\frac{\partial U}{\partial x} = -\frac{dU_s}{dx} f + \frac{U_s}{\ell}\frac{d\ell}{dx}\xi f', \qquad \frac{\partial \overline{uv}}{\partial y} = -\frac{U_s^2}{\ell} g', \tag{4.2.4}$$

where primes denote differentation with respect to ξ. With (4.2.4), (4.1.19) becomes

$$-\frac{U_0 \ell}{U_s^2}\frac{dU_s}{dx} f + \frac{U_0}{U_s}\frac{d\ell}{dx}\xi f' = g'. \tag{4.2.5}$$

If the shapes of f and g are to be universal, so that the normalized profiles of the velocity defect and the Reynolds stress are the same at all x, we must require that the coefficients of f and $\xi f'$ in (4.2.5) be constant. Thus, taking into account that U_0 is a constant, we need

$$\frac{\ell}{U_s^2}\frac{dU_s}{dx} = \text{const}, \quad \frac{1}{U_s}\frac{d\ell}{dx} = \text{const}. \tag{4.2.6}$$

The general solution to the pair (4.2.6) is $\ell \sim x^n$, $U_s \sim x^{n-1}$, so that another relation is needed to make the result determinate. The momentum integral (4.1.26) provides the desired constraint; using (4.2.2), we may rewrite the momentum integral as

$$U_0 U_s \ell \int_{-\infty}^{\infty} f(\xi)\, d\xi - U_s^2 \ell \int_{-\infty}^{\infty} f^2(\xi)\, d\xi = -\frac{M}{\rho}. \tag{4.2.7}$$

The second term in (4.2.7) is of order U_s/U_0 compared to the first. By virtue of (4.1.16), U_s/U_0 is of order ℓ/L, so that the second term in (4.2.7) should be neglected. Substituting for M with (4.1.28), we obtain

$$U_s \ell \int_{-\infty}^{\infty} f(\xi)\, d\xi = -U_0 \theta. \tag{4.2.8}$$

We conclude that the product $U_s \ell$ must be independent of x. If $\ell \sim x^n$ and $U_s \sim x^{n-1}$, we find that $2n-1 = 0$, so that $n = \frac{1}{2}$. Thus, ℓ and U_s are given by

$$U_s = A x^{-1/2}, \quad \ell = B x^{1/2}. \tag{4.2.9}$$

The constants A and B still have to be determined.

A self-preserving solution is possible only if the velocity and length scales behave as stated in (4.2.9). Of course, the fact that such a solution is possible does not guarantee that it occurs in nature. In many problems, possible solutions are not observed because they are not stable and change to a different form when disturbed. We need experimental evidence to determine whether or not the solution (4.2.9) indeed occurs. Experiments with plane turbulent wakes of circular cylinders have shown that the development of ℓ and U_s is well described by (4.2.9) beyond about 80 cylinder diameters. Also, measured mean-velocity profiles agree with (4.2.2) beyond $x = 80d$. However, turbulence intensities and shear stresses do not exhibit self-preservation much before $x = 200d$. In most turbulent flows the mean velocity profile reaches equilibrium long before the turbulence does. Generally, the more complicated the statistical quantity, the longer it takes to reach self-preservation. For example, $\overline{v^3}$ and $\overline{v^4}$ take longer to reach self-preservation than $\overline{v^2}$. However, all measured quantities in wakes are fully self-preserving beyond $x/d = 500$.

The mean-velocity profile If we substitute (4.2.9) into (4.2.5), we obtain

$$\tfrac{1}{2}(U_0 B/A)(\xi f' + f) = g'. \tag{4.2.10}$$

In order to proceed, we need a relation between f and g. If we define an eddy viscosity ν_T by $-\overline{uv} \equiv \nu_T\, \partial U/\partial y$, we can state, by virtue of (4.2.2, 4.2.3),

$$\nu_T = -U_s \ell g/f'. \tag{4.2.11}$$

Thus, we expect $\nu_T/U_s\ell$ to be some function of y/ℓ. Now, g/f' is a symmetric function, so that ν_T is approximately constant near the wake center line. Also, from physical intuition, we expect the turbulence in the wake to be thoroughly mixed, so that the scales of length and velocity should not be functions of the distance from the center line. This again suggests that ν_T may be constant.

It should be noted that (4.2.11) is a consequence of the existence of the single velocity scale U_s and the two length scales y and ℓ. Therefore, (4.2.11) is a consequence of self-preservation; it should not be construed as support for a mixing-length model. The assumption that ν_T is constant is equivalent to assuming that one of the length scales (namely y) is not relevant to ν_T.

Because both g and f' have a zero at the center line, there is some question whether ν_T remains finite as $y \to 0$. This problem is resolved with l'Hôpital's rule, which states that the limit of g/f' as $y \to 0$ is equal to the limit of g'/f''. The latter is finite at $y = 0$.

With these provisions, we proceed on the assumption that ν_T is constant:

$$\nu_T/(U_s\ell) \equiv 1/R_T = -g/f'. \tag{4.2.12}$$

The parameter $R_T \equiv U_s\ell/\nu_T$ is called the *turbulent Reynolds number*; we need experimental data to determine its value. We should keep in mind that (4.2.12) is likely to be valid only near the center line of the wake (because of symmetry); we should expect errors near the edges of the wake.

If we substitute (4.2.12) into (4.2.10), we obtain

$$\alpha(\xi f' + f) + f'' = 0, \tag{4.2.13}$$

in which

$$\alpha = \tfrac{1}{2} R_T U_0 B/A. \tag{4.2.14}$$

The solution of (4.2.13) is

$$f = \exp(-\tfrac{1}{2}\alpha\xi^2). \tag{4.2.15}$$

In accordance with the definition $U_s = \max(U_0 - U)$, we have $f(0) = 1$. We still have not defined ℓ precisely; a convenient definition is to take $\alpha = 1$ so that $f = \exp(-\tfrac{1}{2}) \cong 0.6$ at $\xi = 1$ $(y = \ell)$. The normalized momentum integral then becomes

$$\int_{-\infty}^{+\infty} f(\xi)\, d\xi = (2\pi)^{1/2}. \tag{4.2.16}$$

The observed value of R_T, with U_s and ℓ as previously defined, is 12.5. Substitution of (4.2.16) into (4.2.8) and of (4.2.14) (with $\alpha = 1$) into (4.2.9) then gives, with some algebra,

$$U_s/U_0 = 1.58(\theta/x)^{1/2}, \tag{4.2.17}$$

$$\ell/\theta = 0.252(x/\theta)^{1/2}. \tag{4.2.18}$$

It should be noted that the Reynolds number defined by U_s and ℓ is constant:

$$U_s\ell/\nu = 0.4\, U_0\theta/\nu. \tag{4.2.19}$$

Thus, once turbulent, a plane wake remains turbulent.

The decay laws (4.2.17) and (4.2.18) are similar to those for plane laminar wakes. This is because the momentum deficit, which is proportional to $U_s\ell$, is independent of x, so that both the Reynolds number $U_s\ell/\nu$ and the turbulent Reynolds number $U_s\ell/\nu_T$ are constant.

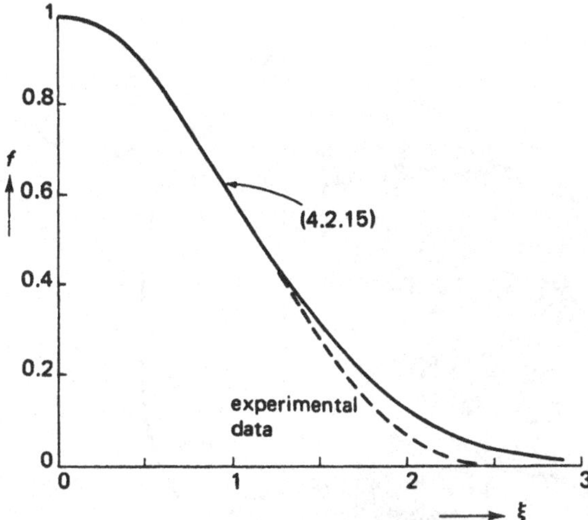

Figure 4.2. The velocity-defect profile of a plane turbulent wake (after Townsend, 1956).

The velocity profile (4.2.15) is in excellent agreement with the observed velocity profiles in wakes for all values of ξ less than 1.3. For larger values of ξ, (4.2.15) has the correct shape, but it predicts somewhat larger values of f than are observed (see Figure 4.2). The deviation is never larger than 5% of U_s. Because the predicted velocity profile (4.2.15) approaches the free-stream velocity U_0 slightly more gradually than the observations indicate, the value of ν_T appropriate for the center of the flow is evidently too large near the edges. A glance at Figure 4.3 makes the main reason for this clear. Within the turbulent part of the flow, average scales of velocity and length do not vary with cross-stream position, because there is thorough mixing from side to side. Here, a constant value of ν_T would be appropriate. Near the edges, however, a point at a fixed distance y spends only a fraction of its time in the turbulent flow. When the point is in the irrotational flow, the Reynolds stress is zero so that the net momentum transport should be multiplied by the relative fraction of time the point is in the turbulent fluid. This fraction is called the *intermittency* γ; the variation of γ is sketched in Figure 4.3. Thus, an expression like $\nu_T = \gamma \nu_{TC}$ (where ν_{TC} is the value appropriate to the center of the wake) would be a better estimate. Indeed, if a velocity profile is computed on this basis, it is found to fit the experimental data extremely well. For many purposes, however, (4.2.15) is sufficiently accurate.

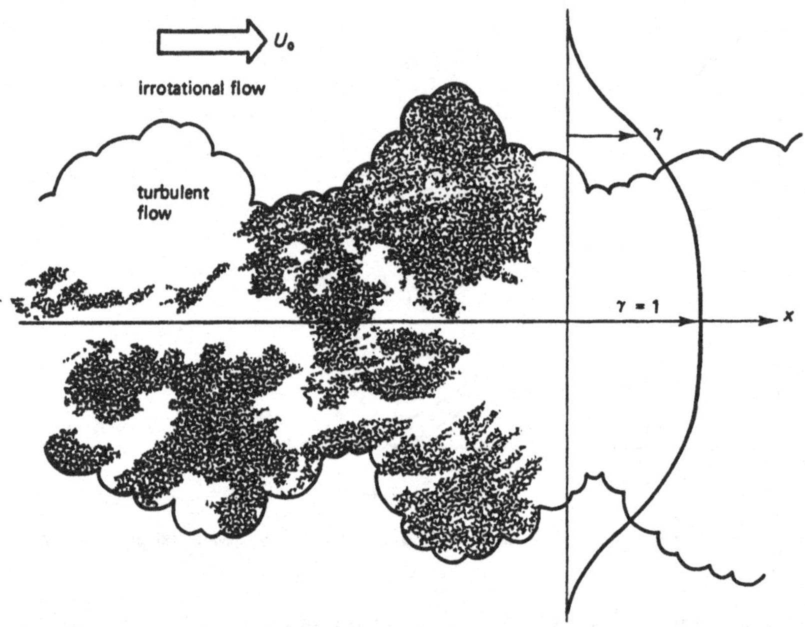

Figure 4.3. Intermittency near the edges of a wake.

Axisymmetric wakes If the foregoing analysis is applied to axisymmetric wakes, there results $U_s \sim x^{-2/3}$, $\ell \sim x^{1/3}$, so that $R_\ell = U_s\ell/\nu \sim x^{-1/3}$. Defining U_s and ℓ in a similar way as before, we obtain $R_T = 14.1$. The structure of the axisymmetric wake is thus not likely to be markedly different from that of the plane wake, with the exception that the Reynolds number of axisymmetric wakes steadily decreases. When R_ℓ is reduced to a value of the order unity, the wake ceases to be turbulent; it develops differently as the residual velocity disturbances decay. This is not a serious practical restriction, however. Let us write

$$U_s/U_0 \sim (\theta/x)^{2/3}, \quad \ell/\theta \sim (x/\theta)^{1/3}, \tag{4.2.20}$$

and let us assume that the coefficients involved are of order unity, as they were for the plane wake. The Reynolds number R_ℓ then varies as

$$R_\ell \sim (U_0\theta/\nu)(\theta/x)^{1/3}, \tag{4.2.21}$$

so that R_ℓ reaches unity when x/θ is of order $(U_0\theta/\nu)^3$. Even for moderate Reynolds numbers this is a large distance.

Scale relations With (4.2.17, 4.2.18), we are in a position to examine quantitatively some of the scale relations in plane wakes. With the help of (4.2.3) and (4.2.12), we may write

$$-\overline{uv} = -U_s^2 f'/R_T. \tag{4.2.22}$$

The Reynolds stress attains a maximum when $\xi = 1$, as differentiation of (4.2.15) (with $\alpha = 1$) shows. This yields

$$(-\overline{uv}/U_s^2)_{max} = (R_T^2 e)^{-1/2} = 0.05. \tag{4.2.23}$$

If the correlation coefficient between u and v is taken to be about 0.4, as it is in most shear flows (see Section 2.2), we obtain as an estimate for the rms velocity fluctuation $\mathscr{u}$ ($\mathscr{u}^2 = \frac{1}{3}\overline{u_i u_i} \cong \overline{u^2} \cong \overline{v^2}$):

$$\mathscr{u} \cong (0.05 U_s^2/0.4)^{1/2} = 0.35 U_s. \tag{4.2.24}$$

The rate at which the wake propagates into the surrounding fluid can be defined as $d\ell/dt = U_0\, d\ell/dx$, which, with (4.2.18) and (4.2.19), becomes

$$d\ell/dt = U_0\, d\ell/dx \cong 0.08\, U_s. \tag{4.2.25}$$

In a self-preserving flow we expect that all velocities are proportional to U_s, so that (4.2.24) and (4.2.25) are not surprising results. However, the values of the coefficients are interesting. The interface in Figure 4.3 propagates into the surrounding irrotational medium because it is contorted by the turbulent eddies. The contortion of the interface is caused by eddies of all scales; on the smallest scales, viscosity acts to propagate vorticity into the irrotational fluid. The net rate of propagation (or *entrainment*, as it is most often called), however, is controlled by the speed at which the contortions with the largest scales move into the surrounding fluid. Evidently, the largest eddies have a characteristic velocity roughly $0.08/0.35 \cong 23\%$ of that of the rms velocity fluctuation $\mathscr{u}$. This is supported by direct measurements; the large eddies contributing most to the entrainment are fairly weak, but have dimensions as large as the flow permits. They are substantially larger than the eddies that contain most of the energy.

A look at time scales is also instructive. A time scale t_p characteristic of the turbulence is given by the total energy $\frac{1}{2}\overline{u_i u_i}$ over the rate of production $-\overline{uv}\,\partial U/\partial y$ (the latter roughly equals the dissipation rate ϵ). With $\overline{u_i u_i} \cong 3\mathscr{u}^2$ and $-\overline{uv} \cong 0.4\mathscr{u}^2$, t_p becomes

$$t_p \equiv \frac{\frac{1}{2}\overline{u_i u_i}}{-\overline{uv}\,\partial U/\partial y} \cong -\frac{3.75\ell}{U_s f'}. \tag{4.2.26}$$

The minimum value of t_p is reached at the maximum of f', which occurs at $\xi = 1$. We obtain

$$t_p \cong 6.2\,\ell/U_s. \tag{4.2.27}$$

On the other hand, a time scale characteristic of the development (the downstream change) of the wake is $t_d = \ell/(d\ell/dt)$, which becomes, on substitution of (4.2.25),

$$t_d \equiv \ell/(d\ell/dt) \cong 12.5\,\ell/U_s. \tag{4.2.28}$$

Hence, the ratio of time scales is about 2:

$$t_d/t_p \cong 2. \tag{4.2.29}$$

The time scale of transfer of energy to small eddies apparently is only about half the time scale of flow development. Clearly, the turbulence can never be in equilibrium because it never has time to adjust to its changing environment. The structure of turbulence in wakes can be self-preserving only because the time scale of the turbulence and that of the flow keep pace with each other as the wake moves downstream.

The turbulent energy budget The equation for the kinetic energy of the turbulence, in an approximation which is consistent with the momentum equation (4.1.19), reads

$$0 = -U_0 \frac{\partial}{\partial x}(\tfrac{1}{2}\overline{q^2}) - \overline{uv}\frac{\partial U}{\partial y} - \frac{\partial}{\partial y}\overline{v\left(\frac{1}{2}q^2 + \frac{p}{\rho}\right)} - \epsilon. \tag{4.2.30}$$

Here, $\overline{q^2} = \overline{u_i u_i}$ is twice the kinetic energy per unit mass. The first term of (4.2.30) is convection of $\frac{1}{2}\overline{q^2}$ by the mean flow. This term is called *advection* in order to distinguish between it and thermal convection. The second term is production, the third is transport by turbulent motion, and the last is dissipation. We designate these terms by the letters *A, P, T,* and *D.*

With a few approximations, the distributions of the terms in (4.2.30) across the plane wake can be computed. We retain the approximation $-\overline{uv} = -U_s^2\, f'/R_T$, which is known to be slightly in error toward the edges of

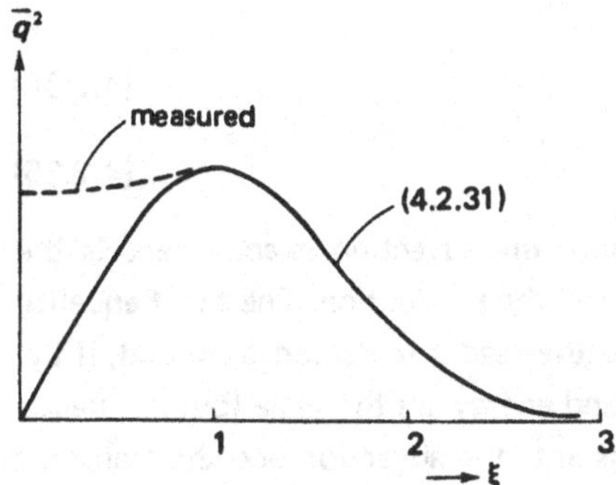

Figure 4.4. Comparison between (4.2.31) and the measured distribution of $\overline{q^2}$ in a plane wake (adapted from Townsend, 1956).

the wake. An expression for $\overline{q^2}$ is also needed. We expect that $\overline{q^2}$ and $-\overline{uv}$ are closely related; let us assume that $-\overline{uv} \cong 0.4\,\overline{q^2}/3$ outboard from the peak in f' (which occurs at $\xi = 1$). Thus for $\xi > 1$, we use

$$\overline{q^2} \cong -7.5\, U_s^2\, f'/R_T. \tag{4.2.31}$$

The region between the center line and $\xi = 1$ has to be dealt with separately, because $\overline{q^2}$ does not vanish at the center line while $-\overline{uv} = 0$ and $f' = 0$ at $\xi = 0$ for reasons of symmetry (Figure 4.4).

For the transport term we use a mixing-length assumption because it also must be self-preserving. Hence, we put

$$\overline{v\left(\frac{1}{2}q^2 + \frac{p}{\rho}\right)} = -\nu_T \frac{\partial}{\partial y}\left(\frac{1}{2}\overline{q^2}\right). \tag{4.2.32}$$

This simple form is adequate for such a crude model. We assume that ν_T is constant, realizing that this assumption is likely to be somewhat in error near the edges of the wake. Further, we take ν_T in (4.2.32) to have the same value as ν_T in (4.2.11), because the transport mechanism is probably similar. We should keep in mind that (4.2.32) cannot be applied to an off-axis peak of $\frac{1}{2}\overline{q^2}$, because we cannot use symmetry to argue for a constant (or even finite) value of ν_T.

With (4.2.31) and (4.2.32), the transport term in (4.2.30) can be expressed in terms of f. Thus we can write all terms except ϵ in terms of f. Using (4.2.15) and (4.2.17, 4.2.18), we obtain

$$\ell R_T\, A/U_s^3 \cong 0.3 f \xi\,(3 - \xi^2), \tag{4.2.33}$$

$$\ell R_T\, P/U_s^3 \cong \xi^2 f^2, \tag{4.2.34}$$

$$\ell R_T\, T/U_s^3 \cong -0.3\, f\xi(3-\xi^2). \tag{4.2.35}$$

We see that, within this approximation, the advection exactly cancels the transport, leaving the dissipation to cancel the production. The exact equality seems hardly accidental. We leave it to the reader to demonstrate that, if the exchange coefficients for momentum and energy are the same (but not necessarily constant) and if $-\overline{uv}/\overline{q^2}$ is constant, the advection and the transport always cancel, except for a term depending on the variation of R_T. The difference between advection and transport becomes smaller as the edge of the wake is approached. Also, production must be relatively small near the edge of the wake because it is quadratic in f.

The overall picture suggested by (4.2.33–4.2.35) is this: in the outer region of the wake (beyond $\xi^2 = 3$) turbulent transport brings kinetic energy from the center of the wake, where it is removed by advection. In other words, the edge of the wake is propagating into the surrounding undisturbed fluid and is blown back by the component of the mean flow normal to the wake boundary. Closer to the center, production becomes important, but it is roughly balanced by dissipation. Inboard of $\xi^2 = 3$, advection deposits kinetic energy, which is removed by transport to the outer edges of the wake. The different terms are sketched in Figure 4.5 with solid lines.

We do not expect dissipation to decrease in the center of the wake. On the contrary, we expect that the dissipation is essentially constant in the turbulent part of the flow because of the thorough mixing from one side of the wake to the other. Hence, the curve representing D should have a shape similar to that of the intermittency γ (Figure 4.3); the dissipation should decrease quite slowly from its value on the axis ($\xi = 0$) to the value $D = -P$ predicted by (4.2.34) near the production peak at $\xi = 1$. This is also sketched in Figure 4.5 with a dashed line.

The expression (4.2.34) for the production, of course, is correct near the center of the wake because $P = 0$ at $\xi = 0$. If advection, which is bringing in turbulent energy, continues to rise as the axis is approached, and if dissipation, which removes energy, does the same, while production falls off sharply, the removal of energy by turbulent transport must decrease near the axis. The decrease is somewhat delayed because the slope of A at $\xi = 1$ is larger than that of D, so that transport must increase for a while. As A and D level off, however, T must decrease. In Figure 4.5 a dashed curve represents this effect.

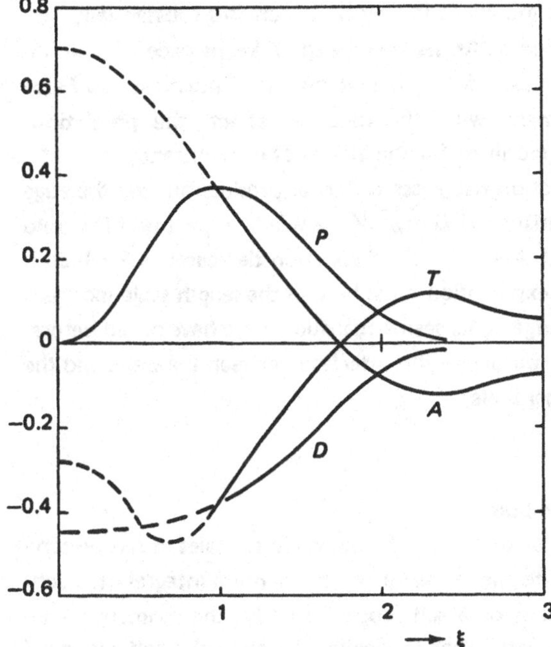

Figure 4.5. The turbulent energy budget of a wake. The solid lines are based on (4.2.33–4.2.35); the dashed lines are extrapolations described in the text.

In the central part of the wake, therefore, the mean-flow transport (advection) deposits turbulent energy, some of which is dissipated locally and some of which is transported toward the outer part of the wake. Most of the energy transported to the outer part of the wake comes from just inboard of the production peak. As an aside, we note that near the center line, gradient-transport (mixing-length) concepts are very poor: there is almost no energy gradient, $\partial(\frac{1}{2}\overline{q^2})/\partial y$, and what little there is has the wrong sign. The energy flux is locally uphill.

The predicted energy budget presented in Figure 4.5 is in good qualitative agreement with the available experimental data. However, the predicted values of advection and transport near the edge of the wake are too small by a factor of about 2. As we saw before, the measured velocity profile in wakes decreases more rapidly than the f calculated on basis of a constant eddy viscosity. Hence, the gradient of the actual f is larger than the gradient of the f that has been used in these predictions (4.2.15). If the measured velocity profile is used to calculate the advection term, it increases substantially and matches the experimental data. As we have seen, T keeps pace with A, independent of what curve is used for f, so that the use of the measured f also brings T in close agreement with the data. In effect, the predictions (4.2.33–4.2.35) should be modified for the effects of intermittency.

The fact that the dissipation decreases as fast as production near the edge of the wake is a little surprising. If $D \sim u^3/\ell$, we would expect that D would be proportional to $(f')^{3/2}/\ell$. Actually, the dissipation decreases as P, which is proportional to $(f')^2$. The explanation must be that the length scale increases as $(f')^{-1/2}$ near the outer edge. This seems realistic; as we have noted before, the eddies responsible for contorting the interface between the wake and the irrotational fluid are of larger scale.

4.3
The wake of a self-propelled body

In order to find the behavior of the length and velocity scales in self-preserving wakes, we were forced to make use of the momentum integral. In a very important practical case, that of a self-propelled body, the momentum integral vanishes. Through its propulsor (propellor, jet engine) a self-propelled body traveling at constant speed adds just enough momentum to cancel the momentum loss due to its drag, so that the wake contains no net momentum deficit. We assume that the body does not operate near an interface of two media, so that no wave drag is involved. Figure 4.6 illustrates this situation. The integral (4.2.8) vanishes identically and the value of n in $\ell \sim x^n$, $U_s \sim x^{n-1}$ remains undetermined.

It is not possible to resolve this problem without making the assumption that ν_T is constant from the beginning of the analysis. In view of the more complex structure of a self-propelled wake, with the secondary extrema of U on either side of the center line, this assumption is even more questionable than it was in the wake with nonzero momentum. For example, at the center line we have $-\overline{uv} = 0$ and $\partial U/\partial y = 0$, so that their ratio is constant because of symmetry. At the secondary extrema, however, symmetry arguments are not applicable, so that there is no reason to expect that $-\overline{uv}$ is zero where $\partial U/\partial y = 0$. All results based on a constant value of ν_T thus have a qualitative significance only. It is particularly important to recognize that the existence of similarity in wakes with finite momentum defect does not depend on the

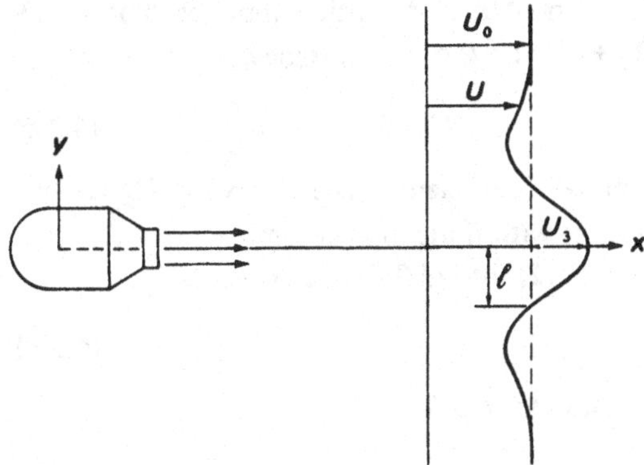

Figure 4.6. The wake of a self-propelled body. The station shown is far downstream of the body.

eddy-viscosity assumption. In the self-propelled wake, however, similarity can be obtained only by assuming that ν_T is independent of y.

Plane wakes Let us consider a plane self-propelled wake. If ν_T is independent of y, we may write the momentum equation (4.1.19) as

$$\frac{\partial}{\partial x}[U_0(U-U_0)] = \nu_T \frac{\partial^2}{\partial y^2}(U-U_0). \tag{4.3.1}$$

Here, the constant U_0 has been subtracted from U for convenience. If we multiply (4.3.1.) by y^n and integrate by parts twice, we obtain

$$\frac{d}{dx}\int_{-\infty}^{\infty} y^n U_0(U-U_0)\,dy = \nu_T n(n-1)\int_{-\infty}^{\infty} y^{n-2}(U-U_0)\,dy. \tag{4.3.2}$$

If we put $n = 2$, the right-hand side of (4.3.2) vanishes, so that we obtain

$$\int_{-\infty}^{\infty} y^2 U_0(U-U_0)\,dy = \text{const.} \tag{4.3.3}$$

If we further assume that the velocity-defect profile is self-preserving, there results

$$U_s\ell^3\int_{-\infty}^{\infty} \xi^2 f(\xi)\,d\xi = \text{const.} \tag{4.3.4}$$

Because self-preservation of the equations of motion requires that $\ell \sim x^n$, $U_s \sim x^{n-1}$ (4.2.6), we obtain $3n + n - 1 = 0$ or $n = \frac{1}{4}$. Hence,

$$U_s = Ax^{-3/4}, \quad \ell = Bx^{1/4}, \tag{4.3.5}$$

where A and B are undetermined coefficients. The decay of U_s is thus substantially faster than in the wake with finite momentum.

If we substitute (4.3.5) and (4.2.12) into (4.2.5), we obtain

$$\alpha(3f + \xi f') + f'' = 0, \tag{4.3.6}$$

where $\alpha = U_0 B R_T/4A$. The solution to (4.3.6) is

$$f = \frac{d^2}{d\xi^2}\left[\exp(-\tfrac{1}{2}\xi^2)\right]. \tag{4.3.7}$$

Here, ℓ has been defined by selecting $\alpha = 1$, as before. The velocity profile (4.3.7) is qualitatively similar to the one sketched in Figure 4.6. No information on the value of R_T in self-propelled wakes is available, although it is not likely to be much different from the value of R_T in ordinary wakes.

From an experimental point of view, it is of interest to ask what would happen if both the self-propelled and the finite-momentum wakes were simultaneously present. Imagine that a slight inaccuracy has been made in satisfying the condition of self-propulsion (zero momentum deficit). The wake then consists of

$$U_0 - U = a\exp(-\tfrac{1}{2}\xi^2) + b\frac{d^2}{d\xi^2}\left[\exp(-\tfrac{1}{2}\xi^2)\right]. \tag{4.3.8}$$

These are the first two terms of a general expansion that could be used for any wake profile (a *Gram-Charlier* expansion). Substitution of (4.3.8) and (4.2.12) in the equation of motion (4.2.5) gives, by equating like powers of ξ,

$$\ell = (2\nu_T x/U_0)^{1/2}, \quad a \propto x^{-1/2}, \quad b \propto x^{-3/2}. \tag{4.3.9}$$

This rather surprising result claims that the presence of a nonzero momentum integral dominates the growth of the length scale and forces quite rapid decay of the self-propelled component of the wake. Consider an attempt to produce a self-propelled wake in the laboratory. If we achieve self-propulsion to the extent that $b/a = 10^2$ at one body diameter (the momentum mismatch then is 1%), it takes only 10^2 body diameters downstream before the self-propelled component is overshadowed by the momentum-deficit component. This may explain why no data on self-preserving, self-propelled wakes are available. The Reynolds number of the self-propelled component of the plane, "mixed" wake varies as x^{-1}, so that this component quickly ceases to be turbulent as it progresses downstream.

Axisymmetric wakes In the case of the axisymmetric wake of a self-propelled body, an analysis similar to that just presented gives $U_s \propto x^{-4/5}$, $\ell \propto x^{1/5}$, so that $R_\ell \propto x^{-3/5}$. In the case of a "mixed" wake with self-propelled and finite-momentum components, the development of the length scale is again forced by the momentum defect, so that $\ell \propto x^{1/3}$. The momentum-defect component then decays as $x^{-2/3}$ and the self-propelled component decays as $x^{-4/3}$. Again, the Reynolds number of the self-propelled component varies as x^{-1}.

The fact that the self-propelled wake decays so much faster than the wake with finite momentum defect has some interesting implications. A maneuvering aircraft or submarine, which is accelerating or decelerating at times, leaves behind it a momentum-defect jet or wake when it is changing speed and a self-propelled wake when it is not. The latter decays much more rapidly. After some time, only the patches of wake representing changes of speed survive.

4.4 Turbulent jets and mixing layers

In jets and mixing layers there are two velocity scales, u and U_s, which are related by $u^2/U_s^2 = \mathcal{O}(\ell/L)$ as given in (4.1.20). It is clear that u/U_s needs to be constant in order to achieve self-preservation. The turbulence must retain the same relative importance as the jet develops; if the relative magnitudes of the turbulence and the mean flow are constantly changing, the flow cannot possibly be self-preserving. Because $u^2/U_s^2 = \mathcal{O}(\ell/L)$, a consequence of u/U_s being constant is that ℓ/L must be constant. Since L is a downstream length scale, we expect that in mixing layers and jets $\ell \propto x$. If ℓ/L is constant, the approximations obtained in Section 4.1 do not improve as x increases. Experiments indicate that $\ell/L \cong 6 \times 10^{-2}$, as was remarked earlier.

Because u is proportional to U_s, either one can be used as a scaling velocity. Let us use U_s, so that we can write

$$U = U_s f(\xi), \tag{4.4.1}$$

$$V = -\int_0^y \frac{\partial U}{\partial x} dy = -\ell \int_0^\xi \left(\frac{dU_s}{dx} f - \frac{U_s}{\ell} \frac{d\ell}{dx} \xi f' \right) d\xi, \tag{4.4.2}$$

$$-\overline{uv} = U_s^2 g(\xi), \tag{4.4.3}$$

$$\xi = y/\ell, \quad \ell = \ell(x), \quad U_s = U_s(x). \tag{4.4.4}$$

Here, as before, primes denote differentiation with respect to ξ; $\xi = 0$ at the center line. We must bear in mind that $g_{max} \neq 1$; instead, we have $g_{max} = u^2/U_s^2 = \mathcal{O}(\ell/L)$.

If we substitute (4.4.1–4.4.3) into (4.1.21), we obtain

$$\frac{\ell}{U_s} \frac{dU_s}{dx} f^2 - \frac{d\ell}{dx} \xi f f' - \frac{\ell}{U_s} \frac{dU_s}{dx} f' \int_0^\xi f \, d\xi + \frac{d\ell}{dx} f' \int_0^\xi \xi f' \, d\xi = g'. \tag{4.4.5}$$

Self-preservation can be obtained only if we require

$$\frac{d\ell}{dx} = A, \quad \frac{\ell}{U_s} \frac{dU_s}{dx} = B, \tag{4.4.6}$$

where A and B are constants. The first of these is not a surprise, because we already knew that $L \sim x$ and that ℓ/L must be constant. The second condition in (4.4.6) can be satisfied by any power law $U_s \propto x^n$, including $n = 0$.

Mixing layers In a mixing layer, the velocity difference U_s is imposed (Figure 4.1) by the external flow. If U_s is constant, (4.4.5) reduces to

$$-\frac{d\ell}{dx} f' \int_0^\xi f \, d\xi = g'. \tag{4.4.7}$$

With the eddy-viscosity assumption (4.2.12), this becomes

$$R_T \frac{d\ell}{dx} f' \int_0^\xi f \, d\xi = f''. \tag{4.4.8}$$

Here, of course, R_T is taken to be constant. It is not possible to obtain a solution of (4.4.8) in closed form. However, for the scale relations this is irrelevant. Let us define ℓ by taking $R_T \, d\ell/dx = 1$, so that all adjustable constants in (4.4.8) are absorbed by ℓ. This corresponds to the normalization used in wakes. The profile predicted by (4.4.8) is in fair agreement with experimental data if

$$R_T = 17.3, \quad \ell = x/17.3 = 5.7 \times 10^{-2}x. \tag{4.4.9}$$

At the edges of the mixing layer there are small discrepancies due to intermittency. The Reynolds number $U_s\ell/\nu$ of mixing layers apparently increases rapidly ($R_\ell \propto x$). Because there is no initial length (such as the jet orifice height or the momentum thickness) in the mixing layer, length scales must be compared with the viscous length ν/U_s. Experiments indicate that the mixing layer becomes self-preserving when $U_s x/\nu > 4 \times 10^5$.

Plane jets In its initial stage of development, a plane jet consists of two plane mixing layers, separated by a core of irrotational flow (Figure 4.7). Some distance after the two mixing layers have merged, the jet becomes a fully developed, self-preserving turbulent flow. The center-line velocity U_s then

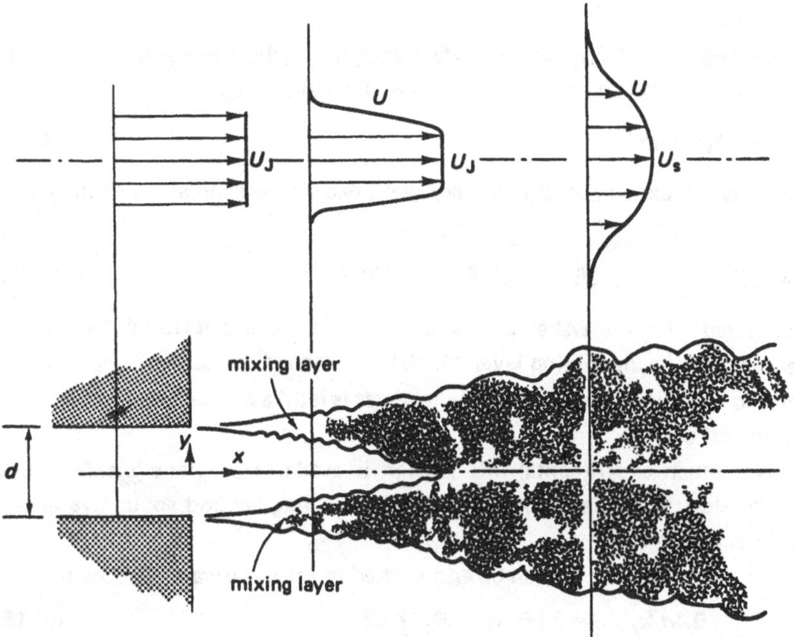

Figure 4.7. A plane turbulent jet. The jet becomes self-preserving some distance after the two mixing layers near the orifice have merged.

varies as x^n ($n \neq 0$), and a momentum integral is needed to determine the power n. If the velocity profile is self-preserving, the momentum integral (4.1.27) becomes

$$\int_{-\infty}^{\infty} U^2\,dy = U_s^2 \ell \int_{-\infty}^{\infty} f(\xi)\,d\xi = U_J^2 d, \tag{4.4.10}$$

where U_J is the initial jet velocity and d is the orifice height (Figure 4.7). We conclude that $2n + 1 = 0$ or $n = -\frac{1}{2}$ in order to make the momentum flux in the jet constant. Thus we obtain, for large enough values of x/d,

$$U_s/U_J = C(x/d)^{-1/2}, \tag{4.4.11}$$

while $\ell = Ax$, as given by (4.4.6). The Reynolds number $R_\ell = U_s\ell/\nu$ increases as $x^{1/2}$, so that the viscous terms become smaller and smaller as x increases. With the use of the eddy-viscosity assumption, (4.4.5) becomes

$$-\frac{1}{2}\frac{d\ell}{dx} R_T \left(f^2 + f' \int_0^\xi f\,d\xi\right) = f''. \tag{4.4.12}$$

If we define ℓ again by taking $d\ell/dx = 2/R_T$ (as in the other cases, this corresponds to $f \cong e^{-1/2}$ at $\xi = 1$), we can solve (4.4.12) to obtain

$$f = \text{sech}^2 (\xi^2/2)^{1/2}. \tag{4.4.13}$$

This fits the experimental data very well, except near the edges of the jet, if we take

$$\ell = 0.078x, \quad R_T = 25.7, \quad U_s/U_J = 2.7(d/x)^{1/2}. \tag{4.4.14}$$

Compared with the wake, the value of R_T in jets is surprisingly large. The value of R_T in the mixing layer (4.4.9) is intermediate between those of the jet and the wake, because the mixing layer is jetlike on one side and wakelike on the other.

Not much experimental information is available on plane jets. Measured mean-velocity profiles appear to be self-preserving beyond about five orifice heights ($x/d > 5$).

The axisymmetric jet can be approached in the same way. We obtain

$$U_s/U_J = 6.4d/x, \quad \ell = 0.067x, \quad R_T = 32. \tag{4.4.15}$$

The Reynolds number $U_s\ell/\nu$ is constant in axisymmetric jets. No measurements have been made beyond 40 orifice diameters. The mean-velocity profile appears to be self-preserving beyond about $x/d = 8$, while the turbulence quantities are still evolving at 40 diameters.

The energy budget in a plane jet If the analysis of Section 4.1 is applied to the turbulent energy budget in a plane jet, we find that to lowest order production balances dissipation. This is too crude; if we want to take advection and transport into account, we have to include terms that are of order $(\ell/L)^{1/2}$ and ℓ/L compared to the leading terms. The full equation reads

$$0 = -U \frac{\partial}{\partial x}(\tfrac{1}{2}\overline{q^2}) - V \frac{\partial}{\partial y}(\tfrac{1}{2}\overline{q^2}) - \overline{uv}\frac{\partial U}{\partial y} - (\overline{u^2} - \overline{v^2})\frac{\partial U}{\partial x}$$

$$-\frac{\partial}{\partial y}\overline{[(\tfrac{1}{2}q^2 + p/\rho)v]} - \epsilon. \tag{4.4.16}$$

We designate the terms by A_1, A_2, P_1, P_2, T, and D. With the same approximations as made in Section 4.2, we can obtain expressions for A_1, A_2, P_1, and T. The only term that presents a problem is P_2, which is a production term caused by normal-stress differences. On grounds of self-preservation we expect that K, defined by

$$\overline{u^2} - \overline{v^2} \equiv K(\overline{u^2} + \overline{v^2}), \tag{4.4.17}$$

is a function of $\xi = y/\ell$ only. The energy in the u component differs from that in the v component because the major production term P_1 feeds energy into $\overline{u^2}$, so that the energy must leak into $\overline{v^2}$ by inertial interaction. The value of the difference depends on the ratio of the supply rate to the leakage rate; this ratio may be expected to be constant because the two rates are determined by the same turbulence dynamics. Hence, we assume that K is not a function of position. Clearly, K is less than unity. If we use (4.2.31), (4.4.17), and the approximate relation $\overline{u^2} + \overline{v^2} \cong \frac{2}{3}q^2$, we can also express P_2 in terms of f.

Even near the edge of the jet ($\xi > 3$), we still have $y/x \ll 1$. Therefore, we do not violate the assumption of a slow evolution, and (4.1.16) remains valid. Approximate expressions for the terms in (4.4.16) near the edge of the jet ($\xi > 1$) are, if we use the mean velocity profile (4.4.13),

$$R_T \ell P_1/U_s^3 = 2f^2, \tag{4.4.18}$$

$$R_T \ell P_2/U_s^3 = 0.28Kf^2, \tag{4.4.19}$$

$$R_T \ell A_1/U_s^3 = -0.58\xi f^2, \tag{4.4.20}$$

$$R_T \ell A_2/U_s^3 = -0.41f, \tag{4.4.21}$$

$$R_T \ell T/U_s^3 = 0.41f. \tag{4.4.22}$$

The dissipation D again can be found by difference. From (4.4.18–4.4.22) it is clear that P_1, P_2, and A_1 all are proportional to f^2, so that near the edge of the jet they become negligible long before A_2 and T, which are proportional to f. As in wakes, we find that A_2 and T have the same numerical coefficient; it can be shown that this is valid for any f if $\overline{uv}/\overline{q^2}$ is constant and if the transport term can be represented by a gradient-transport expression like (4.2.32). Thus, the energy carried by transport from the center of the jet is removed by the second advection term, A_2.

Physically, what is happening is this: near the outer edge of the jet, only one component of the mean velocity, V, is nonzero; it approaches a constant value in the plane jet, thus entraining the fluid surrounding the jet. Because the average boundary of the jet is stationary, turbulent energy must be transported into the "entrainment wind" at just that speed which keeps the average position of the interface stationary. This result is essentially independent of the assumptions embodied in (4.4.18–4.4.22). Because A_2 and T balance, dissipation plays no role. Note that A_1 plays no dominant role here, contrary to the situation in wakes.

Closer to the center line of the jet, the energy budget becomes more complicated. Calculated distributions of the terms in (4.4.16), based on the same approximations that were used for wakes, are presented in Figure 4.8. The mean velocity profile (4.4.13) was used; the second production term (P_2) has not been plotted because it is never larger than -0.003 if K in (4.4.17) is 0.4. The plot shows that A_2 and T balance in the far edge of the jet, as we discussed earlier. Somewhat closer to the center line, the sum of A_1 and A_2 approximately balances T while P and D balance each other, as in wakes. Close to the center line, A_2 becomes negligible and A_1 reverses sign. Also, P_1 must decrease to zero at $\xi = 0$ because $f' = 0$ there, and D levels out near the center line. The energy budget in the center region thus may be expected to be similar to that in the wake (Figure 4.5).

Unfortunately, there are almost no measurements with which this analysis can be compared. Near the edge of a jet, the mean velocity is small, so that the turbulence level, measured as a fraction of the mean velocity, reaches very high values, and reversal of the flow becomes a frequent occurrence. The instruments customarily used to measure turbulence (hot-wire anemometers)

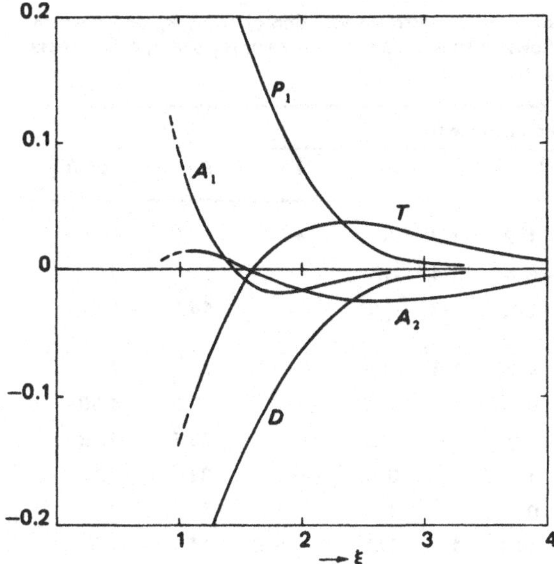

Figure 4.8. Calculated energy budget in the plane jet.

cannot tolerate this situation. However, the agreement between experimental data and predicted values is fairly good in the energy budget of a plane wake, so that we may expect that Figure 4.8, which is based on the same set of assumptions, at least presents a qualitatively correct picture.

4.5
Comparative structure of wakes, jets, and mixing layers

In Table 4.1 are collected the exponents of the power laws describing the downstream variation of U_s, ℓ, and $R_\ell = U_s\ell/\nu$ in the various flows we have examined. Also listed are the exponents, including those of the temperature scale, for buoyant plumes (Section 4.6). The values of R_T of the various flows are also listed.

The large variation in the values of R_T requires some explanation. The definition $R_T = U_s\ell/\nu_T$ uses the velocity scale U_s rather than a velocity scale characteristic of the turbulence. For jets and mixing layers, $u^2/U_s^2 \sim \ell/L$, so that the use of a suitably defined u should substantially reduce the value of R_T. Let us define a velocity scale u_* characteristic for the turbulence by

Table 4.1. Powers of x describing the downstream variation of U_s, ℓ, R_ℓ, and the temperature scale T of free shear flows. Also listed are the values of R_T and $u_* \ell_* / \nu_T$; these parameters are independent of position.

	Powers of x for					
	U_s	ℓ	R_ℓ	T	R_T	$u_* \ell_* / \nu_T$
Plane wake	−1/2	1/2	0	—	12.5	2.75
Self-propelled plane wake	−3/4	1/4	−1/2	—	?	?
Axisymmetric wake	−2/3	1/3	−1/3	—	14.1	2.92
Self-propelled axisymmetric wake	−4/5	1/5	−3/5	—	?	?
Mixing layer	0	1	1	—	17.3	4.00
Plane jet	−1/2	1	1/2	—	25.7	4.18
Axisymmetric jet	−1	1	0	—	32	4.78
Plane plume	0	1	1	−1	?	?
Axisymmetric plume	−1/3	1	2/3	−5/3	14	2.9

$$u_*^2 \equiv \max(-\overline{uv}) = \frac{U_s^2}{R_T}\max(f'). \tag{4.5.1}$$

The maximum value of f' is, of course, different in each case we have discussed. Also, the definition of ℓ varies somewhat from case to case. It would be preferable to use a length scale ℓ_* such that U_s/ℓ_* is the same fraction of the maximum of $\partial U/\partial y$; a convenient number is $e^{1/2}$, because that is the inverse of the maximum slope for plane and axisymmetric wakes. Thus,

$$\max\left(\frac{\partial U}{\partial y}\right) = \frac{U_s}{\ell}\max(f') = \frac{U_s}{\ell_*}e^{-1/2}, \tag{4.5.2}$$

which yields

$$\ell/\ell_* = e^{1/2}\max(f'). \tag{4.5.3}$$

A more meaningful turbulent Reynolds number, which allows us to compare all of the boundary-free shear flows on an equal footing, can now be defined as

$$\frac{u_* \ell_*}{\nu_T} \equiv \left\{\frac{R_T}{e\max(f')}\right\}^{1/2}. \tag{4.5.4}$$

The value of max(f') can be computed from the mean-velocity profile of each flow. If these numbers are substituted into (4.5.4), the values of $U_* \ell_* / \nu_T$ given in Table 4.1 are obtained.

The values of $u_* \ell_* / \nu_T$ clearly separate into two groups, wakes on the one hand, jets and mixing layers on the other hand. Within each group the variations in $u_* \ell_* / \nu_T$ are probably not significant, although there seems to be a consistent tendency for axisymmetric flows to have higher values than plane flows.

The difference between the two groups of flows requires explanation. The only quantity in $u_* \ell_* / \nu_T$ which is open to question is ℓ_*: it is related in a uniform way to the slope of the mean-velocity profile, but we do not know how it is related to the length scale of the turbulent eddies. Suppose that the cross-stream scales of eddies which contribute to the momentum transport in jets and mixing layers are smaller than they are in wakes. We expect that the eddy viscosity $\nu_T \sim u_* \ell_t$, where ℓ_t is a turbulence length scale. The value of $u_* \ell_* / \nu_T$ would then be effectively equal to ℓ_* / ℓ_t. In order to explain the observed difference, the value of ℓ_* / ℓ_t in jets and mixing layers needs to be about 1.5 times the value in wakes. How can we explain this?

The one important way in which jets and mixing layers differ from wakes is that the cross-stream advection term $V\,\partial U/\partial y$ is of the same order as $U\,\partial U/\partial x$ in jets and mixing layers, while the former is negligible compared to the latter in wakes. In jets and mixing layers, therefore, as much momentum is carried by the transverse flow as by the downstream flow. The transverse flow has a strain rate ($\partial V/\partial y$) associated with it, which tends to compress eddies in the cross-stream direction. This may explain why these eddies tend to have smaller length scales than those in wakes. In fact, a crude calculation (Townsend, 1956) indicates that the expected compression factor is about 1.5. This is in agreement with observations on the intermittency γ in axisymmetric jets. The region over which γ decreases from one to zero in jets is much narrower than that in wakes, implying that the large eddies, which are responsible for contorting the interface, are indeed relatively small.

4.6
Thermal plumes

In a medium that expands on heating, a body that is hotter than its surroundings produces an upward jet of heated fluid which is driven by the density difference. The most familiar example is the plume from a cigarette in a quiet room. Atmospheric thermals rising over a surface feature of high temperature and plumes from smokestacks are other common examples. Also, if liquid of a certain density is poured into a liquid of lower density, it forms an upside-down density-driven plume. These flows can be analyzed in the same way as wakes and jets by employing the concept of self-preservation (Zel'dovitch, 1937). In the atmospheric examples we will study, we have to assume that the environment is neutrally stable. The stability plays the role of a streamwise pressure gradient; no self-preserving solutions can be expected if the stability is an arbitrary function of height.

We restrict the analysis to thermal plumes in the atmosphere, in which density differences are created by temperature differences. We use the Boussinesq approximation to the equations of motion, which was introduced in Section 3.4. We recall that in the Boussinesq approximation, the buoyancy term $-g\rho'/\bar{\rho}$ is replaced by $g\vartheta/\Theta_0$, where Θ_0 is the temperature of the adiabatic atmosphere and ϑ is the difference between the actual temperature and Θ_0. The temperature difference ϑ is decomposed into a mean value $\bar{\vartheta}$ and temperature fluctuations θ $(\bar{\theta} \equiv 0)$. If $\bar{\vartheta} = 0$, the atmosphere is neutrally stable. If $\bar{\vartheta}$ increases upward, the atmosphere is stable; if it decreases upward, the atmosphere is unstable.

The Mach number of these plumes is presumed to be low, so that the continuity equation retains its customary form. If the acceleration of gravity points toward the negative x_3 direction, the equations of mean motion and mean temperature difference are

$$U_j \frac{\partial U_i}{\partial x_j} + \frac{\partial}{\partial x_j} \overline{u_i u_j} = -\frac{1}{\rho} \frac{\partial P}{\partial x_i} + \nu \frac{\partial^2 U_i}{\partial x_j \partial x_j} + \frac{g}{\Theta_0} \bar{\vartheta} \delta_{i3}, \tag{4.6.1}$$

$$\frac{\partial U_i}{\partial x_i} = 0, \tag{4.6.2}$$

$$U_j \frac{\partial \bar{\vartheta}}{\partial x_j} + \frac{\partial}{\partial x_j} \overline{\theta u_j} = \gamma \frac{\partial^2 \bar{\vartheta}}{\partial x_j \partial x_j}. \tag{4.6.3}$$

Two-dimensional plumes Let us consider two-dimensional plumes driven by a line source of heat (Figure 4.9). We take the z axis to be vertically upward. The line source is assumed to be parallel to the y axis, so that

$$V = 0, \quad \partial/\partial y = 0. \tag{4.6.4}$$

We assume that the flow in the plume is nearly parallel, just as in ordinary jets; we develop approximate equations of motion based on this premise. Referring to Figure 4.9, we have

$$\partial/\partial z \sim 1/L, \quad \partial/\partial x \sim 1/\ell, \quad \partial W/\partial x \sim U_s/\ell. \tag{4.6.5}$$

Substituting these estimates into (4.62), we obtain for the horizontal velocity component

$$U \sim \ell U_s/L. \tag{4.6.6}$$

We further take the turbulent velocity fluctuations to be of order u, the turbulent temperature fluctuations to be of order t, and ϑ to be of order T. The relations between these scales must be determined in the course of the analysis.

The x-momentum equation is exactly the same as (4.1.7) expressed in the proper coordinate system because the buoyancy term occurs only in the equation for the z momentum. Hence, the orders of magnitude are the same as those given in (4.1.8). Thus, with the provisions expressed in (4.1.10), (4.1.9) holds for plane plumes:

$$\rho\overline{u^2} + P = P_0. \tag{4.6.7}$$

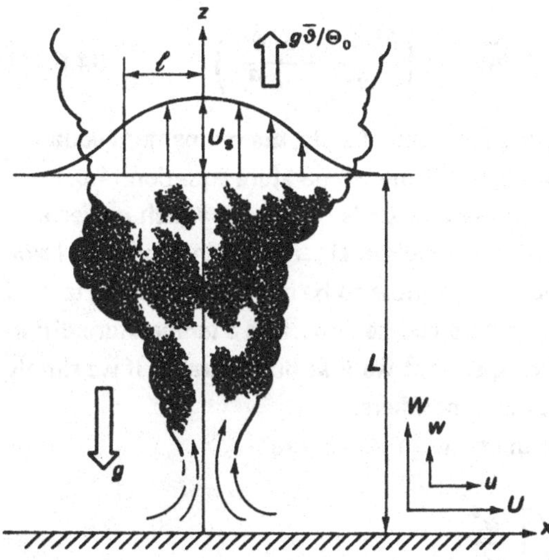

Figure 4.9. Plane thermal plume.

Substitution of (4.6.7) into the z component of (4.6.1) gives

$$U\frac{\partial W}{\partial x}+W\frac{\partial W}{\partial z}+\frac{\partial}{\partial x}(\overline{uw})+\frac{\partial}{\partial z}(\overline{w^2}-\overline{u^2})$$

$$=-\frac{1}{\rho}\frac{dP_0}{dz}+\frac{g}{\Theta_0}\overline{\vartheta}+\nu\left(\frac{\partial^2 W}{\partial x^2}+\frac{\partial^2 W}{\partial z^2}\right). \tag{4.6.8}$$

Far away from the center of the plume, where there is no flow or turbulence, (4.6.8) reduces to

$$0=-\frac{1}{\rho}\frac{\partial P_0}{\partial z}+\frac{g}{\Theta_0}\overline{\vartheta}_0. \tag{4.6.9}$$

Hence, the sum of the pressure term and the buoyancy term in (4.6.8) may be written as

$$-\frac{1}{\rho}\frac{\partial P_0}{\partial z}+\frac{g}{\Theta_0}\overline{\vartheta}=\frac{g}{\Theta_0}(\overline{\vartheta}-\overline{\vartheta_0}). \tag{4.6.10}$$

In plane plumes, the temperature equation reads

$$U\frac{\partial\overline{\vartheta}}{\partial x}+W\frac{\partial\overline{\vartheta}}{\partial z}+\frac{\partial}{\partial x}(\overline{\theta u})+\frac{\partial}{\partial z}(\overline{\theta w})=\gamma\left(\frac{\partial^2\overline{\vartheta}}{\partial x^2}+\frac{\partial^2\overline{\vartheta}}{\partial z^2}\right). \tag{4.6.11}$$

We have to assume that $\overline{\vartheta}_0$ is independent of z. In the momentum equation only the difference $\overline{\vartheta}-\overline{\vartheta}_0$ appears, but if the temperature equation (4.6.11) is written in terms of $\overline{\vartheta}-\overline{\vartheta}_0$, a term $W\,\partial\overline{\vartheta}_0/\partial z$ is generated, which makes self-preservation impossible. We would have additional terms on the right-hand side of (4.6.11), too, but we do not expect those to be dynamically important. If $\overline{\vartheta}_0$ is constant, the mean temperature can be written as a temperature difference everywhere in the equations, so that we lose no generality if we simply put $\overline{\vartheta}_0=0$, which means a neutral atmosphere.

The orders of magnitude of the terms in (4.6.8) are

$$U\frac{\partial W}{\partial x}:\ \frac{U_s^2}{\ell}\frac{\ell}{L}=\left[\frac{U_s^2}{u^2}\frac{\ell}{L}\right]\frac{u^2}{\ell},$$

$$W\frac{\partial W}{\partial z}: \frac{U_s^2}{L} = \left[\frac{U_s^2}{u^2}\frac{\ell}{L}\right]\frac{u^2}{\ell},$$

$$\frac{\partial \overline{uw}}{\partial x}: \frac{u^2}{\ell},$$

$$\frac{\partial(\overline{w^2} - \overline{u^2})}{\partial z}: \frac{u^2}{L} = \frac{\ell}{L}\cdot\frac{u^2}{\ell},$$

$$\frac{g}{\Theta_0}\overline{\vartheta}: \frac{gT}{\Theta_0} = \left[\frac{g T\ell}{\Theta_0 u^2}\right]\frac{u^2}{\ell},$$

$$\nu\frac{\partial^2 W}{\partial x^2}: \nu\frac{U_s}{\ell^2} = \left[R_\ell^{-1}\frac{U_s}{u}\right]\frac{u^2}{\ell},$$

$$\nu\frac{\partial^2 W}{\partial z^2}: \nu\frac{U_s}{L^2} = \left[R_\ell^{-1}\frac{U_s}{u}\left(\frac{\ell}{L}\right)^2\right]\frac{u^2}{\ell}. \tag{4.6.12}$$

In order to have any mean-flow terms at all, we must again assume that

$$u/U_s = \mathcal{O}(\ell/L)^{1/2}. \tag{4.6.13}$$

The scaling thus is the same as in the "mechanical" jet, so that we have, to first order,

$$U\frac{\partial W}{\partial x} + W\frac{\partial W}{\partial z} + \frac{\partial}{\partial x}\overline{uw} = \frac{g}{\Theta_0}\overline{\vartheta}. \tag{4.6.14}$$

Note that the pressure term has been removed from (4.6.11) with (4.6.10) and $\overline{\vartheta}_0 = 0$. The temperature term has been kept in (4.6.11), although we do not know its magnitude yet. If we want thermal effects to be as important as the Reynolds stress, we need

$$gT/\Theta_0 = \mathcal{O}(u^2/\ell). \tag{4.6.15}$$

The orders of magnitude of the various terms in the temperature equation (4.6.11) become

$$U\frac{\partial\bar{\vartheta}}{\partial x}:\ \frac{T}{\ell}U_s\frac{\ell}{L}=\left[\frac{\ell}{L}\frac{T}{t}\frac{U_s}{u}\right]\frac{tu}{\ell},$$

$$W\frac{\partial\bar{\vartheta}}{\partial z}:\ \frac{T}{L}U_s=\left[\frac{\ell}{L}\frac{T}{t}\frac{U_s}{u}\right]\frac{tu}{\ell},$$

$$\frac{\partial}{\partial x}\overline{\theta u}:\frac{tu}{\ell},$$

$$\frac{\partial}{\partial z}\overline{\theta w}:\ \frac{tu}{L}=\frac{\ell}{L}\cdot\frac{tu}{\ell},\tag{4.6.16}$$

$$\gamma\frac{\partial^2\bar{\vartheta}}{\partial x^2}:\ \gamma\frac{T}{\ell^2}=\left[\frac{\gamma}{\nu}\frac{T}{t}\frac{1}{R_\ell}\right]\frac{tu}{\ell},$$

$$\gamma\frac{\partial^2\bar{\vartheta}}{\partial z^2}:\ \gamma\frac{T}{L^2}=\left[\frac{\gamma}{\nu}\frac{T}{t}\frac{1}{R_\ell}\left(\frac{\ell}{L}\right)^2\right]\frac{tu}{\ell}.$$

In order to have a term which is of the same order as the third, we must require that

$$\ell TU_s=\mathcal{O}(L\,tu).\tag{4.6.17}$$

If the molecular diffusion terms in (4.6.12) and (4.6.16) are to be of the same order as the neglected turbulent transport terms, we need

$$\frac{U_s}{u}R_\ell^{-1}=\mathcal{O}(\ell/L),\quad \frac{\gamma}{\nu}\frac{T}{t}R_\ell^{-1}=\mathcal{O}(\ell/L).\tag{4.6.18}$$

With the aid of (4.6.13) and (4.6.17), these conditions reduce to

$$R_\ell^{-1}=\mathcal{O}(\ell/L)^{3/2},\quad (\gamma/\nu)R_\ell^{-1}=\mathcal{O}(\ell/L)^{3/2}.\tag{4.6.19}$$

In gases, $\gamma/\nu\cong 1$, so that the provisions (4.6.19) are equally stringent. Of course, if R_ℓ is larger than $(L/\ell)^{3/2}$, the molecular terms in (4.6.12, 4.6.16) are even smaller than the neglected transport terms.

With these provisions, the temperature equation reduces to

$$U\frac{\partial\bar{\vartheta}}{\partial x}+W\frac{\partial\bar{\vartheta}}{\partial z}+\frac{\partial}{\partial x}\overline{\theta u}=0.\tag{4.6.20}$$

The combination of (4.6.13) and (4.6.17) gives

$$t/T = \mathcal{O}(u/U_s) = \mathcal{O}(\ell/L)^{1/2}. \tag{4.6.21}$$

Self-preservation In order to have self-preservation, t/T and u/U_s need to be constant, because the temperature fluctuations θ should play the same relative role in the mean-temperature field $\overline{\vartheta}$ at all z, and the velocity fluctuations u,w should have the same relative importance in the mean-velocity field U,W at all z. We conclude from (4.6.21) that these requirements are consistent with the approximations developed so far if ℓ/L is a constant. Because z is the only possible choice for L, thermal plumes grow linearly ($\ell \propto z$), just like jets.

Because t/T and u/U_s are constant, we can use T and U_s as scales of temperature and velocity. The assumption of self-preservation then can be expressed as

$$W = U_s f(x/\ell) = U_s f(\xi),$$

$$U = -\ell \int_0^{\xi} \left(\frac{dU_s}{dz} f - \frac{U_s}{\ell} \frac{d\ell}{dx} \xi f' \right) d\xi, \tag{4.6.22}$$

$$-\overline{uw} = U_s^2 g(\xi), \quad -\overline{\theta u} = T U_s h(\xi), \quad \overline{\vartheta} = TF(\xi),$$

where $\ell = \ell(z)$, $U_s = U_s(z)$, $T = T(z)$, and $\xi = x/\ell$. If (4.6.22) is substituted into (4.6.14) and (4.6.20), there results

$$\frac{\ell}{U_s}\frac{dU_s}{dz} f' \int_0^{\xi} f\, d\xi + \frac{d\ell}{dz} f' \int_0^{\xi} \xi f'\, d\xi + \frac{\ell}{U_s}\frac{dU_s}{dz} f^2 - \frac{d\ell}{dz} \xi f f' - g' = \frac{g}{\Theta_0}\frac{T\ell}{U_s^2} F, \tag{4.6.23}$$

$$-\frac{\ell}{U_s}\frac{dU_s}{dz} F' \int_0^{\xi} \xi f\, d\xi + \frac{d\ell}{dz} F' \int_0^{\xi} \xi f'\, d\xi + \frac{\ell}{T}\frac{dT}{dz} fF - \frac{d\ell}{dz} \xi f F' = h'. \tag{4.6.24}$$

Here the primes denote differentiation with respect to ξ. If we are to obtain self-preservation, the coefficients in (4.6.23, 4.6.24) must be constant:

$$\frac{\ell}{U_s}\frac{dU_s}{dz} = c_1, \quad \frac{d\ell}{dz} = c_2, \quad \frac{\ell}{T}\frac{dT}{dz} = c_3, \quad \frac{g}{\Theta_0}\frac{T\ell}{U_s^2} = c_4. \tag{4.6.25}$$

We clearly need linear growth of the plume; that is, $\ell = c_2 z$. The first and third relations in (4.6.25) only state that U_s and T must be powers of z. If $U_s \propto z^n$ and $T \propto z^m$, the fourth relation in (4.6.25) gives $m + 1 = 2n$, so that

$$U_s = Az^n, \quad T = Bz^{2n-1}. \tag{4.6.26}$$

We obviously need a constraint similar to a momentum integral. However, momentum is not conserved in a plume because the potential energy represented by the buoyancy is being converted into kinetic energy, so that the momentum is continually increasing. Instead, an integral related to the amount of heat added per unit time is constant.

The heat-flux integral Let us take (4.6.20) and rewrite it, with the help of the continuity equation, as

$$\frac{\partial}{\partial x}(\overline{\vartheta}\, U) + \frac{\partial}{\partial z}(\overline{\vartheta} W) + \frac{\partial}{\partial x}(\overline{\theta u}) = 0. \tag{4.6.27}$$

This may be integrated with respect to x, which yields

$$\int_{-\infty}^{\infty} \overline{\vartheta} W \, dx = \text{const} = \frac{H}{\rho c_p}. \tag{4.6.28}$$

The constant may be identified as $H/\rho c_p$, where H is the total heat flux in the plume, because $\rho c_p\, \overline{\vartheta}$ is the amount of heat per unit volume and $W\,dx$ is the volume flux per unit depth. Substituting the first and last of (4.6.22) into (4.6.28), we obtain

$$\ell T U_s \int_{-\infty}^{\infty} fF \, d\xi = \frac{H}{\rho c_p}. \tag{4.6.29}$$

Therefore, with (4.6.26) and $\ell = c_2 z$, we find

$$U_s = \text{const}, \quad T = Bz^{-1}. \tag{4.6.30}$$

If exactly the same reasoning is applied to axisymmetric plumes, we obtain

$$\ell \propto z, \quad U_s \propto z^{-1/3}, \quad T \propto z^{-5/3}. \tag{4.6.31}$$

Further results Let us return to the equations for the plane plume. Because $dU_s/dz = 0$ by virtue of (4.6.30), several terms in (4.6.23) and (4.6.24) are zero. With a little manipulation, the equation of motion reduces to

$$-\frac{d\ell}{dz}f'\int_0^\xi f\,d\xi - g' = \frac{g}{\Theta_0}\frac{T\ell}{U_s^2}F. \tag{4.6.32}$$

The simplified temperature equation can be integrated once, to yield

$$-\frac{d\ell}{dz}F\int_0^\xi f\,d\xi = h. \tag{4.6.33}$$

The presence of $d\ell/dz \sim u^2/U_s^2$ is due to the use of $-\overline{uw} = U_s^2\, g$ rather than $u^2 g$: g is not of order one, but of order $d\ell/dz$.

The set (4.6.32, 4.6.33) can be solved only if the turbulent transport of momentum and heat is represented by mixing-length expressions. The eddy viscosity ν_T and the eddy thermal diffusivity γ_T may be assumed to be constant. The turbulent Prandtl number γ_T/ν_T may be taken to be equal to one, because the horizontal temperature transport depends mainly on the temperature fluctuations produced by the horizontal temperature gradient, so that temperature transport is governed by the same mechanism as momentum transport. As in "mechanical" jets, ℓ may be defined by putting $d\ell/dz \equiv 1/R_T$. No experimental data on R_T in plane plumes are available, but in axisymmetric plumes the value of R_T is about 14, with $d\ell/dz \cong 1/R_T$ if ℓ is taken as the value of x where $f \cong \exp(-\frac{1}{2})$ (Rouse, Yih, and Humphreys, 1952). This value is comparable to that in wakes, but it is substantially smaller than that in mechanical jets (Table 4.1). The entrainment wind apparently does not reduce the size of eddies in plumes. This is due to the stable temperature gradient near the plume, which compresses eddies vertically and expands them horizontally. We leave it to the reader to convince himself that this effect quantitatively tends to balance the horizontal compression caused by the entrainment wind during the life of a rising eddy.

If mixing-length expressions for $\overline{uw}$ and $\overline{\theta w}$ are substituted into (4.6.32) and (4.6.33), there results

$$-f'\int_0^\xi f\,d\xi - f'' = \frac{g}{\Theta_0}\frac{T\ell R_T}{U_s^2}F, \tag{4.6.34}$$

$$-F\int_0^\xi f\,d\xi = F'. \tag{4.6.35}$$

These equations incorporate the assumptions $d\ell/dz = 1/R_T$ and $\gamma_T = \nu_T$. At the center line of the plume, $F = 1$ *and* $f = 1$ by definition. If the shape of f may be approximated by $\exp(-\frac{1}{2}\xi^2)$, $f'' = -1$ at $\xi = 0$. At the center line, the first term of (4.6.34) vanishes, so that we obtain

$$\frac{g}{\Theta_0}\frac{T\ell R_T}{U_s^2} = 1. \tag{4.6.36}$$

The integral in (4.6.29) is about one if $F \cong f \cong \exp(-\frac{1}{2}\xi^2)$. Therefore, we obtain the approximate relation

$$\ell T U_s \cong H/\rho c_p. \tag{4.6.37}$$

From (4.6.36), (4.6.37), and $\ell = z/R_T$, we obtain

$$U_s^3 \cong \frac{gR_T H}{\Theta_0 \rho c_p}, \tag{4.6.38}$$

$$T \cong \frac{R_T H}{\rho c_p U_s z}. \tag{4.6.39}$$

With $R_T = 14$, T and U_s can be determined if the heat flux is known.

Problems

4.1 Consider an axisymmetric jet that issues from an orifice of diameter d with a velocity U_0. The ambient fluid is not at rest but moves in the same direction as the jet, with a velocity $0.1U_0$. Describe the early and late stages of development of this jet.

4.2 A very long cylinder (diameter 1 mm) is placed perpendicular to a steady airstream whose velocity is 10 m/sec. The cylinder is heated electrically; the power input is 100 watts per meter span. At what distance downstream is the rms temperature fluctuation in the wake of the cylinder reduced to 1°C? Assume that the distribution of the mean temperature difference in the wake is similar to the distribution of the mean velocity defect. For air at room temperature and pressure, $\rho = 1.25$ kg/m^3, $c_p \cong 10^3$ joule/kg°C.

4.3 A Boeing 747 taxies away from the airport gate. The pilot applies a thrust of 10,000 lb (5×10^4 newton) per engine; the engines are at a height of about 4 m above the ground. The jet exhaust is initially hot, but it rapidly cools through mixing with the ambient air. For the purposes of this problem, the initial jet velocity may be taken as the one that produces the correct amount of thrust at ambient density through the 1-m-diam engine exhaust. How far behind the engine must a 2-m-tall man stand to be reasonably sure that he will not encounter gusts (mean plus fluctuating velocities) greater than 10 m/sec? As a rule of thumb, you may assume that the probability of encountering a velocity fluctuation greater than three times the rms value is negligible.

4.4 Fresh cooling water from a nuclear power station at a river mouth is pumped out to sea in a large pipe and released at the bottom to avoid thermal pollution. Assuming that the cooling water rises as an axisymmetric density-driven plume, at what depth must the cooling water be released to avoid raising the temperature in the first 30 m below the surface by more than 1°C? The volume flow of cooling water is 10 m^3/sec; the temperature and density at the point of release are 100°C and 0.96kg/m^3, respectively. At 5°C, the density of fresh water is 1 kg/m^3, and the density of sea water is 1.03 kg/m^3.

5

WALL-BOUNDED SHEAR FLOWS

Boundary-layer flows are more complicated than flows in free shear layers because the presence of a solid wall imposes constraints that are absent in wakes and jets. The most obvious constraint is that the viscosity of the fluid, no matter how small it is, enforces the no-slip condition: the velocity of the fluid at a solid surface must be equal to the velocity of the surface. This viscous constraint gives rise to a viscosity-dominated characteristic length, which is of order ν/w if w is characteristic of the level of turbulent velocity fluctuations. At large Reynolds numbers, the boundary-layer thickness δ is much larger than ν/w, so that we have to deal with two different length scales simultaneously. This problem will be thoroughly discussed for turbulent flow in channels and pipes. After the consequences of the presence of more than one length scale are fully understood, turbulent boundary layers in the atmosphere and turbulent boundary layers in pressure gradients will be studied.

5.1
The problem of multiple scales

It is instructive to take a preliminary look at the problem of multiple scales. We do so in a qualitative way, leaving the analytical details for Section 5.2. The solid wall may be smooth or rough, so that we have a small viscous length ν/w or a characteristic height k of the roughness elements in addition to the boundary-layer thickness δ. Because δ is generally much larger than ν/w and/or k, we expect that the latter do not influence the entire flow. Instead, we expect that these small length scales control the dynamics of the flow only in some narrow region in the immediate vicinity of the surface. This region, called the wall layer or surface layer, has an asymptotic behavior in the limit as $\delta w/\nu \to \infty$ or $\delta/k \to \infty$, which is quite distinct from the overall development of the boundary layer. Therefore, we must treat boundary layers in a piecemeal fashion by first dealing with the surface layer and the rest of the flow (which is called the outer layer) separately and then reconciling these partial descriptions with appropriate asymptotic methods.

As in boundary-free shear flows, a comprehensive analysis of boundary-layer flows can be performed only if the downstream evolution is slow. If L is a streamwise length scale, we need to require that $\delta/L \ll 1$ in order to make sure that only the local scales δ, ν/w, and w are relevant in the dimensional analysis.

Inertial sublayer There exists a close analogy between the spatial structure of turbulent boundary layers and the spectral structure of turbulence. At sufficiently large Reynolds numbers, the overall dynamics of turbulent boundary layers is independent of viscosity, just as the large-scale spectral dynamics of turbulence is. In the wall layer of a turbulent boundary layer, viscosity generates a "sink" for momentum, much like the dissipative sink for kinetic energy at the small-scale end of the turbulence spectrum. In particular, the asymptotic rules governing the link between the large-scale description and the small-scale description lead to the closely related concepts of an inertial subrange in the turbulence energy spectrum (see Chapter 8) and an inertial sublayer in wall-bounded shear flows. In the literature, the inertial sublayer is called the logarithmic region because its mean-velocity profile is logarithmic, as we shall see later.

A preview of the concept of an inertial sublayer is in order. If the length-scale ratio $\delta w/\nu$ is large enough, it should be possible to find a range of distances y from the surface such that $yw/\nu \gg 1$ and $y/\delta \ll 1$ simultaneously. In this region, the length scale ν/w is presumably too small to control the dynamics of the flow, and the length scale δ is presumably too large to be effective. If this occurs, the distance y itself is the only relevant length.

A graphical representation of the situation is given in Figure 5.1. If w is representative of the turbulence intensity and if no other characteristic velocities occur in the problem, the mean-velocity gradient $\partial U/\partial y$ can depend on w and y only in the following way:

$$\partial U/\partial y = cw/y. \tag{5.1.1}$$

This integrates to

$$U/w = c \ln y + d. \tag{5.1.2}$$

Under the assumptions already stated, (5.1.1) is a dimensional necessity, so that we may expect to find a logarithmic velocity profile wherever $yw/\nu \gg 1$ and $y/\delta \ll 1$ (see also Section 2.5).

In most boundary-layer flows, the velocity scale w is not known a priori. It turns out that (5.1.2) is a crucial link in the determination of the dependence of w on the independent variables of the problem.

Velocity-defect law As in wake flow, the scaling length for most of the boundary layer (with exclusion of the surface layer) is the thickness δ. This is

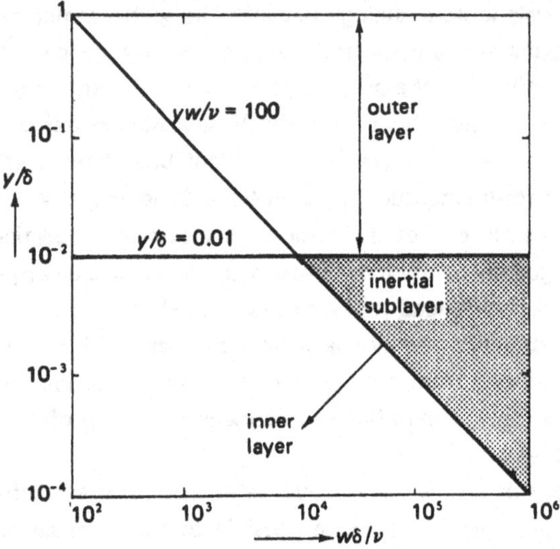

Figure 5.1. An inertial sublayer can exist only if the Reynolds number is large enough. For illustrative purposes, $yw/\nu >> 1$ and $y/\delta << 1$ have been interpreted as $yw/\nu > 100$, $y/\delta < 0.01$. For many practical applications, the limits do not need to be as strict as this.

the appropriate length because the large eddies in the flow have sizes comparable to δ. If the turbulence in a boundary layer is driven by Reynolds stresses, the mean-velocity gradient $\partial U/\partial y$, which is the reciprocal of a "transverse" time scale for the mean flow, has to be of order w/δ if w is the scaling velocity for the Reynolds stress. This argument does not apply to the flow near the surface, because the length scale is different there. The differential similarity law

$$\partial U/\partial y = (w/\delta)\, f(y/\delta) \tag{5.1.3}$$

thus has to be integrated from outside the boundary layer toward the wall in order to obtain a similarity law for U. The result is

$$U - U_0 = -(w/\delta) \int_y^\infty f(y/\delta)\, dy = wF(y/\delta), \tag{5.1.4}$$

where U_0 is the velocity outside the boundary layer. We find later in this chapter that self-preservation can be obtained only if $w/U_0 << 1$. However, a velocity defect $(U_0 - U)$ of order w can never meet the no-slip condition $U_0 - U = U_0$ at the surface if $w/U_0 << 1$. This indicates that a dynamically distinct surface layer with very steep velocity gradients must exist in order to satisfy the boundary condition. If the velocity and length scales in the surface layer are w and ν/w, respectively, the velocity gradients must be of order w^2/ν; hence, they are very large compared to the velocity gradients in the outer layer (which are of order w/δ) if $w\delta/\nu$ is large enough.

5.2
Turbulent flows in pipes and channels

The equations of motion for turbulent flows in pipes and in channels with parallel walls are relatively simple, because the geometry prohibits the continuing growth of their thickness. If the pipe or channel is long enough, the velocity profile has to become independent of the downstream distance x. As a result, the nonlinear inertia terms $U_j \partial U_i / \partial x_j$ are suppressed. This simplifies the theoretical analysis considerably and separates the surface layer–outer layer problem from the problems associated with the downstream development in other wall-bounded shear flows.

Channel flow We consider turbulent flow of an incompressible fluid between two parallel plates separated at a distance $2h$. The plates are assumed infinitely long and wide; they are at rest with respect to the coordinate system used. A definition sketch is given in Figure 5.2. The mean flow is assumed to be in the x,y plane and steady, and all derivatives of mean quantities normal to that plane are assumed to be zero. All derivatives with respect to x are also assumed to be zero, except for the pressure gradient dP/dx, which drives the

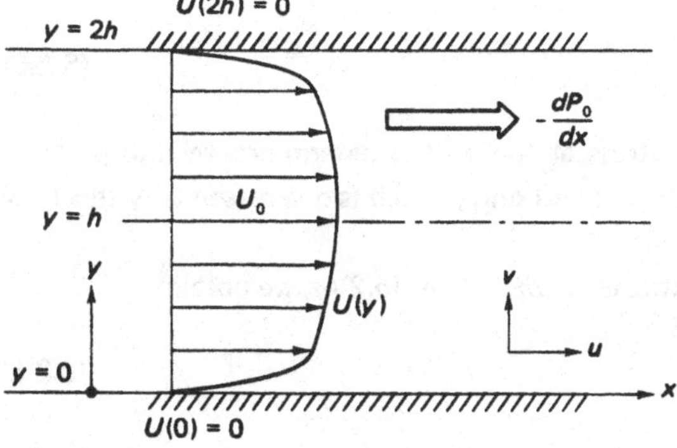

Figure 5.2. Definition sketch for flow between plane parallel walls.

flow against the shear stresses at the two walls. The continuity equation requires that the y component of the mean velocity is zero everywhere if it is zero at both walls.

The relevant equations of motion for the mean flow are

$$0 = -\frac{1}{\rho}\frac{\partial P}{\partial x} - \frac{d}{dy}\overline{uv} + \nu \frac{d^2 U}{dy^2}, \tag{5.2.1}$$

$$0 = -\frac{1}{\rho}\frac{\partial P}{\partial y} - \frac{d}{dy}\overline{v^2}. \tag{5.2.2}$$

Integration of (5.2.2) yields

$$P/\rho + \overline{v^2} = P_0/\rho, \tag{5.2.3}$$

where P_0 is a function of x only. Because $\overline{v^2}$ is independent of x (by assumption), $\partial P/\partial x$ is equal to dP_0/dx. Both of these gradients should be independent of x to avoid streamwise acceleration of the flow. Therefore, (5.2.1) can be integrated from $y = 0$ upward, to yield

$$0 = -\frac{y}{\rho}\frac{dP_0}{dx} - \overline{uv} + \nu \frac{dU}{dy} - u_*^2. \tag{5.2.4}$$

As in Section 2.5, the stress at the surface has been defined as ρu_*^2; the velocity u_* is called the *friction velocity*. The turbulent velocity fluctuations have to satisfy the no-slip condition, so that the Reynolds stress is zero at the surface. The surface stress is thus purely viscous stress.

At the center of the channel ($y = h$), the shear stress ($-\rho\overline{uv} + \mu\, dU/dy$) must be zero for reasons of symmetry. Hence, if $y = h$, (5.2.4) reads

$$u_*^2 = -\frac{h}{\rho}\frac{dP_0}{dx}. \tag{5.2.5}$$

In this problem the shear stress at the wall is determined by the pressure gradient and the width of the channel only, which is one reason why this flow is less complicated than others.

If we use (5.2.5) to substitute for dP_0/dx in (5.2.4), we obtain

$$-\overline{uv} + \nu \frac{dU}{dy} = u_*^2\left(1 - \frac{y}{h}\right). \tag{5.2.6}$$

Contemplating possible nondimensional forms of (5.2.6), we conclude that u_*^2 is the proper scaling factor for $-\overline{uv}$ because we expect the viscous stress to be small at large Reynolds numbers. Also, the experience gained in the study of wakes suggests that $\partial U/\partial y$ should be scaled with u_*/h, because the turbulent scales of velocity and length presumably are u_* and h. Thus, we should write

$$-\frac{\overline{uv}}{u_*^2}+\frac{\nu}{u_*h}\frac{d(U/u_*)}{d(y/h)}=1-\frac{y}{h}. \tag{5.2.7}$$

If the Reynolds number $R_*=u_*h/\nu$ is large, this particular nondimensional form suppresses the viscous stress. Because the stress at the surface is purely viscous, (5.2.7) cannot be valid near the wall in the limit as $R_*\rightarrow\infty$. In the immediate vicinity of the wall, therefore, another nondimensional form of (5.2.6) must be found; it should be selected in such a way that the viscous term does not become small at large Reynolds numbers. From (5.2.7) we conclude that this can be done by absorbing R_* in the scale for y. The resulting equation is

$$-\frac{\overline{uv}}{u_*^2}+\frac{d(U/u_*)}{d(yu_*/\nu)}=1-\frac{\nu}{hu_*}\frac{yu_*}{\nu}. \tag{5.2.8}$$

It is clear that this nondimensionalization tends to suppress the change of stress in the y direction if $R_*=u_*h/\nu\rightarrow\infty$.

For convenience, let us define

$$y_+\equiv yu_*/\nu, \quad \eta\equiv y/h. \tag{5.2.9}$$

Equations (5.2.7) and (5.2.8) then can be written as

$$-\frac{\overline{uv}}{u_*^2}+R_*^{-1}\frac{d}{d\eta}\left(\frac{U}{u_*}\right)=1-\eta, \tag{5.2.10}$$

$$-\frac{\overline{uv}}{u_*^2}+\frac{d}{dy_+}\left(\frac{U}{u_*}\right)=1-R_*^{-1}y_+. \tag{5.2.11}$$

We are looking for asymptotic solutions of these equations in the limit as $R_*\rightarrow\infty$. From (5.2.10, 5.2.11) it is evident that these solutions depend on our point of view: for all but very small values of η we expect the viscous stress to be negligibly small, and at finite values of y_+ (which correspond to very small values of η) we expect that viscous stresses are important and that the total stress is approximately constant. The region of viscous effects must be confined to the immediate vicinity of the wall, since only there can we expect the local Reynolds numbers Uy/ν and y_+ to be so small that turbulence cannot sustain itself.

In the limit as $R_* \to \infty$, but with η remaining of order one, (5.2.10) reduces to

$$-\overline{uv}/u_*^2 = 1 - \eta. \tag{5.2.12}$$

This equation cannot represent conditions as $\eta \to 0$, which corresponds to finite values of y_+. We call the part of the flow governed by (5.2.12) the *core region* (the name "outer layer" is not appropriate in channel flow).

In the limit as $R_* \to \infty$, but with y_+ remaining of order one, (5.2.11) becomes

$$-\frac{\overline{uv}}{u_*^2} + \frac{d(U/u_*)}{d(yu_*/\nu)} = 1. \tag{5.2.13}$$

This equation cannot represent reality if $y_+ \to \infty$, which corresponds to finite values of η. The part of the flow governed by (5.2.13) is called the *surface layer*.

The surface layer on a smooth wall We now restrict ourselves momentarily to flow over smooth surfaces, so that the roughness height k does not occur as an additional parameter. The flow in the surface layer is governed by (5.2.13), which is free of explicit dependence on parameters. If the surface is smooth, no additional parameters occur in the boundary conditions on (5.2.13), so that we may expect the solution of (5.2.13) to be

$$U/u_* = f(y_+), \tag{5.2.14}$$

$$-\overline{uv}/u_*^2 = g(y_+). \tag{5.2.15}$$

These relations are called the *law of the wall*. The only boundary conditions that the system (5.2.13, 5.2.14, 5.2.15) needs to satisfy at this point are $f(0) = 0, g(0) = 0$. The similarity expressions (5.2.14, 5.2.15) may not be valid if $y_+ \to \infty$, unless that limit is approached rather carefully. The shapes of f and g have been determined experimentally, but we prefer not to discuss the experimental evidence before we have taken a look at the other side of the coin.

The core region In the core region, all we have is a statement, (5.2.12), on the Reynolds stress. The momentum equation thus gives no explicit information on U itself. Let us look at an equation in which U does occur explicitly. Such an equation is the turbulent energy budget, which in this channel-flow geometry is

$$-\overline{uv}\frac{dU}{dy} = \epsilon + \frac{d}{dy}\left(\frac{1}{\rho}\overline{vp} + \tfrac{1}{2}\overline{q^2 v}\right). \tag{5.2.16}$$

In (5.2.16), ϵ stands for the viscous dissipation of the turbulent kinetic energy $\frac{1}{2}\overline{q^2}$; viscous transport of $\frac{1}{2}\overline{q^2}$ has been neglected (see Chapter 3). Referring back to (5.2.12), we see that the Reynolds stress $-\overline{uv}$ is of order u_*^2 for all finite values of η. Since the turbulent energy is generated by this stress, we expect q^2 and p/ρ to be of order u_*^2, too. We have seen before that the large eddies in turbulent flows scale with the cross-stream dimensions for the flow. Hence, the terms on the right-hand side of (5.2.16) must be of order u_*^3/h. Since the Reynolds stress is of order u_*^2, we conclude that dU/dy is of order u_*/h. If we stay well above the surface layer, so that no other characteristic lengths can complicate the picture, we can state without any loss of generality that

$$\frac{dU}{dy} = \frac{u_*}{h} \cdot \frac{dF}{d\eta}, \tag{5.2.17}$$

with the understanding that $dF/d\eta$, which is the derivative of some unknown function F, is of order one. Because h is not an appropriate length scale near the surface, (5.2.17) has to be integrated from the center of the channel ($\eta = 1$) toward the wall. This results in

$$(U - U_0)/u_* = F(\eta), \tag{5.2.18}$$

where U_0 is the mean velocity at the center of the channel. We see that the appropriate similarity law for the core region is a *velocity-defect law*. Of course, (5.2.18) is not applicable as $\eta \to 0$.

Inertial sublayer A two-layer description as developed here requires special attention in the region where the two descriptions merge into each other. The existence of a *region of overlap* or *matched layer* is possible only if the limits $y_+ \to \infty$ and $\eta \to 0$ can be taken simultaneously. In Section 5.1 it was demonstrated that this is possible if the Reynolds number is large enough (see Figure 5.1). More specifically, if $y_+ = cR_*^\alpha$, then $\eta = cR_*^{\alpha-1}$, so that $y_+ \to \infty$ and $\eta \to 0$ simultaneously if $0 < \alpha < 1$. This is called an *intermediate limit process*; it corresponds to travel toward the right on a straight line with slope $\alpha - 1$ in the plot given in Figure 5.3.

The process of obtaining the proper limiting behavior of the law of the wall and the velocity-defect law is called *asymptotic matching*. Formally, matching requires that the intermediate limits of the functions involved be equal for any α in the interval $0 < \alpha < 1$. However, in this particular case no such elegance is needed. Since we have demonstrated that an intermediate limit process is possible, we can now assume that the surface layer and the wall layer can be matched. It is most convenient to match the velocity gradients of the wall layer and the core region. According to (5.2.14), the velocity gradient in the surface layer is given by

$$\frac{dU}{dy} = \frac{u_*^2}{\nu} \frac{df}{dy_+}. \tag{5.2.19}$$

In the core region, (5.2.17) must be valid. Equating (5.2.17) and (5.2.19) and

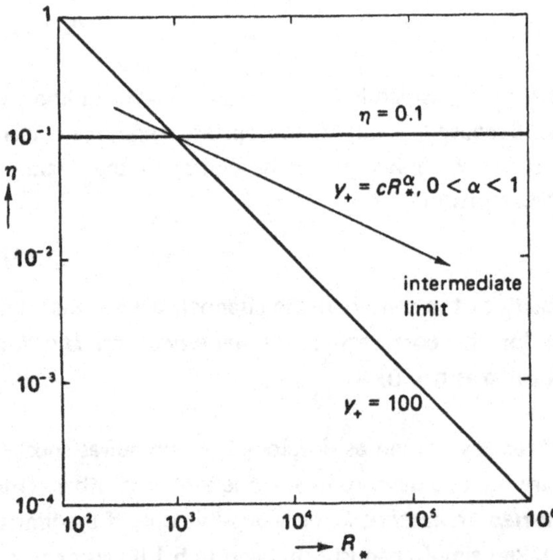

Figure 5.3. An intermediate limit process in which $y_+ \to \infty$ and $\eta \to 0$ simultaneously.

keeping in mind that we are considering some limit process in which $y_+ \to \infty$ and $\eta \to 0$ simultaneously, we obtain

$$\frac{u_*}{h}\frac{dF}{d\eta} = \frac{u_*^2}{\nu}\frac{df}{dy_+}. \tag{5.2.20}$$

On multiplication by y/u_*, this becomes

$$\eta\frac{dF}{d\eta} = y_+\frac{df}{dy_+} = \frac{1}{\kappa}. \tag{5.2.21}$$

The left-hand side of (5.2.21) can be a function only of η and the right-hand side can be a function only of y_+, because neither F nor f depends on any parameters. Thus, in the inertial sublayer both sides of (5.2.21) must be equal to the same universal constant. If the constant is denoted by $1/\kappa$, (5.2.21) can be integrated to yield

$$F(\eta) = \frac{1}{\kappa}\ln\eta + \text{const}, \tag{5.2.22}$$

$$f(y_+) = \frac{1}{\kappa}\ln y_+ + \text{const}. \tag{5.2.23}$$

Both of these are valid only if $\eta \ll 1$ and $y_+ \gg 1$.

The chain of arguments leading to (5.2.22, 5.2.23) was developed by Clark B. Millikan, who presented it at the Fifth International Congress of Applied Mechanics (Millikan, 1939). At that time, the formal theory of singular-perturbation problems was unknown; not until the decade 1950–1960 was a rational theory of multiple length-scale problems developed by Kaplun, Lagerstrom, Cole, and others (see Cole, 1968). The constant κ in (5.2.22, 5.2.23) is called von Kármán's constant, because Th. von Kármán was one of the first to derive the logarithmic velocity profile from similarity arguments (von Kármán, 1930).

The logarithmic velocity profile in the inertial sublayer is one of the major landmarks in turbulence theory. With analytical tools of a rather general nature a very specific result has been obtained, even though the equations of motion cannot be solved in general.

In this flow, matching of the Reynolds stress is straightforward. According to (5.2.12), $-\overline{uv}/u_*^2 \to 1$ if $\eta \to 0$. According to (5.2.13) and (5.2.21), for $y_+ \to \infty$,

$$-\overline{uv}/u_*^2 = 1 - 1/\kappa y_+, \tag{5.2.24}$$

so that

$$-\overline{uv}/u_*^2 \to 1 \text{ if } y_+ \to \infty. \tag{5.2.25}$$

The inertial sublayer thus is a region of approximately constant Reynolds stress. From (5.2.24) it is also clear that the viscous stress (which is proportional to the second term in (5.2.24)) is very small compared to the Reynolds stress if $y_+ \gg 1$. The matched layer is called *inertial sublayer* because of this absence of local viscous effects.

Logarithmic friction law If (5.2.18) and (5.2.14) are substituted into (5.2.22) and (5.2.23), respectively, there results

$$\frac{U - U_0}{u_*} = \frac{1}{\kappa} \ln \eta + b, \tag{5.2.26}$$

$$\frac{U}{u_*} = \frac{1}{\kappa} \ln y_+ + a. \tag{5.2.27}$$

These expressions are valid only in the inertial sublayer. The constants a and b must be finite; they cannot depend on the Reynolds number $R_* = u_* h/\nu$ because f and F are independent of R_*. It follows from (5.2.26) and (5.2.27) that

$$\frac{U_0}{u_*} = \frac{1}{\kappa} \ln R_* + a - b, \tag{5.2.28}$$

because (5.2.26) and (5.2.27) must be valid simultaneously in the inertial sublayer. This relation is called the *logarithmic friction law*; it determines U_0 if the pressure gradient and the channel width are known.

Turbulent pipe flow Axisymmetric parallel flow in a circular pipe of constant diameter is of greater practical importance than plane channel flow. The geometry of pipe flow is sketched in Figure 5.4. We assume that the flow is *fully developed*, that is, independent of x. The origin of the y coordinate is put at the inner surface. This would be very inconvenient for most purposes, but it is convenient here, because we only need the mean-flow equation, which in these coordinates becomes

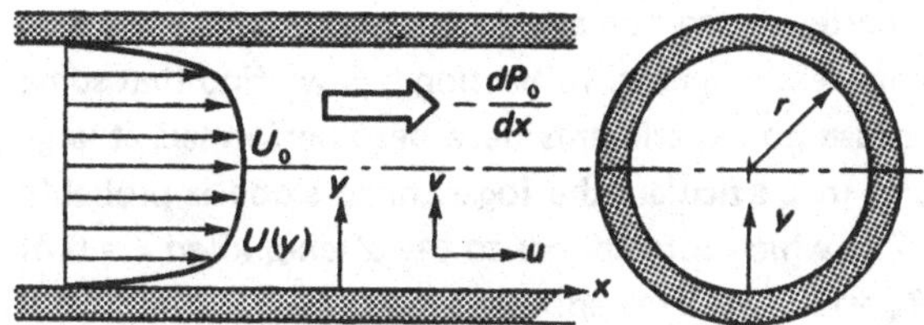

Figure 5.4. Definition sketch for pipe flow.

$$-\overline{uv} + \nu \frac{dU}{dy} = \tfrac{1}{2}\,(y - r)\,\frac{dP_0}{\rho dx}. \tag{5.2.29}$$

The derivation of (5.2.29) is left as an exercise for the reader. The momentum integral in fully developed pipe flow is, if the wall stress is again denoted by ρu_*^2,

$$2\pi r u_*^2 = \pi r^2 \frac{dP_0}{\rho dx}. \tag{5.2.30}$$

The momentum equation thus becomes

$$-\overline{uv} + \nu \frac{dU}{dy} = u_*^2 \left(1 - \frac{y}{r}\right), \tag{5.2.31}$$

which is identical to (5.2.6) if r is replaced by h. All of the conclusions obtained for channel flow thus apply equally to pipe flow. The shape of $F(\eta)$, where η now is defined as y/r, may be different from the shape of F in plane channel flow because of different geometrical constraints. However, the shape of $f(y_+)$ should be identical to that in plane channel flow, because the curvature of the wall is nearly zero if seen from points close enough to the surface to make y_+ finite.

Experimental data on pipe flow For turbulent flows in pipes with smooth walls, the logarithmic velocity profile and the logarithmic friction law are well represented by

$$U/u_* = 2.5 \ln y_+ + 5, \tag{5.2.32}$$

$$(U - U_0)/u_* = 2.5 \ln \eta - 1, \tag{5.2.33}$$

$$U_0/u_* = 2.5 \ln R_* + 6. \tag{5.2.34}$$

There is considerable scatter in the numerical constants; the values given represent averages over many experiments. In Section 5.4, we find that some of the "scatter" arises because no experiments have been performed at large enough Reynolds numbers. In particular, the logarithmic slope is probably very nearly 3 (instead of 2.5, which corresponds to the often-quoted $\kappa = 0.4$) if the Reynolds number $R_* = ru_*/\nu$ is large enough.

A volume-flow velocity U_b ("bulk" velocity) can be defined by

$$\pi r^2 U_b = \int_0^r 2\pi(r - y)U\,dy. \tag{5.2.35}$$

A fairly crude, but frequently used, approximation to the relation between U_b/u_* and R_* is

$$U_b/u_* = 2.5 \ln R_* + 1.5. \tag{5.2.36}$$

This relation has an interesting application. The local velocity $U(y)$ is equal to U_b at some point in the flow. If (5.2.32) and (5.2.36) are valid at that point, this occurs when

$$2.5 \ln r/y = 3.5, \tag{5.2.37}$$

which yields $y/r = \frac{1}{4}$. It so happens that in pipe flow the velocity profile follows (5.2.32) closely up to and somewhat beyond $\eta = \frac{1}{4}$, even though this is well outside the reach of the inertial sublayer. Thus, the volume flow through a smooth pipe can be determined simply by putting a small total-head probe at $\eta = \frac{1}{4}$ and drilling a static-pressure tap in the wall at the same value of x as that of the tip of the total-head tube. This is called a *quarter-radius probe*.

The viscous sublayer We now want to consider the law of the wall, (5.2.14, 5.2.15), in more detail. The first issue to be considered is whether or not the Reynolds stress can contribute to the stress at small values of y_+. At the surface itself, all of the stress is viscous stress. However, if the surface is rough and if $y = 0$ is taken at the mean height of the roughness elements, the shear stress at $y = 0$, as distinct from the shear stress at the surface, can be borne partly by the Reynolds stress if the roughness elements are large enough. We return to this issue later; for the moment, we restrict the discussion to flow over smooth surfaces.

It is useful to look at the problem from the point of view of the turbulence, rather than the mean flow, and to look from the inertial sublayer downward toward the wall. In the inertial sublayer, the Reynolds stress is approximately equal to ρu_*^2, and the mean-velocity gradient is given by $u_*/\kappa y$. Hence, the turbulence production rate $-\overline{uv}\, dU/dy$ is equal to $u_*^3/\kappa y$. If turbulence production is mainly balanced by the viscous dissipation ϵ (experiments have shown that this is a fairly accurate statement in the inertial sublayer), we have

$$\epsilon \cong u_*^3/\kappa y. \tag{5.2.38}$$

The Kolmogorov microscale η (not easily confused with $\eta = y/r$ in this context) thus varies with y according to

$$\eta = \left(\frac{\nu^3}{\epsilon}\right)^{1/4} \cong \left(\frac{\kappa y \nu^3}{u_*^3}\right)^{1/4} = \kappa^{1/4} y\, y_+^{-3/4}. \tag{5.2.39}$$

The integral scale (ℓ) of the turbulence, on the other hand, must be of order y because the largest eddies should scale with the distance from the wall. In the inertial sublayer, $\partial U/\partial y = u_*/\kappa y$, so that $\ell \cong \kappa y$ is a suitable estimate. Nondimensionally, we obtain

$$\eta_+ \equiv \eta u_*/\nu \cong (\kappa y_+)^{1/4}, \tag{5.2.40}$$

$$\ell_+ \equiv \ell u_*/\nu \cong \kappa y_+. \tag{5.2.41}$$

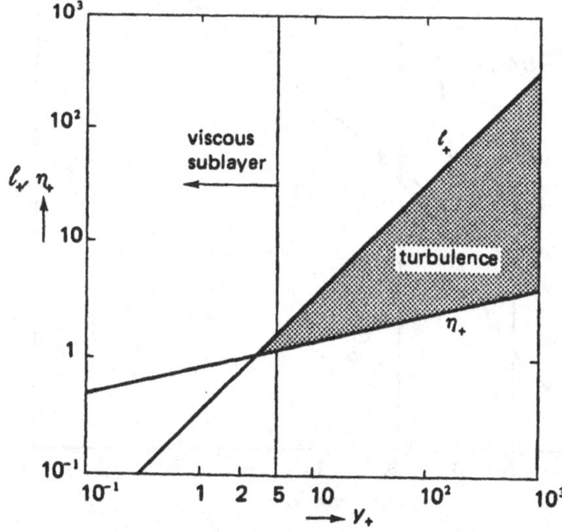

Figure 5.5. The variation of ℓ and η near the surface.

These relations are plotted in Figure 5.5; they show that the integral scale becomes smaller than the Kolmogorov microscale if y_+ is small. This is impossible, so that we must conclude that the turbulence cannot sustain itself and cannot generate Reynolds stresses if y_+ is small. Experimental evidence has shown that the Reynolds stress remains a small fraction of u_*^2 up to about $y_+ = 5$. This region is called the *viscous sublayer*. In the viscous sublayer, the flow is not steady, but the velocity fluctuations do not contribute much to the total stress because of the overwhelming effects of the viscosity. In some of the literature, the viscous sublayer is called the *laminar sublayer*; this name, however, is misleading because it suggests that no velocity fluctuations are present. In the viscous sublayer, the velocity profile must be linear ($U/u_* = y_+$), as indicated by the solution of (5.2.13) when $-\overline{uv}$ is neglected.

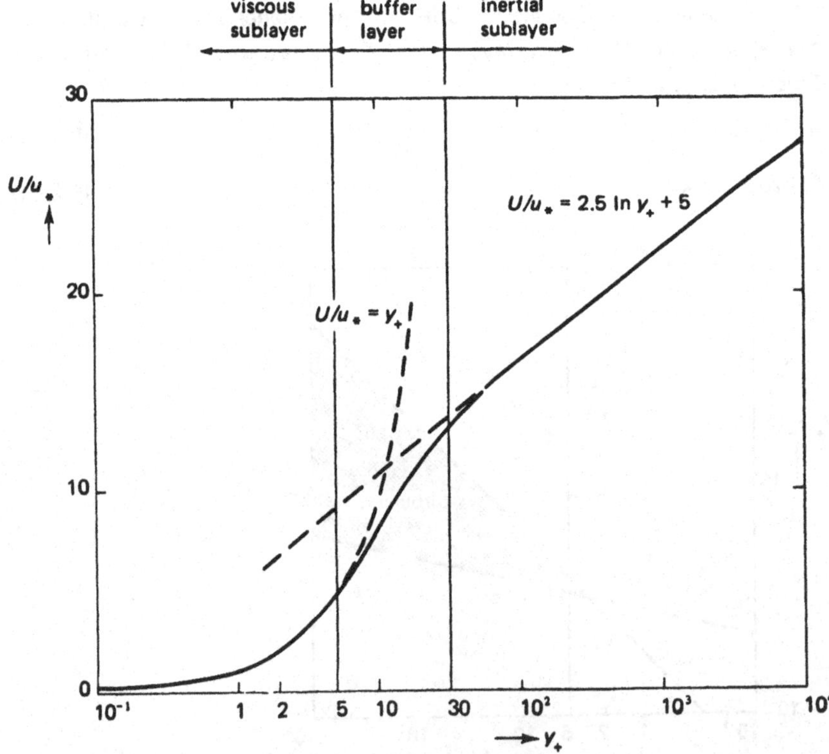

Figure 5.6. The law of the wall.

Experimental data on the law of the wall The velocity profile in the surface layer must satisfy $f = y_+$ for small y_+ and the logarithmic law (5.2.32) at large y_+. Experimentally obtained velocity profiles have the shape given in Figure 5.6. Another useful plot is the distribution of stresses. According to (5.2.13), the sum of the (nondimensionalized) viscous and Reynolds stresses must be equal to one throughout the surface layer. The two curves are sketched in Figure 5.7. The region where neither one of the stresses can be neglected is sometimes called the *buffer layer*. In many engineering calculations, the buffer layer is disposed of by linking the linear velocity profile in the viscous sublayer to the logarithmic velocity profile in the inertial sublayer. This causes an abrupt change from purely viscous stress to purely turbulent stress at $y_+ = 11$ approximately. The buffer layer is the site of vigorous turbulence dynamics, because the turbulent energy production rate $g\, df/dy_+$ reaches a maximum of $\frac{1}{4}$ at the value of y_+ where the Reynolds stress is equal to the viscous stress ($g = df/dy_+ = \frac{1}{2}$). This occurs at $y_+ = 12$ approximately, as is shown in Figure 5.7.

A few approximate numbers on the turbulence intensity in the surface layer may be useful. If the rms value of a variable is denoted by a prime, the

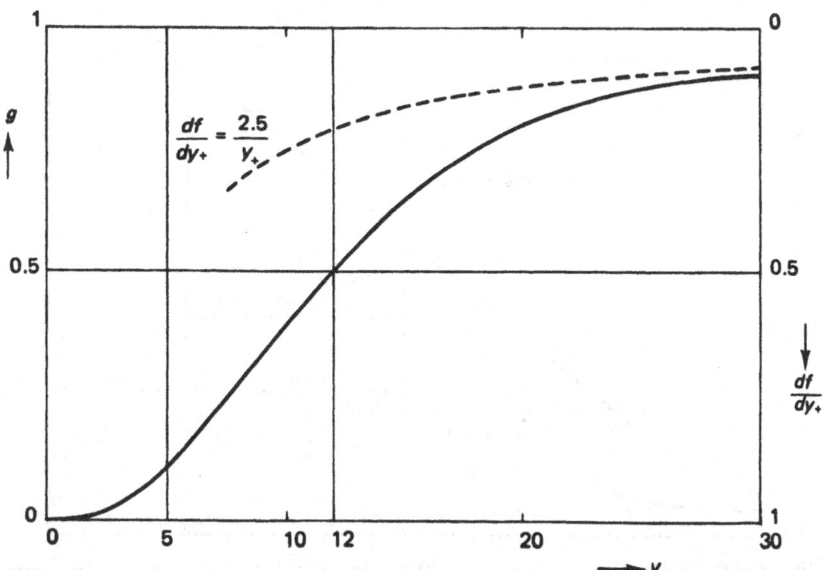

Figure 5.7. Distribution of Reynolds stress $g = -\overline{uv}/u_*^2$ and of viscous stress df/dy_+ in the surface layer (adapted from Hinze, 1959).

following relations hold in the inertial sublayer: $u' \cong 2u_*$, $v' \cong 0.8u_*$, $w' \cong 1.4u_*$, $\frac{1}{2}q^2 \cong 3.5u_*^2$, $-\overline{uv} \cong u_*^2 \cong 0.4\,u'v'$. The u component is largest because the turbulence-production mechanism favors it; the distribution of energy among the components is performed by nonlinear interaction.

Experimental data on the velocity-defect law A plot of the velocity-defect law is presented in Figure 5.8. In pipe flow, the logarithmic velocity profile (5.2.33) happens to represent the actual velocity profile fairly well all through the pipe, which is often quite convenient in engineering applications. The difference between the actual velocity profile in the core region and the logarithmic law, normalized by u_*, is called the *wake function* $W(\eta)$:

$$W(\eta) = 1 - 2.5 \ln \eta + F(\eta). \tag{5.2.42}$$

The wake function happens to be approximately sinusoidal in many wall-bounded flows; in this particular case, $W(\eta)$ is fairly well represented by

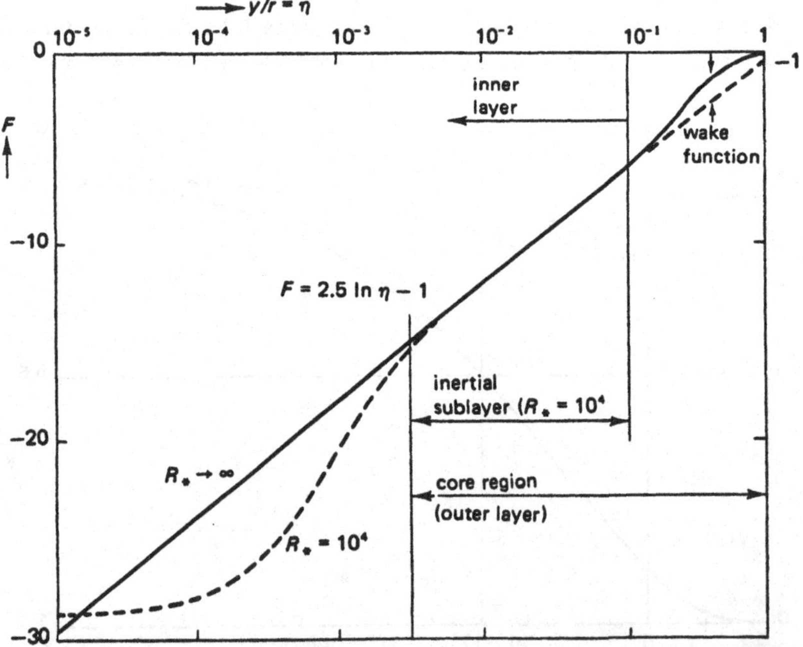

Figure 5.8. The velocity-defect law in pipe flow. The dashed curve on the left represents the actual velocity profile in the wall layer for $R_* = 10^4$. The width of the inertial sublayer increases with R_*.

$$W = \tfrac{1}{2}\,[\sin \pi\,(\eta - \tfrac{1}{2}) + 1]\,. \qquad (5.2.43)$$

The amplitude of the sine wave is equal to $\frac{1}{2}$ in this case, but in boundary layers with opposing pressure gradients W can become quite large. The wake function W may be represented by a universal shape function $\mathscr{F}$ multiplied by a numerical constant that depends on the conditions of the flow. This representation is called the *law of the wake* because $\mathscr{F}$ is similar to the shape of the velocity-defect profile in wakes (Coles, 1956).

The turbulence intensity drops slowly if one goes from the surface toward the center. In the core region, a crude approximation to the experimental data is $u' = v' = w' \cong 0.8\,u_*$. The fluctuating velocity component v has a nearly constant amplitude across the pipe.

The flow of energy The surface layer is a "sink" for momentum, and therefore also for kinetic energy associated with the mean flow. Mean-flow kinetic energy transferred into the surface layer by Reynolds stresses is converted into turbulent kinetic energy (turbulence production) and into heat (viscous dissipation). If we integrate the transport term $\partial(\overline{uv}\,U)/\partial y$ between the surface and a value of y near the outer edge of the inertial sublayer, we conclude that the total loss of energy in that region is of order $\rho U_0 u_*^2$ per unit area and time, because U is fairly close to U_0 at the edge of the inertial sublayer. The direct loss to viscous dissipation occurs primarily in the viscous sublayer, because $\partial U/\partial y$ has a sharp peak at the surface. This loss is of order ρu_*^3: $\mu(\partial U/\partial y)^2$ is of order $\rho u_*^4/\nu$ in the viscous sublayer, but this loss is concentrated in a region whose height is only of order ν/u_*. Most of the mean-flow kinetic energy transported into the surface layer is thus used for the maintenance of turbulent kinetic energy.

In the core region, on the other hand, the Reynolds stress is of order ρu_*^2 and dU/dy is of order u_*/r. Integrating over the entire core region, the turbulence production per unit area and time in the core region is of order ρu_*^3. We conclude that most of the turbulence production occurs in the surface layer. The surface layer is the source of most of the turbulent energy. This conclusion must be viewed with caution, though, because the rate of dissipation of turbulent energy is also high in the surface layer.

The main function of the core region is not turbulence production, but transport of mean-flow kinetic energy into the surface layer. In the core of the pipe, the pressure gradient performs work at a rate of roughly $\rho u_*^2 U_0/r$

per unit volume and time. This energy input is carried off by the Reynolds stress to the surface layer, where it is converted into turbulent kinetic energy.

Flow over rough surfaces If the surface of a pipe or channel is rough, the arguments leading to the law of the wall require some modification. If the ratio k/r (k is an rms roughness height, say) is small enough, the roughness does not affect the velocity-defect law.

A definition sketch of flow over a rough surface is given in Figure 5.9. If $y = 0$ at the average vertical position at the surface, the velocity at $y = 0$ cannot be defined for a substantial fraction of the streamwise distance. As discussed earlier, the no-slip condition has to be satisfied at the surface, but the mean velocity obtained by averaging the instantaneous velocity at $y = 0$ over time and over all intervals Δx where the surface is below $y = 0$ need not be zero.

The surface layer over a rough wall has two characteristic lengths, k and ν/u_*, whose ratio is the roughness Reynolds number $R_k = ku_*/\nu$. We thus expect a law of the wall which can be written as

$$\frac{U}{u_*} = f_1(y_+, R_k), \tag{5.2.44}$$

or

$$\frac{U}{u_*} = f_2(y/k, R_k). \tag{5.2.45}$$

These expressions must be matched with the velocity-defect law. Because the latter is independent of roughness as long as $k/r \ll 1$ and because the matching is performed on the velocity derivative, the effects of roughness on

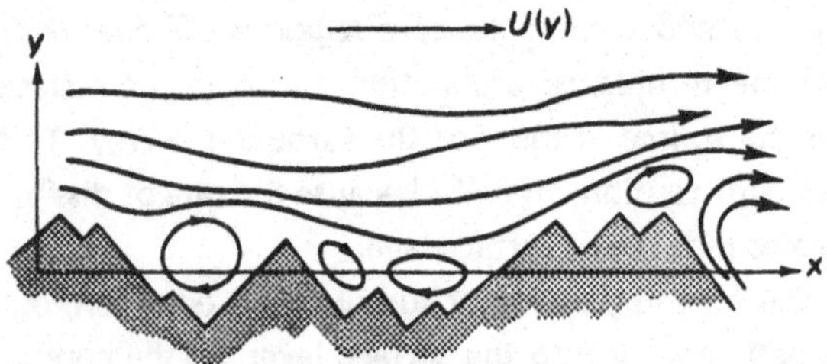

Figure 5.9. Flow over a rough surface.

the logarithmic velocity profile in the inertial sublayer can appear only as an additive function of the parameter:

$$\frac{U}{u_*} = \frac{1}{\kappa} \ln y_+ + f_3(R_k), \tag{5.2.46}$$

$$\frac{U}{u_*} = \frac{1}{\kappa} \ln \frac{y}{k} + f_4(R_k). \tag{5.2.47}$$

In the limit as $R_k \to 0$, f_3 has to become equal to 5, as comparison with (5.2.32) indicates. It turns out that roughness has no effect on (5.2.46) as long as $R_k < 5$, because the roughness elements are then submerged in the viscous sublayer in which no Reynolds stresses can be generated, however much the flow is disturbed.

For large values of R_k, a suitable nondimensional form of the equation of motion (5.2.31) is

$$-\frac{\overline{uv}}{u_*^2} + R_k^{-1} \frac{d(U/u_*)}{d(y/k)} = 1 - \frac{y}{k}\frac{k}{r}. \tag{5.2.48}$$

This shows that the viscous stress is very small at values of y/k of order one if $R_k \to \infty$. It should be noted that k/r must remain small, or else a distinct surface layer cannot exist. From (5.2.48) we conclude that $f_4(R_k)$ in (5.2.47) should be independent of R_k if it is large enough. This indeed occurs in practice for values of R_k above 30. The physical concept here is that roughness elements with large R_k generate turbulent wakes, which are responsible for essentially inviscid drag on the surface. For values of R_k between 5 and 30, the additive constant in the logarithmic part of the velocity profile (5.2.44, 5.2.45) depends on R_k.

The rough-wall velocity profile becomes, in the limit as $R_k \to \infty$,

$$\frac{U}{u_*} = \frac{1}{\kappa} \ln \frac{y}{k} + \text{const.} \tag{5.2.49}$$

Often, the position $y = 0$ is not known accurately enough to bother with the additive constant; instead, it is absorbed in the definition of k. Also, the logarithmic profile is often assumed to be valid all the way down to $y/k = 1$ (which makes $U = 0$ at $y/k = 1$ if the additive constant is ignored), even though its derivation was based on the limit process $y/k \to \infty$. The friction law corresponding to (5.2.49) is

$$\frac{U_0}{u_*} = \frac{1}{\kappa} \ln \frac{r}{k} + \text{const.} \tag{5.2.50}$$

5.3
Planetary boundary layers

The geostrophic wind The flow of air over the surface of the earth is affected by the Coriolis force that arises in any coordinate system that is rotating with respect to an inertial frame of reference. Under favorable conditions, the flow outside the boundary layer at the earth's surface is approximately steady, horizontal, and homogeneous in horizontal planes. In that case, the equations of motion reduce to a simple balance between pressure gradient and Coriolis forces. In the coordinate system of Figure 5.10, which is a Cartesian frame whose *x*, *y* plane is normal to the local vertical at latitude ϕ, this geostrophic balance is

$$-fV_g = -\frac{1}{\rho}\frac{\partial P}{\partial x}, \tag{5.3.1}$$

$$fU_g = -\frac{1}{\rho}\frac{\partial P}{\partial y}. \tag{5.3.2}$$

In these expressions U_g and V_g are the *x* and *y* components of the *geostrophic wind*, whose modulus is $G = (U_g^2 + V_g^2)^{1/2}$. The parameter *f*, which may be taken to be constant if the flow covers only a small range of latitudes

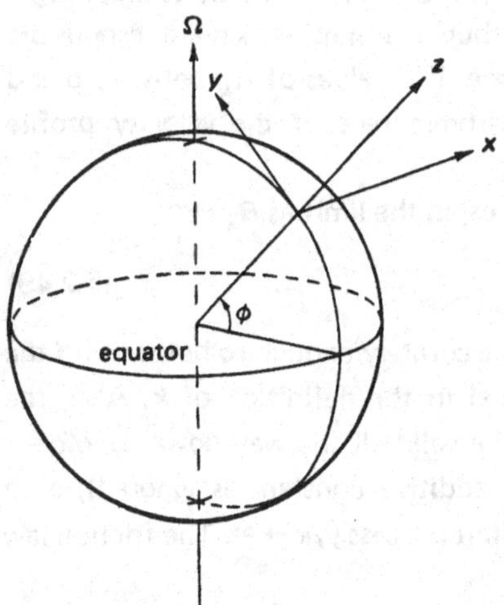

Figure 5.10. Coordinate system for planetary boundary layers.

ϕ, is equal to twice the z component of the angular velocity Ω at latitude ϕ:

$$f = 2\,\Omega \sin \phi. \tag{5.3.3}$$

This is called the *Coriolis parameter*. Its value is approximately $10^{-4}\,\text{sec}^{-1}$ at $\phi = 40°$.

The Ekman layer The geostrophic wind does not meet the no-slip condition at the surface, so that a boundary layer must exist. If the flow in the boundary layer is steady and homogeneous in horizontal planes, the equations of motion for this planetary boundary layer, or *Ekman layer*, become

$$-f(V - V_g) = \frac{d}{dz}(-\overline{uw}), \tag{5.3.4}$$

$$f(U - U_g) = \frac{d}{dz}(-\overline{vw}). \tag{5.3.5}$$

Here, (5.3.1, 5.3.2) have been used to substitute for the pressure gradient. Also, it has been assumed that the roughness Reynolds number is so large that viscous stresses can be neglected. It is convenient to assume that the stress at the surface (ρu_*^2, by definition) has no y component, so that, for $z \to 0$,

$$-\overline{uw} = u_*^2, \qquad -\overline{vw} = 0. \tag{5.3.6}$$

The velocity-defect law The equations of motion show quite clearly that a velocity-defect law is called for. We assume that u_* is the only characteristic velocity; this restricts us to flows in which no appreciable heat transfer occurs, because heat flux in a flow exposed to gravity may cause additional turbulence or may suppress turbulence, depending on its direction (see Chapter 3). The Reynolds stresses are presumably of order u_*^2, but the height h of the Ekman layer is unknown. A tentative nondimensional form of (5.3.4, 5.3.5) is thus

$$-\frac{fh}{u_*}\left(\frac{V - V_g}{u_*}\right) = \frac{d}{d(z/h)}\left(-\frac{\overline{uw}}{u_*^2}\right), \tag{5.3.7}$$

$$\frac{fh}{u_*}\left(\frac{U - U_g}{u_*}\right) = \frac{d}{d(z/h)}\left(-\frac{\overline{vw}}{u_*^2}\right). \tag{5.3.8}$$

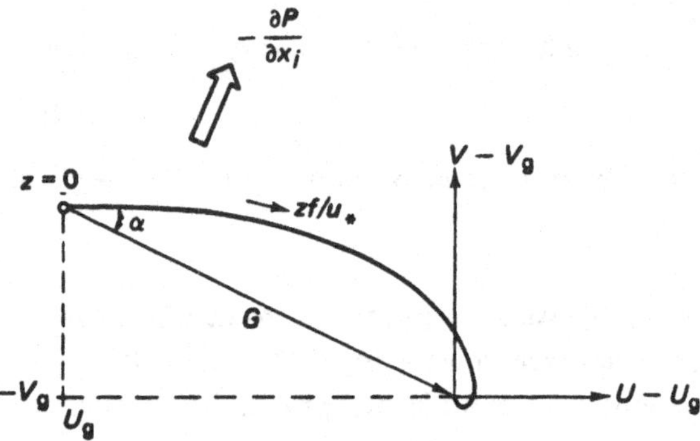

Figure 5.11. The Ekman velocity-defect spiral. The nondimensional height zf/u_* increases toward the right on the curve.

We are at liberty to select h for maximum convenience. If we choose

$$h = cu_*/f, \tag{5.3.9}$$

where c is some constant of order unity, the equations of motion become independent of any parameters, because all possible dependence has been absorbed by careful scaling. Thus, we expect that the velocity-defect law for Ekman layers should be (Blackadar and Tennekes, 1968)

$$(U - U_g)/u_* = F_u(zf/u_*), \tag{5.3.10}$$

$$(V - V_g)/u_* = F_v(zf/u_*). \tag{5.3.11}$$

Figure 5.11 shows a polar velocity plot of the experimentally observed defect law (5.3.10, 5.3.11) for the velocity vector in the Northern Hemisphere. The pressure-gradient vector is normal to the geostrophic wind, as (5.3.1) and (5.3.2) show. The Ekman spiral is located to the left of the geostrophic wind vector, because the Coriolis force in the boundary layer, where velocities are generally smaller than G, is insufficient to balance the pressure gradient. The angle between the surface wind (which is, as we shall see, parallel to the surface stress, that is, in the positive x direction) and the pressure gradient is thus less than 90°, so that the Ekman spiral rotates clockwise with increasing z.

The surface layer The Ekman spiral equations (5.3.10, 5.3.11) are not valid near the surface, because h is not the relevant length scale there. If the surface is rough, with a roughness height z_0 such that $z_0 u_*/\nu \gg 1$, the relevant nondimensional form of (5.3.4, 5.3.5) is

$$-\frac{fz_0}{u_*^2}(V - V_g) = \frac{d(-\overline{uw}/u_*^2)}{d(z/z_0)}, \quad (5.3.12)$$

$$\frac{fz_0}{u_*^2}(U - U_g) = \frac{d(-\overline{vw}/u_*^2)}{d(z/z_0)}. \quad (5.3.13)$$

The left-hand sides of (5.3.12, 5.3.13) can be at most of order fz_0G/u_*^2. If we use (5.3.9), this can be written as z_0G/hu_*. For typical conditions in the atmosphere, $G/u_* \cong 30$, $h \cong 1{,}000$ m, $z_0 \cong 0.01$ m, so that $z_0G/hu_* \cong 3 \times 10^{-4}$. This is very small indeed. We shall neglect the left-hand sides of (5.3.12, 5.3.13); we shall shortly see that this is justified under the limit process which is involved. The surface layer, to first approximation, is thus a constant-stress layer which does not feel the turning effects of the Coriolis force. Because the stress at the surface has been assumed to have no y component, the wind in the surface also has no y component. The law of the wall must read

$$V/u_* = 0, \quad (5.3.14)$$

$$U/u_* = f_u(z/z_0). \quad (5.3.15)$$

These relations show that near the surface the wind is in the positive x direction, so that the Ekman spiral in Figure 5.11 must depart to the right horizontally from $U = 0$, $V = 0$. Also, because the spiral rotates clockwise, $V_g < 0$.

The logarithmic wind profile The law of the wall (5.3.14, 5.3.15) must be matched to the velocity-defect law (5.3.10, 5.3.11). This yields, where the usual procedures have been followed (Blackadar and Tennekes, 1968),

$$\frac{V_g}{u_*} = -F_v(0) = -A, \quad (5.3.16)$$

$$\frac{U}{u_*} = \frac{1}{\kappa}\ln\left(\frac{z}{z_0}\right) + B, \quad (5.3.17)$$

$$\frac{U - U_g}{u_*} = \frac{1}{\kappa}\ln\left(\frac{zf}{u_*}\right) + C, \quad (5.3.18)$$

$$\frac{U_g}{u_*} = \frac{1}{\kappa}\ln\left(\frac{u_*}{fz_0}\right) + B - C. \quad (5.3.19)$$

Here, (5.3.17) and (5.3.18) are valid only in the region where $z/z_0 \gg 1$ and $zf/u_* \ll 1$ simultaneously. The parameter u_*/fz_0 functions like a Reynolds number for the turbulent flow over smooth surfaces; it is called the *friction Rossby number*. The relations given above are asymptotic approximations, valid only for large enough friction Rossby numbers. From (5.3.16) and (5.3.19) we conclude that $fz_0 G/u_*^2 \cong fz_0 U_g/u_*^2 \cong (fz_0/u_*) \ln(u_*/fz_0)$, so that $fz_0 G/u_*^2 \to 0$ as $u_*/fz_0 \to \infty$. The approximation involved in obtaining the law of the wall is indeed valid asymptotically.

The angle α between the wind in the surface layer and the geostrophic wind is given by (see Figure 5.11)

$$\tan \alpha = -V_g/U_g = Au_*/U_g = A/[(1/\kappa) \ln (u_*/fz_0) + B - C]. \tag{5.3.20}$$

Measurements suggest that $A \cong 12$, $C \cong 4$. The value of B is often set at zero, with a consequent minor change in z_0. If $B = 0$ and if z_0 and $1/\kappa$ are known, (5.3.17) can be used to determine the friction velocity u_* from a wind profile near the surface. This is a common practice because direct measurements of stress are quite difficult.

Ekman layers in the ocean The turbulent boundary layer near the surface of a body of water exposed to wind stresses is similar to the Ekman layer in the atmosphere, except for the boundary conditions. If there is no current at great depth and if pressure gradients may be neglected, the water current at the surface makes an angle α, given by the equivalent of (5.3.20), with respect to the stress at the surface. The polar plot of water currents in the Northern Hemisphere is given in Figure 5.12. The formal analysis of the problem is left to the reader.

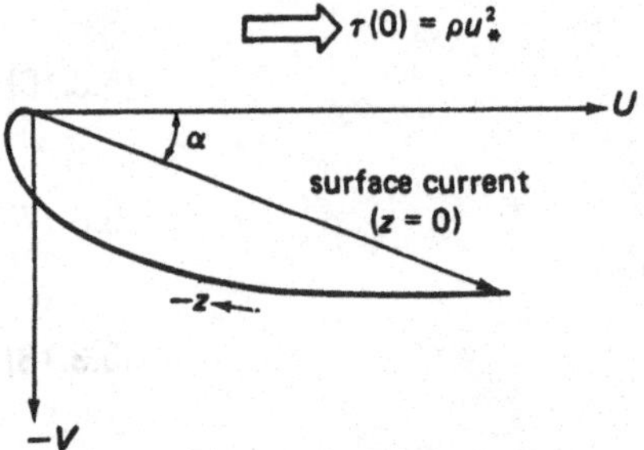

Figure 5.12. Ekman layer near the surface of the ocean (Northern Hemisphere).

5.4
The effects of a pressure gradient on the flow in surface layers

So far, we have encountered only surface layers in which the characteristic velocity is the friction velocity u_*. However, there exist conditions under which u_* is not appropriate. An interesting case is the surface layer of a boundary layer in which the stress at the wall is kept equal to zero for a considerable downstream distance by a carefully chosen distribution of an opposing pressure gradient. In engineering terms, such a boundary layer is on the verge of *separation*. Normally, this requires a rather large opposing pressure gradient, because Reynolds stresses can transfer momentum rapidly enough to prevent excessive deceleration in moderate pressure gradients.

The equations of motion, for steady two-dimensional flow, read

$$\frac{\partial U}{\partial x} + \frac{\partial V}{\partial y} = 0, \tag{5.4.1}$$

$$U\frac{\partial U}{\partial x} + V\frac{\partial U}{\partial y} = -\frac{1}{\rho}\frac{\partial P}{\partial x} - \frac{\partial}{\partial y}(\overline{uv}) - \frac{\partial}{\partial x}(\overline{u^2}) + \nu\frac{\partial^2 U}{\partial y^2} + \nu\frac{\partial^2 U}{\partial x^2}, \tag{5.4.2}$$

$$U\frac{\partial V}{\partial x} + V\frac{\partial V}{\partial y} = -\frac{1}{\rho}\frac{\partial P}{\partial y} - \frac{\partial}{\partial y}(\overline{v^2}) - \frac{\partial}{\partial x}(\overline{uv}) + \nu\frac{\partial^2 V}{\partial y^2} + \nu\frac{\partial^2 V}{\partial x^2}. \tag{5.4.3}$$

We use a coordinate system with a solid wall at $y = 0$. The mean flow in the half-plane $y \geqslant 0$ is in the positive x direction; the pressure gradient $\partial P/\partial x$ is positive. If the characteristic velocity in the surface layer is w, the length scale must be ν/w in order to preserve the viscous-shear stress in (5.4.2). We assume that U, u, and v scale with w, because no self-preservation can exist if the mean flow and the turbulence scale in different ways. The downstream length scale is L; we assume that $Lw/\nu >> 1$.

With $\partial U/\partial x \sim w/L$ and $\partial V/\partial y \sim Vw/\nu$, the continuity equation (5.4.1) gives $V \sim \nu/L$. The left-hand side of the y-momentum equation (5.4.3) is then of order $\nu w/L^2$. The orders of magnitude of the turbulence terms in (5.4.3) are

$$\partial(\overline{v^2})/\partial y = \mathcal{O}(w^3/\nu), \quad \partial(\overline{uv})/\partial x = \mathcal{O}(w^2/L); \tag{5.4.4}$$

the viscous terms in (5.4.3) are of order

$$\nu\,\partial^2 V/\partial y^2 = \mathcal{O}(w^2/L), \quad \nu\,\partial^2 V/\partial x^2 = \mathcal{O}(\nu^2/L^3). \tag{5.4.5}$$

Because $Lw/\nu >> 1$, the major turbulence term, $\partial(\overline{v^2})/\partial y$, must be balanced by $\partial P/\rho\partial y$ to first order. Integration of this simplified equation with respect to y and differentiation with respect to x yields the familiar equation

$$\frac{1}{\rho}\frac{\partial P}{\partial x}+\frac{\partial \overline{v^2}}{\partial x}=\frac{1}{\rho}\frac{dP_0}{dx}. \qquad (5.4.6)$$

Here, P_0 is the pressure at the surface ($y=0$), which, of course, is not a function of y.

The various terms of (5.4.2) now may be estimated as follows:

$$\begin{aligned}
&(U\,\partial U/\partial x + V\,\partial U/\partial y) = \mathcal{O}(w^2/L),\\
&(1/\rho)(\partial P/\partial x - dP_0/dx) = \mathcal{O}(w^2/L),\\
&\partial(\overline{uv})/\partial y = \mathcal{O}(w^3/\nu), \quad \nu\,\partial^2 U/\partial y^2 = \mathcal{O}(w^3/\nu),\\
&\partial\,\overline{u^2}/\partial x = \mathcal{O}(w^2/L), \quad \nu\,\partial^2 U/\partial x^2 = \mathcal{O}(\nu w/L^2).
\end{aligned} \qquad (5.4.7)$$

If $wL/\nu \gg 1$, only the shear-stress terms survive, while $\partial P/\partial x$ may be approximated by dP_0/dx. The approximate equation of motion is thus

$$\frac{\partial}{\partial y}\left(-\overline{uv}+\nu\frac{\partial U}{\partial y}\right)=\frac{1}{\rho}\frac{dP_0}{dx}. \qquad (5.4.8)$$

Because P_0 is independent of y, this integrates to

$$-\overline{uv}+\nu\frac{\partial U}{\partial y}=\frac{y}{\rho}\frac{dP_0}{dx}. \qquad (5.4.9)$$

Here, we have put the stress at the wall equal to zero, because that is the special case we want to consider. The pressure gradient now plays the role of an independent parameter, much like ρu_*^2 is treated as an independent parameter in other surface layers. Because we are considering a surface layer, the boundary-layer thickness δ and the downstream scale L are not relevant, so that a characteristic velocity has to be constructed with dP_0/dx and ν (the surface is smooth). The only possible choice is

$$u_p^3=\frac{\nu}{\rho}\frac{dP_0}{dx}. \qquad (5.4.10)$$

The only parameter-free nondimensional form of (5.4.9) is

$$-\frac{\overline{uv}}{u_p^2}+\frac{\partial(U/u_p)}{\partial(yu_p/\nu)}=\frac{yu_p}{\nu}. \qquad (5.4.11)$$

This equation has only one characteristic velocity (u_p) and one characteristic length, and its boundary conditions are homogeneous (both U and the stress are zero at $y = 0$). Its solution must be a law of the wall of the form

$$U/u_p = f(yu_p/\nu), \tag{5.4.12}$$

$$-\overline{uv}/u_p^2 = g(yu_p/\nu). \tag{5.4.13}$$

The derivation of the corresponding velocity-defect law would carry us too far from the problem at hand. In first approximation, the flow in the outer part of these boundary layers is probably a pure "wake flow" in the sense that a wake function $W(y/\delta)$ like the one defined in (5.2.43), but with a peak-to-peak amplitude U_0, gives a good description of the first-order flow. At finite Reynolds numbers this wake flow is modified by a velocity-defect law that matches the law of the wall (5.4.12).

The mere existence of a velocity-defect law is all that needs to be assumed to predict that, at large yu_p/ν,

$$U/u_p = \alpha \ln yu_p/\nu + \beta. \tag{5.4.14}$$

This statement is supported by the observation that $\partial U/\partial y$ must be of order u_p/y if u_p is the only velocity scale in the problem and if $y >> \nu/u_p$. Experiments with a flow with zero wall stress were performed by Stratford (1959); his results suggest that $\alpha \cong 5$, $\beta \cong 8$. A sketch of the velocity profile is given in Figure 5.13.

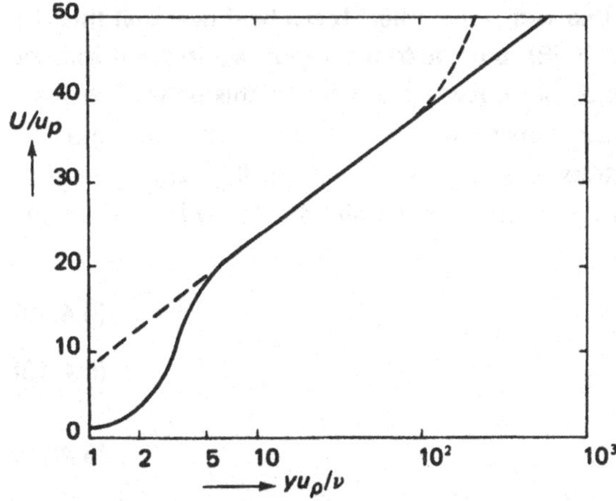

Figure 5.13. The surface layer in a flow with zero wall stress. The dashed curve at the right gives the velocity profile in the outer layer as it begins to deviate from the logarithm (based on data by Stratford, 1959).

A second-order correction to pipe flow The results obtained above suggest that it should be possible to estimate the effect of the pressure gradient on flows in other surface layers. Generally, pressure gradients are associated with acceleration or deceleration of the mean flow, so that their effects may be inseparable from nonlinear inertia effects. However, in pipe flows the inertia terms in the equation of motion vanish because of downstream homogeneity. Let us recall (5.2.29) and (5.2.30):

$$-\overline{uv} + \nu \frac{dU}{dy} = \frac{1}{2}(y - r)\frac{dP_0}{\rho dx}, \tag{5.4.15}$$

$$\frac{r}{\rho}\frac{dP_0}{dx} = -2u_*^2. \tag{5.4.16}$$

If we substitute for $r\, dP_0/dx$ with (5.4.16), the equation of motion becomes

$$-\overline{uv} + \nu \frac{dU}{dy} = u_*^2 + \frac{1}{2}\frac{y}{\rho}\frac{dP_0}{dx}. \tag{5.4.17}$$

The second term on the right-hand side of (5.4.17) is small in the surface layer, so that it is commonly neglected (see Section 5.2). In this particular case, there is no need to do so if we are willing to exploit the results obtained for the surface layer with zero wall stress.

We will think of the wall-layer flow and stress as consisting of two parts which add without interacting with each other. It can be shown that this is a valid procedure (Tennekes, 1968), but the formal proof requires multivariate asymptotic techniques, which are outside the scope of this book. The first-order flow and stress are associated with the constant stress ρu_*^2, and the second-order flow and stress are related to the small stress correction $\frac{1}{2} y\, dP_0/dx$. With these assumptions, we obtain the following system of equations:

$$U = U_1 + U_2, \tag{5.4.18}$$

$$-\overline{uv} = -(\overline{uv})_1 - (\overline{uv})_2, \tag{5.4.19}$$

$$-(\overline{uv})_1 + \nu \frac{dU_1}{dy} = u_*^2, \tag{5.4.20}$$

$$-(\overline{uv})_2 + \nu \frac{dU_2}{dy} = \frac{1}{2}\frac{y}{\rho}\frac{dP_0}{dx}. \tag{5.4.21}$$

The solution of (5.4.20) is the familiar law of the wall

$$U_1/u_* = f(yu_*/\nu), \tag{5.4.22}$$

which, at large $y_+ = yu_*/\nu$, behaves as

$$U_1/u_* = \frac{1}{\kappa} \ln y_+ + C. \tag{5.4.23}$$

The solution of (5.4.21) must be similar to the solution of (5.4.9), that is, it must be a law of the wall like (5.4.12). However, in pipe flow the pressure gradient is negative, so that U_2 is presumably also negative. The appropriate velocity scale for U_2 is u_{p2}, which is defined as

$$u_{p2}^3 = -\frac{\nu}{2\rho}\frac{dP_0}{dx}. \tag{5.4.24}$$

In this way, $u_{p2} > 0$. Nondimensionalized with u_{p2} and ν, (5.4.21) becomes

$$\frac{-(\overline{uv})_2}{u_{p2}^2} + \frac{d(U_2/u_{p2})}{d(yu_{p2}/\nu)} = -\frac{yu_{p2}}{\nu}. \tag{5.4.25}$$

This is identical to (5.4.11), except for the sign reversal in the total stress. The solution of (5.4.25) is thus identical to the solution of (5.4.11), except for a change of sign. This yields the counterpart of (5.4.12):

$$U_2/u_{p2} = -f(yu_{p2}/\nu). \tag{5.4.26}$$

In particular, for large yu_{p2}/ν,

$$U_2/u_{p2} = -\alpha \ln (yu_{p2}/\nu) - \beta. \tag{5.4.27}$$

According to (5.4.16) and (5.4.24), u_{p2} and u_* are related to each other by

$$u_{p2}^3 = u_*^3 \frac{\nu}{u_* r} = u_*^3 R_*^{-1}. \tag{5.4.28}$$

Hence, (5.4.27) can be written as

$$U_2/u_* = -\alpha R_*^{-1/3} \ln y_+ + h(R_*^{-1/3}), \tag{5.4.29}$$

where $h(R_*^{-1/3})$ contains all additive constants.

The slope of the logarithmic velocity profile There is too much experimental scatter in pipe-flow data to allow for a verification of all aspects of (5.4.29).

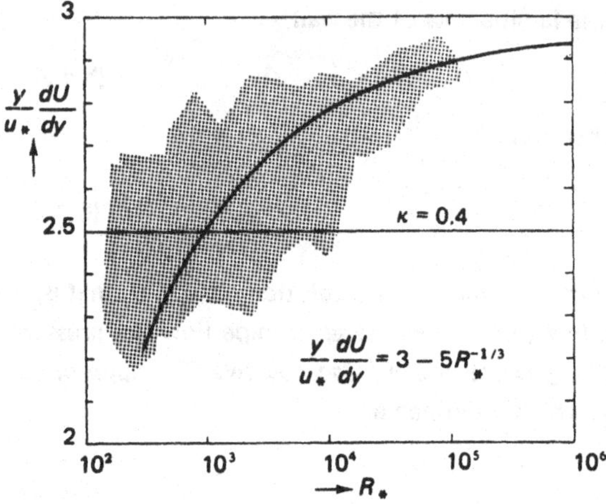

Figure 5.14. The slope of the velocity profile in pipe flow. The experimental scatter is indicated by the shaded area (adapted from Tennekes, 1968).

One major consequence of (5.4.29) is that the slope of the logarithmic velocity profile, in a region where $u_{p2}y/\nu \gg 1$, $u_* y/\nu \gg 1$, but $y/r \ll 1$, is a function of the Reynolds number R_*:

$$\frac{y}{u_*}\frac{dU}{dy} = \frac{y}{u_*}\left(\frac{dU_1}{dy} + \frac{dU_2}{dy}\right) = \frac{1}{\kappa} - \alpha R_*^{-1/3}. \tag{5.4.30}$$

The correction term is appreciable: if $R_* = 1{,}000$ and $\alpha \cong 5$, $\alpha R_*^{-1/3} \cong 0.5$, which is 20% of the value $1/\kappa = 2.5$ that is most often used. The asymptotic value of $1/\kappa$ must thus be about equal to 3. Experimental data (Figure 5.14) show that the trend predicted by (5.4.30) indeed exists.

The characteristic length for the second-order flow is ν/u_{p2}, which is larger than ν/u_* by a factor $R_*^{1/3}$. Therefore, the inertial sublayer of the second-order flow begins at a value of y much larger than the lower edge of the first-order inertial sublayer. It is instructive to look at this problem graphically. Figure 5.15 shows that the second-order inertial sublayer is substantially narrower than the first-order one. The limit lines in the figure are more or less arbitrary, but Figures 5.6 and 5.13 suggest that the respective flows are nearly inviscid for $y_+ > 30$ and $yu_{p2}/\nu > 10$, respectively.

If the asymptotic value of $1/\kappa$ is approximately 3 and if $\alpha \cong 5$, it takes an experiment at $R_* \cong 5 \times 10^6$ (which corresponds to $U_b r/\nu \cong 2 \times 10^8$) to

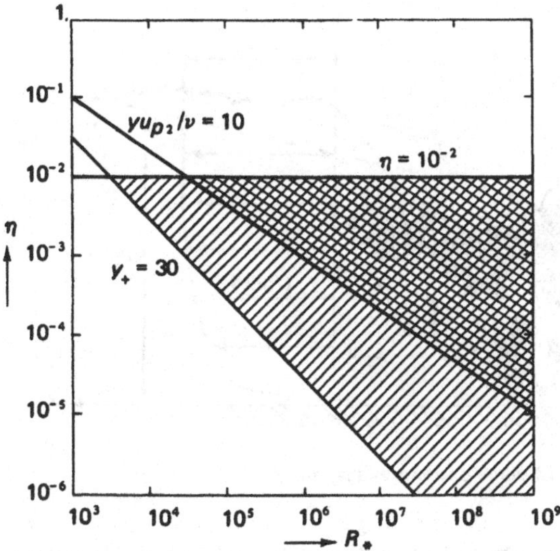

Figure 5.15. First-order and second-order inertial sublayers in pipe flow. Equation (5.4.30) should be valid in the crosshatched area.

determine $1/\kappa$ within 1% error. An experiment set up near a hydraulic power plant with a pipe of 4 m radius and water flowing at a velocity U_b = 50 m/sec would do the job. The pipe would have to be at least 1,000 m long in order to make sure that downstream homogeneity is achieved near the exit.

5.5
The downstream development of turbulent boundary layers

The thickness of boundary layers flowing over solid surfaces generally increases in the downstream direction, because the loss of momentum at the wall is diffused either by viscosity (molecular mixing) or by turbulent mixing. The growth of turbulent boundary layers, of course, is generally quite rapid compared to the growth of laminar boundary layers.

A general treatment of boundary-layer development under arbitrary boundary conditions is out of the question, because the equations of motion cannot be solved in general. Engineers who have to predict the development of a turbulent boundary layer on a wing or a ship's hull, say, use semiempirical techniques, such as described by Schlichting (1960). Here, we concentrate on a family of turbulent boundary layers in steady, plane flow in which

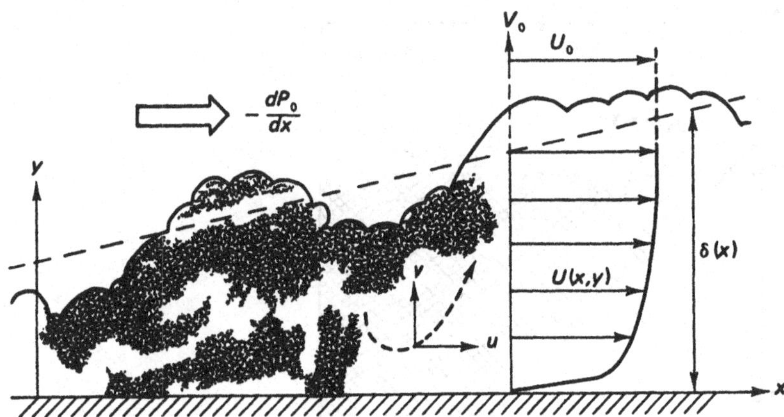

Figure 5.16. Definition sketch of plane boundary-layer flow.

the downstream pressure distribution is adjusted in such a way that their velocity profiles, if nondimensionalized with an appropriate velocity-defect law, are independent of the Reynolds number and of the downstream distance x. Such boundary layers are called *equilibrium layers*; they are equivalent to the Falkner-Skan family of laminar boundary layers.

We consider steady, incompressible, plane flows over a smooth surface without heat or mass transfer. A definition sketch is given in Figure 5.16; the equations of motion are

$$U\frac{\partial U}{\partial x}+V\frac{\partial U}{\partial y}=-\frac{1}{\rho}\frac{\partial P}{\partial x}+\frac{\partial}{\partial y}\left(-\overline{uv}+\nu\frac{\partial U}{\partial y}\right)+\frac{\partial}{\partial x}\left(-\overline{u^2}+\nu\frac{\partial U}{\partial x}\right), \tag{5.5.1}$$

$$U\frac{\partial V}{\partial x}+V\frac{\partial V}{\partial y}=-\frac{1}{\rho}\frac{\partial P}{\partial y}+\frac{\partial}{\partial y}\left(-\overline{v^2}+\nu\frac{\partial V}{\partial y}\right)+\frac{\partial}{\partial x}\left(-\overline{uv}+\nu\frac{\partial V}{\partial x}\right), \tag{5.5.2}$$

$$\frac{\partial U}{\partial x}+\frac{\partial V}{\partial y}=0. \tag{5.5.3}$$

The flow outside the boundary layer is assumed to be irrotational:

$$\frac{\partial U_0}{\partial y}-\frac{\partial V_0}{\partial x}=0. \tag{5.5.4}$$

A length scale L, associated with the rate of change of U_0 downstream, is defined by

$$\frac{1}{L} \equiv \left| \frac{1}{U_0} \frac{\partial U_0}{\partial x} \right| . \tag{5.5.5}$$

If the flow outside the boundary layer is uniform, $L \to \infty$. In that case, the distance x from a suitably defined origin is the appropriate length scale; the procedures used to obtain approximate equations of motion in that case are identical to those used for turbulent wakes. It turns out that the limit $L/x \to \infty$ does not cause any change or singularity in the first-order equations of motion, so that we can conveniently ignore the special case $L/x \to \infty$ in the analysis to follow.

We look for solutions to the equations of motion that satisfy a velocity-defect law,

$$(U - U_0)/u_* = F(y/\delta), \tag{5.5.6}$$

in such a way that F is independent of the downstream distance x. In other words, we are looking for *self-preserving* flows. Of course, the self-preserving solutions should be asymptotically independent of Reynolds number, so that they can describe an entire family of flows, in which a suitably nondimensionalized pressure gradient is the only parameter. The velocity-defect law (5.5.6) is not expected to be valid in the surface layer, so that the latter must be treated separately.

From the experience gained with pipe flow we can safely assume that $u_*/U_0 \ll 1$ if the Reynolds number $\delta u_*/\nu$ is made sufficiently large. This implies that the shear stress at the wall, ρu_*^2, is very small compared to ρU_0^2. We also assume that the boundary layer grows fairly slowly: $\delta/L \ll 1$ and $\delta/x \ll 1$. All of these assumptions will have to be justified a posteriori.

The potential flow The flow outside the boundary layer is governed by

$$U_0 \frac{\partial U_0}{\partial x} + V_0 \frac{\partial U_0}{\partial y} = -\frac{1}{\rho} \frac{\partial P_0}{\partial x}, \tag{5.5.7}$$

$$U_0 \frac{\partial V_0}{\partial x} + V_0 \frac{\partial V_0}{\partial y} = -\frac{1}{\rho} \frac{\partial P_0}{\partial y}, \tag{5.5.8}$$

together with the appropriate continuity equation and (5.5.4).

With (5.5.6), the continuity equation (5.5.3) may be written as

$$\frac{\partial V}{\partial y} = -\frac{\partial U_0}{\partial x} - \frac{\partial}{\partial x}(u_* F). \tag{5.5.9}$$

Now, the length scale associated with changes in the potential flow is L, so that $\partial U_0/\partial x$ is essentially constant over a distance δ if $\delta/L \ll 1$. Treating $\partial U_0/\partial x$ as a constant, we obtain by integrating (5.5.9) from $y = 0$ to $y = \delta$:

$$V_0(\delta) = -\delta \frac{\partial U_0}{\partial x} - \frac{d}{dx}(u_*\delta) \int_0^\delta F\, d\eta. \tag{5.5.10}$$

Here, $V_0(\delta)$ is the value of V_0 just outside the boundary layer. If the integral in (5.5.10) is finite and if $u_* \ll U_0$, (5.5.10) may be approximated by

$$V_0(\delta) = -\delta \frac{\partial U_0}{\partial x}. \tag{5.5.11}$$

This equation is not valid if $\partial U_0/\partial x$ is very small, as it would be if the pressure gradient $\partial P_0/\partial x$ were small. In that case, the approximations developed for turbulent wakes should be used.

Differentiating the condition of zero vorticity (5.5.4) with respect to x, we estimate

$$\frac{L}{U_0}\frac{\partial}{\partial y}\left(\frac{\partial U_0}{\partial x}\right) = \frac{L}{U_0}\frac{\partial^2 V_0}{\partial x^2} = \mathcal{O}\left(\frac{\delta}{L^2}\right). \tag{5.5.12}$$

This shows that the relative change of $\partial U_0/\partial x$ over a distance δ is of order $(\delta/L)^2$, so that $\partial U_0/\partial x$ can indeed be treated as a constant as far as the boundary layer is concerned.

From (5.5.4) and (5.5.11) we find that $\partial U_0/\partial y = \partial V_0/\partial x = \mathcal{O}(\delta U_0/L^2)$. We can now estimate the left-hand side terms of (5.5.7, 5.5.8) just outside the boundary layer. The result is

$$U_0 \frac{\partial U_0}{\partial x} = \mathcal{O}\left(\frac{U_0^2}{L}\right), \quad V_0 \frac{\partial U_0}{\partial y} = \mathcal{O}\left(\frac{\delta^2 U_0^2}{L^3}\right), \tag{5.5.13}$$

$$U_0 \frac{\partial V_0}{\partial x} = \mathcal{O}\left(\frac{\delta U_0^2}{L^2}\right), \quad V_0 \frac{\partial V_0}{\partial y} = \mathcal{O}\left(\frac{\delta U_0^2}{L^2}\right). \tag{5.5.14}$$

If $\delta/L \ll 1$, $\partial P_0/\partial y \ll \partial P_0/\partial x$, because both terms of (5.5.8) are of the same order and both are a factor δ/L smaller than the dominant term of (5.5.7). This implies that the entire equation for V_0 is dynamically insignificant. The second term on the left-hand side of (5.5.7) is of order δ^2/L^2 compared to the first. If $\delta/L \ll 1$, the equations for the inviscid flow above the boundary layer may thus be approximated by the single equation

$$U_0 \frac{dU_0}{dx} = -\frac{1}{\rho} \frac{dP_0}{dx}. \tag{5.5.15}$$

No partial derivatives are needed to this approximation, because U_0 and P_0 are essentially independent of y as far as the boundary layer is concerned.

The pressure inside the boundary layer We now estimate the order of magnitude of all terms in (5.5.2). If the Reynolds number is large enough, viscous stresses are small compared to Reynolds stresses, so that we may write

$$U \frac{\partial V}{\partial x} + V \frac{\partial V}{\partial y} = -\frac{1}{\rho} \frac{\partial P}{\partial y} - \frac{\partial \overline{v^2}}{\partial y} - \frac{\partial \overline{uv}}{\partial x}. \tag{5.5.16}$$

Since the velocity defect is small, U is of order U_0. The order of magnitude of V is $V_0(\delta)$, which is equal to $\delta U_0/L$. Thus, $U\,\partial V/\partial x = \mathcal{O}(\delta U_0^2/L^2)$. The gradient $\partial V/\delta y = \mathcal{O}(V_0/\delta)$, so that $V\,\partial V/\partial y = \mathcal{O}(\delta U_0^2/L^2)$. The Reynolds stress terms are $\partial \overline{v^2}/\partial y = \mathcal{O}(u_*^2/\delta)$ and $\partial \overline{uv}/\partial x = \mathcal{O}(u_*^2/L)$. The last two estimates are based on the assumption that the stress is of order ρu_*^2 throughout the boundary layer, so that u_* is the relevant velocity scale for the turbulent motion. This assumption is not valid if the pressure gradient causes separation, as we have seen in Section 5.4.

The second Reynolds-stress term in (5.5.16) may be neglected compared to the first. An approximate integral of (5.5.16) then reads

$$\frac{P_0}{\rho} - \frac{P}{\rho} = \overline{v^2} - \int_y^\delta \left(U \frac{\partial V}{\partial x} + V \frac{\partial V}{\partial y} \right) dy. \tag{5.5.17}$$

The first term on the right-hand side of (5.5.17) is of order u_*^2 and the integral is of order $(\delta U_0/L)^2$. These two terms are of the same order of magnitude if γ, defined by

$$\gamma \equiv u_* L/U_0 \delta, \tag{5.5.18}$$

is finite. This amounts to $u_*/U_0 = \mathcal{O}(\delta/L)$, which is similar to the scale relation used in wakes. We assume, subject to later verification, that γ indeed is of order one. Differentiating (5.5.17), we obtain

$$\frac{1}{\rho} \frac{\partial P}{\partial x} - \frac{1}{\rho} \frac{dP_0}{dx} = \mathcal{O}\left(\frac{u_*^2}{L} \right). \tag{5.5.19}$$

The boundary-layer equation With these results, the boundary-layer approximation to (5.5.1) can be obtained. The approximation has to be performed rather carefully, because (5.5.1) is dominated by $U\,\partial U/\partial x$ and $\partial P/\rho\partial x$, both of which are of order U_0^2/L. We are looking for flows which satisfy the velocity-defect law (5.5.6), so that the following decomposition is useful:

$$U\frac{\partial U}{\partial x} = U_0\frac{dU_0}{dx} + U_0\frac{\partial}{\partial x}(U-U_0) + (U-U_0)\frac{dU_0}{dx} + (U-U_0)\frac{\partial}{\partial x}(U-U_0). \tag{5.5.20}$$

The first term on the right-hand side of (5.5.20) cancels the pressure gradient by virtue of (5.5.15). If F is finite, the next two terms are of order U_0u_*/L; it is clear that these should be retained. However, terms of order u_*^2/L can be neglected. The difference between $\partial P/\rho\partial x$ and $dP_0/\rho dx$ is of order (u_*^2/L), as (5.5.19) shows, so that $\partial P/\partial x$ can be replaced by dP_0/dx. The stress term $\partial\overline{u^2}/\partial x = \mathcal{O}(u_*^2/L)$ can be neglected for the same reason. The viscous term $\nu\partial^2U/\partial x^2 = \mathcal{O}(\nu U_0/L^2) = \mathcal{O}(u_*^2/L)(\nu/u_*\delta)$ if $U_0/u_* = \mathcal{O}(L/\delta)$ (5.5.18), so that it also can be neglected if $u_*\delta/\nu$ is not small.

On basis of the results obtained so far, (5.5.1) may be approximated by

$$U_0\frac{\partial}{\partial x}(U-U_0) + (U-U_0)\frac{dU_0}{dx} + (U-U_0)\frac{\partial}{\partial x}(U-U_0) + V\frac{\partial}{\partial y}(U-U_0$$

$$= \frac{\partial}{\partial y}\left(-\overline{uv} + \nu\frac{\partial U}{\partial y}\right). \tag{5.5.21}$$

The last term on the left-hand side could be written in terms of the velocity defect because U_0 is independent of y. The assumption that the velocity defect $(U_0 - U)$ is of order u_* has not yet been applied to the left-hand side of this equation. Because the velocity defect is not small in the surface layer and because the surface stress is purely viscous, further simplification of (5.5.21) should be delayed until a momentum integral has been obtained.

Before we do this, let us look at the orders of magnitude of the various terms in (5.5.21). The Reynolds-stress term in (5.5.21) is of order u_*^2/δ. The first and second terms on the left-hand side of (5.5.21) are of order U_0u_*/L. The Reynolds-stress term, therefore, is of order $u_*L/\delta U_0 = \gamma$ compared to the major inertia terms. Three limit processes are possible. If $\gamma\to 0$ as $\delta u_*/\nu\to\infty$, Reynolds stresses are negligible. This corresponds to situations with extremely rapid acceleration or deceleration of the flow; the particular limit process involved is sometimes used to compute the initial reaction of turbulent boundary layers to very rapid changes in pressure. If $\gamma \rightarrow \infty$, the inertia terms are small compared to the Reynolds stresses. Physically speaking, this is an impossible situation; it corresponds to a Reynolds stress which is independent of y, and therefore equal to zero (because the stress must be zero outside the boundary layer).

The *distinguished limit* is clearly the case in which γ remains finite, no matter how large the Reynolds number is. This is a significant conclusion, because it implies that equilibrium flows can be obtained only if the ratio of the turbulence time scale δ/u_* to the flow time scale L/U_0 is finite and remains constant as the boundary layer develops. In other words, the boundary-layer turbulence has to keep pace with the flow.

Let us return to the momentum integral. Rearranging (5.5.21) with help of the continuity equation (5.5.3), we obtain

$$\frac{\partial}{\partial x}[U(U-U_0)] + \frac{\partial}{\partial y}[V(U-U_0)] + (U-U_0)\frac{dU_0}{dx} = \frac{\partial}{\partial y}\left(-\overline{uv} + \nu\frac{\partial U}{\partial y}\right). \tag{5.5.22}$$

Integration of (5.5.22) yields

$$-\frac{d}{dx}\int_0^\infty U(U-U_0)\,dy - \frac{dU_0}{dx}\int_0^\infty (U-U_0)\,dy = u_*^2. \tag{5.5.23}$$

As before, the stress at the surface is defined as ρu_*^2. Outside the boundary layer, the stress and the velocity defect are zero. The exact location of the upper limit of the integrals in (5.5.23) is immaterial; the infinity symbol is merely used for convenience.

A *normalized boundary-layer thickness* Δ may be defined by

$$\Delta u_* \equiv \int_0^\infty (U_0 - U)\,dy. \tag{5.5.24}$$

If $U_0 - U$ is of order u_* through most of the boundary layer, Δ and δ are of the same order of magnitude. Using (5.5.24), we can write the first integral in (5.5.23) as

$$-\int_0^\infty U(U-U_0)\,dy = U_0 u_* \Delta - \int_0^\infty (U-U_0)^2\,dy. \tag{5.5.25}$$

If the velocity defect is small, the last integral in (5.5.25) is of order $u_*^2\Delta$. In the surface layer, however, $U - U_0 \sim U_0$, so that the contribution to the last integral made in the surface layer is of order $U_0^2\nu/u_*$ (the thickness of the surface layer is of order ν/u_*). Therefore, it is of order $(U_0/u_*)^2\ (\nu/u_*\Delta)$ relative to the contribution made by the rest of the boundary layer. Because we expect that $U_0/u_* \to \infty$ much slower than $\Delta u_*/\nu$, this contribution can be neglected. Finally, because $u_*/U_0 << 1$, only the first term on the right-hand side of (5.5.25) needs to be retained in first approximation. Therefore, (5.5.23) may be approximated by

$$\frac{d}{dx}(\Delta u_* U_0) + \Delta u_* \frac{dU_0}{dx} = u_*^2. \tag{5.5.26}$$

We can now return to (5.5.21). Outside the surface layer, the viscous term can be neglected if $\delta u_*/\nu$ is large. The third inertia term is of order u_*^2/L if the velocity defect is of order u_*, so that it is small compared to the leading terms. The cross-stream velocity component V occurs in (5.2.21); if the analysis leading from (5.5.9) to (5.5.11) is repeated with an arbitrary upper limit of integration, there results $V = -y\, dU_0/dx$ with a correction term that can be neglected if $u_*/U_0 << 1$. The equation of motion for the outer layer thus becomes

$$U_0 \frac{\partial}{\partial x}(U - U_0) + (U - U_0)\frac{dU_0}{dx} - y\frac{dU_0}{dx}\frac{\partial}{\partial y}(U - U_0) = -\frac{\partial \overline{uv}}{\partial y}. \tag{5.5.27}$$

This equation is linear in the velocity defect $U_0 - U$; it is called the *linearized boundary-layer equation.*

Equilibrium flow We want to find solutions to (5.5.27) which satisfy

$$(U - U_0)/u_* = F(\eta), \tag{5.5.28}$$

$$-\overline{uv}/u_*^2 = G(\eta), \tag{5.5.29}$$

where

$$\eta = y/\Delta. \tag{5.5.30}$$

The normalized boundary-layer thickness Δ has been used here for convenience. Substitution of (5.5.28) and (5.5.29) into (5.5.27) yields

$$\frac{\Delta}{u_*^2}\frac{d}{dx}(U_0 u_*)F - \frac{1}{u_*}\frac{d}{dx}(\Delta U_0)\,\eta\frac{dF}{d\eta} = \frac{dG}{d\eta}. \tag{5.5.31}$$

If the coefficients in this equation can be made independent of x, the equation of motion allows self-preserving solutions. However, further analysis of (5.5.31) cannot proceed until the equations for the wall layer have been examined.

The flow in the wall layer Let us consider the equation for U in the immediate vicinity of the surface. We expect U to be of order u_*, so that $U\,\partial U/\partial x$ is of order u_*^2/L. Also, $\partial U/\partial x = \mathcal{O}(u_*/L)$, so that $V = \mathcal{O}(\nu/L)$ if ν/u_* is the length scale for the wall layer. Hence, $V\,\partial U/\partial y = \mathcal{O}(u_*^2/L)$. The pressure-gradient term is of order U_0^2/L, so that the inertia terms should be neglected if $u_*/U_0 \ll 1$. The length scale in the wall layer is ν/u_* in order to keep Reynolds stress and viscous stress of the same order of magnitude. The leading stress terms in the equation for U are $\partial\overline{uv}/\partial y$ and $\nu\,\partial^2 U/\partial y^2$; they are of order u_*^3/ν. The pressure gradient is of order $\nu U_0^2/Lu_*^3$ compared to the other stress terms. Now,

$$\frac{\nu U_0^2}{Lu_*^3} = \frac{U_0\delta}{u_*L}\cdot\frac{U_0}{u_*}\cdot\frac{\nu}{u_*\delta}. \tag{5.5.32}$$

The first factor on the right-hand side of (5.5.32) is $1/\gamma$. Because we decided not to deal with very rapidly accelerating or decelerating flows, γ is finite. As was stated before, $U_0/u_* \to \infty$ rather slowly compared to $\delta u_*/\nu$. Therefore, the pressure gradient is small compared to the principal Reynolds-stress and viscous-stress terms in the inner layer. The equation of motion reduces to

$$0 = \frac{\partial}{\partial y}\left(-\overline{uv} + \nu\frac{\partial U}{\partial y}\right). \tag{5.5.33}$$

It can be seen intuitively that this approximation is correct. Throughout the analysis, it has been assumed that the velocity defect is of order u_* and that Reynolds stresses are of order ρu_*^2. These assumptions can be valid only if no other characteristic velocity is relevant. In conditions where the pressure gradient might generate a new velocity scale (like the one used in Section 5.4), the obvious requirement is that it be small compared to the friction velocity u_*.

The law of the wall Equation (5.5.33) defines a constant-stress layer with wall stress ρu_*^2. The nature of the solutions of (5.5.33) has been studied in Section 5.2; we recall that

$$U/u_* = f(yu_*/\nu), \tag{5.5.34}$$

$$-\overline{uv}/u_*^2 = g(yu_*/\nu). \tag{5.5.35}$$

To first approximation the flow in the wall layer is independent of the pressure gradient. This result was first discovered in experiments made by Ludwieg and Tillmann (1949).

The logarithmic friction law We assume that solutions to the equation of motion for the outer layer which satisfy the velocity-defect law (5.5.28) do exist. If this is the case, the law of the wall (5.5.34) must be matched to the velocity-defect law (5.5.28) through a logarithmic velocity profile. The logarithmic velocity profile gives a logarithmic friction law, which may be written as

$$\frac{U_0}{u_*} = \frac{1}{\kappa} \ln \frac{\Delta u_*}{\nu} + A. \tag{5.5.36}$$

The additive constant A can be a function of a pressure-gradient parameter. For later use, a differentiated form of (5.5.36) is given. A convenient form is

$$\left(1 + \frac{u_*}{\kappa U_0}\right) \frac{d}{dx}\left(\frac{U_0}{u_*}\right) = \frac{1}{\kappa\Delta}\frac{d\Delta}{dx} + \frac{1}{\kappa U_0}\frac{dU_0}{dx}. \tag{5.5.37}$$

The pressure-gradient parameter We now determine under what conditions self-preserving solutions to (5.5.31) may be expected. The coefficients occurring in (5.5.31) may be expanded as follows:

$$\frac{\Delta}{u_*^2}\frac{d}{dx}(U_0 u_*) = 2\frac{\Delta}{u_*}\frac{dU_0}{dx} + \frac{\Delta U_0^2}{u_*^2}\frac{d}{dx}\left(\frac{u_*}{U_0}\right), \tag{5.5.38}$$

$$\frac{1}{u_*}\frac{d}{dx}(\Delta U_0) = \frac{U_0}{u_*^2}\frac{d}{dx}(\Delta u_*) - \frac{\Delta U_0^2}{u_*^2}\frac{d}{dx}\left(\frac{u_*}{U_0}\right). \tag{5.5.39}$$

The momentum integral (5.5.26) may be rearranged to read

$$\frac{U_0}{u_*^2}\frac{d}{dx}(\Delta u_*) = 1 - 2\frac{\Delta}{u_*}\frac{dU_0}{dx}. \tag{5.5.40}$$

Substitution of the differential friction law (5.5.37) into (5.5.38, 5.5.39) shows that the last terms of (5.5.38) and (5.5.39) are small compared to the others if $u_*/U_0 \ll 1$. Inspection of the set (5.5.38, 5.5.39, 5.5.40) then indicates that a convenient pressure-gradient parameter is Π, defined by

$$\Pi \equiv -\frac{\Delta}{u_*}\frac{dU_0}{dx}. \tag{5.5.41}$$

This result is not surprising: self-preservation can be obtained only if the ratio of the time scales $(dU_0/dx)^{-1}$ and Δ/u_* is a constant (see the discussion following (5.5.21) and (5.5.56)).

In terms of Π, (5.5.31) and (5.5.40) read

$$-2\Pi F - (1 + 2\Pi)\,\eta\frac{dF}{d\eta} = \frac{dG}{d\eta}, \tag{5.5.42}$$

$$\frac{U_0}{u_*^2}\frac{d}{dx}(\Delta u_*) = 1 + 2\Pi. \tag{5.5.43}$$

The system (5.5.36, 5.5.42, 5.5.43) is subject to a normalization condition imposed by the definitions (5.5.24) and (5.5.28) of Δ and F, respectively. The normalization condition is

$$\int_0^\infty F\,d\eta = -1. \tag{5.5.44}$$

The boundary conditions imposed on (5.5.42) are

$$F \to 0,\; G \to 0 \quad \text{for } \eta \to \infty, \tag{5.5.45}$$

$$G \to 1 \quad \text{for } \eta \to 0, \tag{5.5.46}$$

$$\eta\, dF/d\eta \to 1/\kappa \quad \text{for } \eta \to 0. \tag{5.5.47}$$

The system of equations (5.5.36, 5.5.42–5.5.47) is independent of x if Π is a constant. Therefore, we may expect self-preserving boundary-layer flows in pressure distributions that make Π independent of x. The problem defined by (5.5.42–5.5.47) is also independent of the Reynolds number, so that the solutions $F(\eta)$, $G(\eta)$ exhibit asymptotic invariance (Reynolds-number similarity). Therefore, boundary layers in which Π is constant are *equilibrium layers*; their velocity profiles are self-preserving and the velocity profiles of two different boundary layers at the same value of Π are identical, even if their Reynolds numbers are not the same. Of course, all of these statements are only valid asymptotically as $\Delta u_*/\nu \to \infty$.

These conclusions were first obtained by F. H. Clauser (1956). Clauser performed a series of experiments in which the pressure distribution was carefully adjusted in order to obtain downstream invariance of the velocity-defect function $F(\eta)$. His experiments showed that the pressure distribution

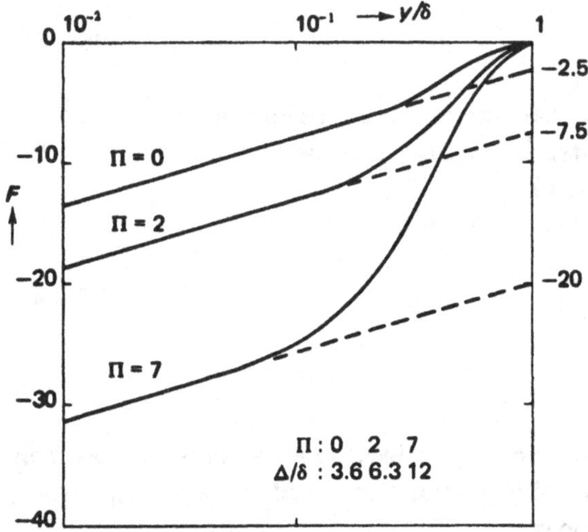

Figure 5.17. Velocity-defect profiles at different values of Π. The scaling length for y is the boundary-layer thickness δ, defined as the value of y where $F = -0.1$ (based on data by Clauser, 1956).

was well represented by a constant value of Π. The significance of Π was discovered from an ad hoc argument involving the relative contributions of the wall stress and the pressure gradient to the rate of increase of momentum deficit in the boundary layer.

Some of the velocity profiles obtained by Clauser and others for different values of Π are shown in Figure 5.17. The additive constant in the logarithmic part of F increases rapidly with Π; the amplitude of the wake function $W(\eta)$, which is the difference between F and its logarithmic part, therefore also increases with Π. In the limit $\Pi \to \infty$, the velocity profile may be a pure wake function.

Free-stream velocity distributions The equations governing the downstream development of equilibrium layers (5.5.41, 5.5.43, 5.5.36) are

$$\frac{\Delta}{u_*}\frac{dU_0}{dx} = -\Pi, \quad \frac{U_0}{u_*^2}\frac{d}{dx}(\Delta u_*) = 1 + 2\Pi, \quad \frac{U_0}{u_*} = \frac{1}{\kappa}\ln\frac{\Delta u_*}{\nu} + A(\Pi).$$

No general solution to the set (5.5.36, 5.5.41, 5.5.43) is known. Approximate solutions, however, can easily be obtained if the very slow change of U_0/u_* with respect to x is exploited. If the range of values of x for which an approximate solution is desired is fairly small, it may be assumed that u_*/U_0 is equal to its value at the beginning of the interval ($x = x_i$). If we put

$$\frac{u_*}{U_0} = \left(\frac{u_*}{U_0}\right)_i = \beta_i, \tag{5.5.48}$$

we may replace (5.5.41) and (5.5.43) by

$$\frac{\Delta}{U_0}\frac{dU_0}{dx} = -\beta_i\,\Pi, \tag{5.5.49}$$

$$\frac{1}{U_0}\frac{d}{dx}(\Delta U_0) = (1 + 2\Pi)\,\beta_i. \tag{5.5.50}$$

In this approximation, the logarithmic friction law has to be ignored. The solution of (5.5.49, 5.5.50) was first given by A. A. Townsend (1956); it reads

$$\frac{\Delta}{\Delta_i} = 1 + \tilde{\gamma}_i\left(\frac{x}{x_i} - 1\right), \tag{5.5.51}$$

$$\frac{U_0}{U_{0i}} = \left[1 + \tilde{\gamma}_i\left(\frac{x}{x_i} - 1\right)\right]^{\alpha}, \tag{5.5.52}$$

where

$$\alpha = -\frac{\Pi}{1 + 3\Pi}, \tag{5.5.53}$$

and

$$\tilde{\gamma}_i = (1 + 3\Pi)\,\beta_i x_i/\Delta_i. \tag{5.5.54}$$

The coefficient $\tilde{\gamma}_i$ is of order $u_* x/\Delta U_0$, so that it is similar to the time-scale ratio γ defined in (5.5.18). The length scale L, defined in (5.5.5) has the value $\Delta_i/\beta_i\Pi$ at $x = x_i$, so that γ_i may be written as

$$\tilde{\gamma}_i = \left(\frac{1 + 3\Pi}{\Pi}\right)\frac{x_i}{L_i}. \tag{5.5.55}$$

The time-scale ratio γ, on the other hand, is given by

$$\gamma = \frac{\Delta}{\delta}\,\Pi^{-1}. \tag{5.5.56}$$

The singularity of (5.5.56) in the limit as $\Pi \to 0$ is due to the particular way in which L is defined. If $\Pi = 0$, (5.5.50) yields $d\Delta/dx = \beta_i$, which corresponds to finite values of $\Delta U_0/xu_*$, so that again the ratio of time scales is finite. It should be noted that Δ/δ is always finite if δ is defined as the value of y where F is some small number (say 0.1).

It is clear that (5.5.51) and (5.5.52) are singular if $\Pi \to \infty$. This singularity has physical significance, because it represents flows that are approaching separation. According to (5.5.52) and (5.5.53), this occurs if $U_0 \propto x^{-1/3}$. In experimental practice, no steady, stable flows at $\Pi > 10$ can be obtained. Equation (5.5.51) also shows that equilibrium layers become thicker more rapidly at large positive values of Π. It should be noticed that all boundary layers grow linearly in x if u_*/U_0 is assumed to be constant. For large values of $(x - x_i)$, the slow decrease of u_*/U_0 with Δ (and thus with x) takes effect; the boundary-layer thickness then increases roughly as $\delta \propto x/\ln x$.

Boundary layers in zero pressure gradient A somewhat more detailed discussion of the case $\Pi = 0$ (corresponding to constant U_0) is in order. If the pressure gradient is zero, (5.5.42), (5.5.43), and (5.5.36) become

$$-\eta \frac{dF}{d\eta} = \frac{dG}{d\eta}, \tag{5.5.57}$$

$$\frac{U_0}{u_*^2} \frac{d}{dx} (\Delta u_*) = 1, \tag{5.5.58}$$

$$\frac{U_0}{u_*} = \frac{1}{\kappa} \ln \frac{\Delta u_*}{\nu} + A(0). \tag{5.5.59}$$

The short-range growth of Δ may be approximated by

$$\frac{d\Delta}{dx} = \beta_i, \tag{5.5.60}$$

where β_i is the value of u_*/U_0 at x_i. In the case $\Pi = 0$, $\Delta/\delta \cong 3.6$ if δ is defined as the value of y where $F = -0.1$.

It is worthwhile to consider the entrainment of fluid outside the boundary layer by the turbulent motion at the edge of the boundary layer. The continuity equation may be integrated to yield

$$V_0 = -\int_0^\infty \frac{\partial U}{\partial x} dy = -\frac{d}{dx} \left(u_* \Delta \int_0^\infty F \, d\eta \right). \tag{5.5.61}$$

Since the integral in (5.5.61) is equal to -1 by virtue of (5.5.44), we may write for the slope α_0 of the mean streamlines at the edge of the boundary layer

$$\alpha_0 \cong \tan \alpha_0 = \frac{V_0}{U_0} = \frac{1}{U_0}\frac{d}{dx}(\Delta u_*) = \frac{d\delta^*}{dx}, \tag{5.5.62}$$

where δ^* is the *displacement thickness* (5.5.66). By substitution of (5.5.58) we find that

$$\alpha_0 = (u_*/U_0)^2. \tag{5.5.63}$$

The average slope α_δ of the edge of the boundary layer is $d\delta/dx \cong 0.28\ d\Delta/dx$ if $\Delta/\delta \cong 3.6$. From (5.5.58) and (5.5.37) we conclude that

$$\alpha_\delta = \frac{d\delta}{dx} \cong \frac{0.28}{U_0/u_* - 1/\kappa}. \tag{5.5.64}$$

If $u_*/U_0 \ll 1$, $\alpha_\delta \gg \alpha_0$. A few numbers may be helpful. If $U_0/u_* = 30$, $\alpha_0 \cong 0.064°$ and $\alpha_\delta \cong 0.57°$. If $U_0/u_* = 20$, $\alpha_0 \cong 0.14°$ and $\alpha_\delta \cong 0.92°$. Figure 5.18 illustrates the situation. The entrainment process is believed to be maintained by large-eddy motions like those sketched in the figure. These

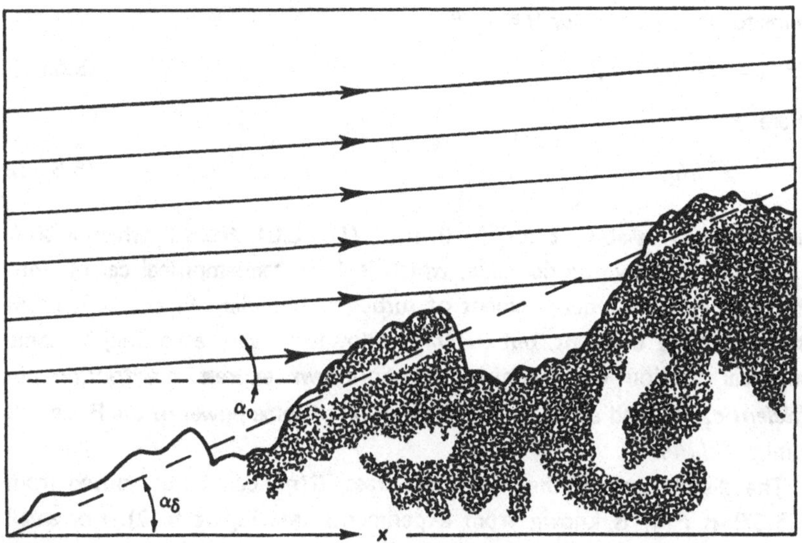

Figure 5.18. Entrainment by a boundary layer in zero pressure gradient. The mean streamlines do not represent the actual flow pattern over the interface shown.

eddies continually distort the interface between the turbulent and nonturbulent fluid and may on occasion engulf parts of the nonturbulent fluid. The entrainment velocity is about $0.28u_*$ if $1/\kappa$ is neglected compared to U_0/u_* and if α_0 is neglected compared to α_δ. The interface between the turbulent boundary layer and the potential flow is quite sharp; its characteristic thickness is believed to be of order ν/u_*, which is comparable to the thickness of the viscous sublayer (Corrsin and Kistler, 1954).

The momentum integral (5.5.58) is associated with the linearized equations of motion. This implies that the momentum thickness θ, defined by

$$U_0^2\theta \equiv \int_0^\infty U(U_0 - U)\,dy, \tag{5.5.65}$$

has been assumed to be equal to the displacement thickness δ^*, defined by

$$U_0\delta^* \equiv \int_0^\infty (U_0 - U)\,dy. \tag{5.5.66}$$

This approximation, of course, is consistent with the assumption that the velocity defect $U_0 - U$ is small compared to U_0. Experiments have shown that the velocity-defect law is satisfied rather accurately even if the velocity defect is not small. Substitution of (5.5.28) into the definitions of δ^* and θ yields for the *shape factor* $H \equiv \delta^*/\theta$

$$H = (1 - C u_*/U_0)^{-1}, \tag{5.5.67}$$

where

$$C = \int_0^\infty F^2\,d\eta. \tag{5.5.68}$$

The value of C is about 6 for $\Pi = 0$. If $u_*/U_0 = 0.04$, $H \cong 1.3$, which is 30% larger than the asymptotic value, which is 1. In semiempirical calculations of the downstream development of turbulent boundary layers, H is often assumed to be constant, but u_*/U_0 is allowed to vary according to some empirical friction law (empirical friction laws express the *friction coefficient* c_f, defined as $2u_*^2/U_0^2$, as a function of some power of the Reynolds number $\theta U_0/\nu$).

The distribution of the Reynolds stress, $G(\eta)$, can be computed from (5.5.57) if $F(\eta)$ is known from experiments (see Figure 5.17). For small values of η, F is logarithmic, so that (5.5.57) gives

$$dG/d\eta = -1/\kappa \quad \text{for } \eta \to 0. \tag{5.5.69}$$

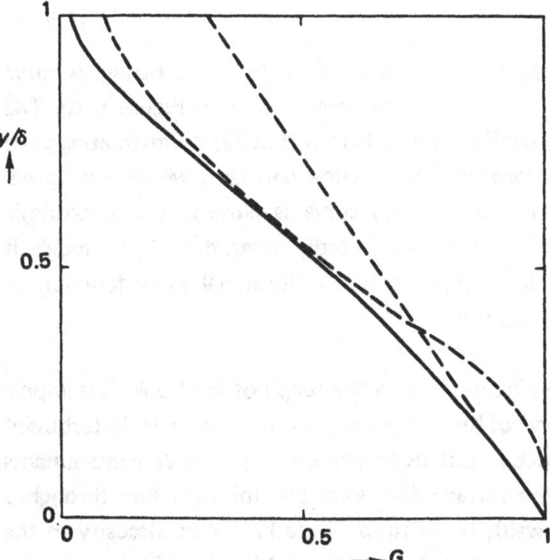

Figure 5.19. The Reynolds-stress distribution for $\Pi = 0$. The solid line is computed with (5.5.57) and F given by Figure 5.17. The straight dashed line is (5.5.70), with $1/\kappa = 2.5$, $\Delta/\delta = 3.6$. The curved dashed line is (5.5.72), with $K = 1/60$.

Since $G \to 1$ if $\eta \to 0$, (5.5.69) may be integrated to yield

$$G(\eta) = 1 - \eta/\kappa. \tag{5.5.70}$$

This expression is valid only near the surface. Figure 5.19 gives a sketch of the distribution of $G(\eta)$.

Equation (5.5.57) relates the velocity profile to the stress profile. So far, we have avoided any assumptions on the relation between stress and velocity gradient. With similarity arguments and asymptotic rules, we have resolved all of the essential features of boundary-layer flows without ever solving the equations of motion. If we want to solve equations like (5.5.57), we need a constitutive relation to link the stress to the velocity gradient. A simple constitutive relation is

$$G = K\, dF/d\eta, \tag{5.5.71}$$

where K is an eddy viscosity, nondimensionalized with u_* and Δ. If K is independent of η, (5.5.57) and (5.5.71) can easily be solved for the stress G. The result is

$$G(\eta) = \exp(-\eta^2/2K). \tag{5.5.72}$$

The value of K, of course, has to be determined by curve fitting. A curve according to (5.5.72), with $K = \frac{1}{60}$, has been drawn in Figure 5.19. The velocity distribution $F(\eta)$ can be obtained from (5.5.72) by integrating once more. This introduces an arbitrary integration constant, which can be adjusted in such a way that the resulting curve is close to the logarithmic velocity profile at small values of η. This is hardly worth the effort, though; if an analytical expression for $F(\eta)$ is desired, a sinusoidal wake function of suitable amplitude does just as well.

Transport of scalar contaminants Within the scope of this book, it is impossible to discuss the transport of heat or other scalar contaminants in turbulent boundary layers in any detail. Let us briefly consider passive contaminants that are released from the surface (for example, the heat flux through a boundary layer on a hot wall). If the ratio of the kinematic viscosity to the diffusivity of the contaminant is near unity, the distribution of the contaminant is similar to the distribution of the mean-velocity defect; the rate of spread of contaminant in the y direction is the same as the rate of growth of the boundary layer. The rate of transfer of contaminant away from the surface is coupled to the stress at the surface. In the case of temperature, the transfer law reads

$$\frac{\Theta_w - \Theta_0}{\theta_*} = \frac{1}{\kappa} \ln \frac{\Delta u_*}{\nu} + \text{const}, \tag{5.5.73}$$

where

$$\theta_* \equiv H/\rho c_p u_*. \tag{5.5.74}$$

In these expressions, it has been assumed that the thermal diffusivity is equal to ν. The rate of heat transfer from the surface, H, can be computed if the temperatures at the surface (Θ_w) and outside the boundary layer (Θ_0), as well as u_* and Δ, are known.

If the diffusivities for the scalar and for momentum are not the same, the thickness of the viscous (momentum) sublayer and of the molecular diffusion layer of the scalar near the surface are not the same. The transfer of scalar contaminants through the boundary layer then becomes a very complicated problem. A case in point is heat transfer in turbulent flow of liquid mercury. In mercury at room temperature, the thermal diffusivity (γ) is 35 times as large as the kinematic viscosity. If the transport of heat by turbulent motion is represented by an eddy diffusivity γ_e, which is about $u_*\Delta/60$, the ratio γ_e/γ becomes equal to one for $u_*\Delta/\nu \sim 2{,}000$. At moderate Reynolds numbers like this, much of the heat transfer is caused by molecular motion, even though nearly all of the momentum transfer is caused by the turbulent motion. Effectively, the molecular diffusion layer extends through the entire momentum boundary layer.

Problems

5.1 Consider fully developed turbulent flow in a two-dimensional diffuser with plane walls. Estimate the opening angle of the diffuser for which the downstream pressure gradient is equal to zero.

5.2 Describe the radial distribution of the circulation and of the mean tangential velocity in a turbulent line vortex. The circulation outside the turbulent vortex is constant; it has a value Γ_0. This is an inner-outer layer problem. The inner core of the vortex is in solid-body rotation; it has negligible Reynolds stresses. For the equations of motion in cylindrical coordinates, see Batchelor (1967) or other texts.

5.3 Estimate the volume flow in the Gulf Stream. This flow is due to the flow in the Ekman layer of the North Atlantic Ocean. Assume that the Ekman layer is driven by westerly winds across the Atlantic at middle latitudes. The wind speeds are of order 10 m/sec. What is the direction of the volume flux in the Ekman layer?

5.4 Experiments have shown that small amounts of high molecular weight, linear polymers added to water can cause a substantial drag reduction in turbulent pipe and boundary-layer flow of water. No satisfactory explanation of this phenomenon has been found, but an appreciation for the order of magnitude of this effect can be obtained by assuming that the polymer solution doubles the viscosity experienced by the turbulence without changing the viscosity experienced by the mean flow. Obtain an estimate for the drag reduction on basis of this assumption. An analysis of the effects of polymers on Figure 5.5 is helpful.

5.5 Write an equation for the kinetic energy $\frac{1}{2}U^2$ of the mean velocity in fully developed turbulent flow in a plane channel. Sketch the distributions of all terms across the channel. Use the data in Figures 5.6, 5.7, and 5.8 whenever needed to obtain reasonable accuracy. The energy exchange between the core region and the wall layers is of particular interest. Interpret your results carefully.

5.6 Repeat the analysis of Problem 5.5 for a turbulent boundary layer over a plane wall without pressure gradient.

5.7 From the data in Section 5.5, obtain an approximate friction law of the type $c_f = \alpha R_\theta^{-\beta}$ ($c_f = 2(u_*/U_0)^2$, $R_\theta = \theta U_0/\nu$, θ is the momentum thickness) for turbulent boundary layers in zero pressure gradient. Integrate the momentum integral equation ($c_f = 2\, d\theta/dx$ if $dP/dx = 0$) to obtain an approximate drag formula for a plate of length L.

6

THE STATISTICAL DESCRIPTION OF TURBULENCE

Up to now, we have considered only average values of fluctuating quantities, such as U and $-\overline{uv}$. It is just as important to our understanding of turbulence to examine how fluctuations are distributed around an average value and how adjacent fluctuations (next to each other in time or space) are related. The study of distributions around a mean value requires the introduction of the probability density and its Fourier transform, the characteristic function. The study of the relation between neighboring fluctuations calls for the introduction of the autocorrelation and its Fourier transform, the energy spectrum. This chapter is devoted to the development of these mathematical tools; in the following two chapters, they are used in the study of turbulent transport ("diffusion") and of spectral dynamics. One other tool needed in the study of turbulent transport is the central limit theorem, which makes predictions about the shape of the probability density of certain quantities. The central limit theorem is introduced and discussed at the end of this chapter.

6.1
The probability density

We restrict the discussion to fluctuating quantities that are statistically steady, so that their mean values are not functions of time. Only under this condition does the idea of a time average make sense. A statistically steady function is called *stationary*; an example of a stationary function is given in Figure 6.1. The fluctuating $\tilde{u}(t)$ might be the streamwise velocity component measured in a wind tunnel behind a grid. We are interested in measuring the relative amount of time that $\tilde{u}(t)$ spends at various levels. We could get a crude idea of this by displaying $\tilde{u}(t)$ on the y axis of an oscilloscope, with a rapid sweep on the x axis. A time exposure would have a variable density, proportional to the time spent at each value of y. A more accurate measurement can be obtained by the use of a gating circuit, which turns on when the signal $\tilde{u}(t)$ is between two adjacent levels. In Figure 6.1 the levels are placed fairly close together in terms of the width of $\tilde{u}(t)$. The output of the gating circuit is shown below $\tilde{u}(t)$. If this is averaged, we obtain the percentage of time spent by $\tilde{u}(t)$ between the two levels. Adjusting the electronic "window" successively to different heights, we obtain a function similar to the one shown to the right of $\tilde{u}(t)$ in Figure 6.1.

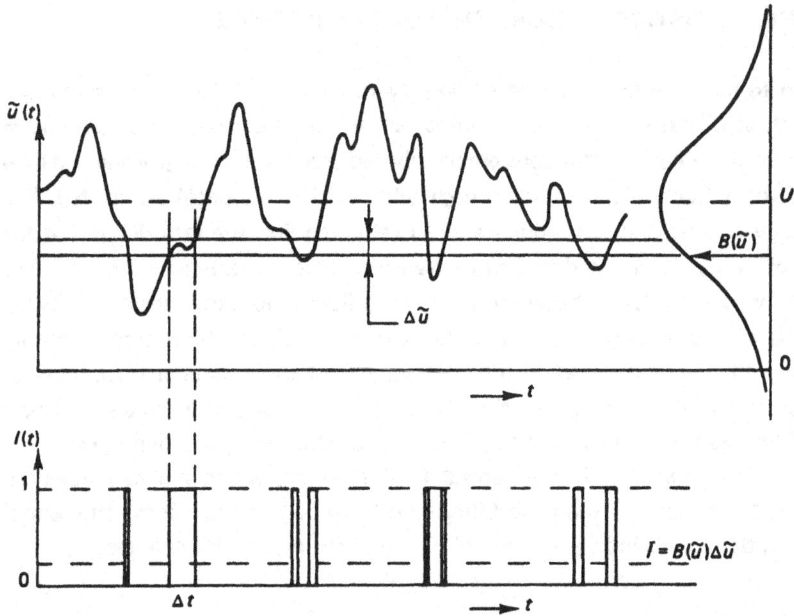

Figure 6.1. Measurement of the probability density of a stationary function. The function I (t) is the discriminator output.

We expect that the averaged output of the gating circuit is proportional to the window width $\Delta\tilde{u}$, so that it is convenient to define a quantity $B(\tilde{u})$ by

$$B(\tilde{u})\,\Delta\tilde{u} \equiv \lim_{T\to\infty} \frac{1}{T}\Sigma(\Delta t). \tag{6.1.1}$$

The function $B(\tilde{u})$ is called a *probability density*; the probability of finding $\tilde{u}(t)$ between $\tilde{u}$ and $\tilde{u}+\Delta\tilde{u}$ is equal to the proportion of time spent there. Because $B(\tilde{u})$ represents a fraction of time, it is always positive, while the sum of the values of $B(\tilde{u})$ for all $\tilde{u}$ must be equal to one:

$$B(\tilde{u}) \geqslant 0, \quad \int_{-\infty}^{\infty} B(\tilde{u})\,d\tilde{u} = 1. \tag{6.1.2}$$

The shape of $B(\tilde{u})$ sketched in Figure 6.1 is typical of probability densities measured in turbulence. Many other shapes are possible; the probability density of a sine wave is sketched in Figure 6.2. This curve is zero beyond ± 1, because the sine wave has unit amplitude. Near ± 1, the slope goes to zero, so that the sine wave spends most time there, making the values of $B(\tilde{u})$ near ± 1 very large.

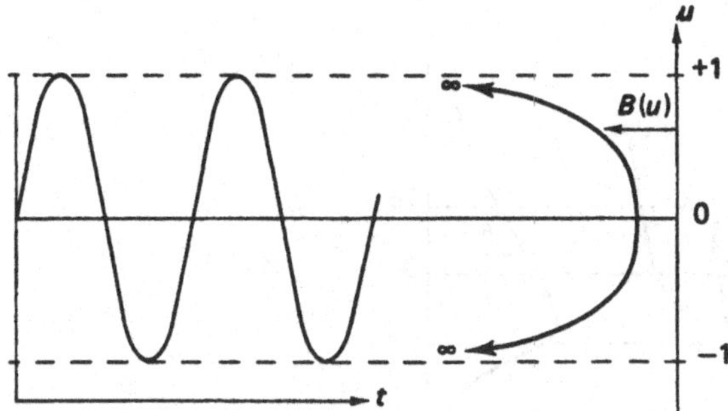

Figure 6.2. The probability density distribution of a sine wave.

We may express the averages we are familiar with in terms of $B(\tilde{u})$. Suppose we wish to average some function $f(\tilde{u})$. The time average

$$\bar{f} = \lim_{T\to\infty} \frac{1}{T}\int_{t_0}^{t_0+T} f(\tilde{u})\, dt \tag{6.1.3}$$

can be formed by adding all of the time intervals between t_0 and $t_0 + T$ during which $\tilde{u}(t)$ is between $\tilde{u}$ and $\tilde{u} + \Delta\tilde{u}$, multiplying this by $f(\tilde{u})$, and summing over all levels. The proportion of time spent between $\tilde{u}$ and $\tilde{u} + \Delta\tilde{u}$ is equal to $B(\tilde{u})\, \Delta\tilde{u}$, so that we can write

$$\bar{f} = \lim_{T\to\infty} \frac{1}{T}\int_{t_0}^{t_0+T} f(\tilde{u})\, dt = \int_{-\infty}^{\infty} f(\tilde{u})B(\tilde{u})\, d\tilde{u}. \tag{6.1.4}$$

The mean values of the various powers of $\tilde{u}$ are called *moments*. The first moment is the familiar mean value, which is defined by

$$U \equiv \int_{-\infty}^{\infty} \tilde{u}B(\tilde{u})\, d\tilde{u}. \tag{6.1.5}$$

In experimental work, the mean value is always subtracted from the fluctuating function $\tilde{u}(t)$. As in Chapter 2, we denote the fluctuations by u, so that $u = \tilde{u} - U$ and $\bar{u} = 0$. We then have $B(\tilde{u}) = B(U + u)$, so that it is convenient to use a probability density $B(u)$, which is obtained by shifting $B(\tilde{u})$ over a distance U along the $\tilde{u}$ axis. The moments formed with u^n and $B(u)$ are called *central moments*. The first central moment, of course, is zero.

The mean-square departure σ^2 from the mean value U is called the *variance*, or second (central) moment. It is defined by

$$\sigma^2 \equiv \overline{u^2} = \int_{-\infty}^{\infty} u^2 B(\tilde{u})\, d\tilde{u} = \int_{-\infty}^{\infty} u^2 B(u)\, du. \tag{6.1.6}$$

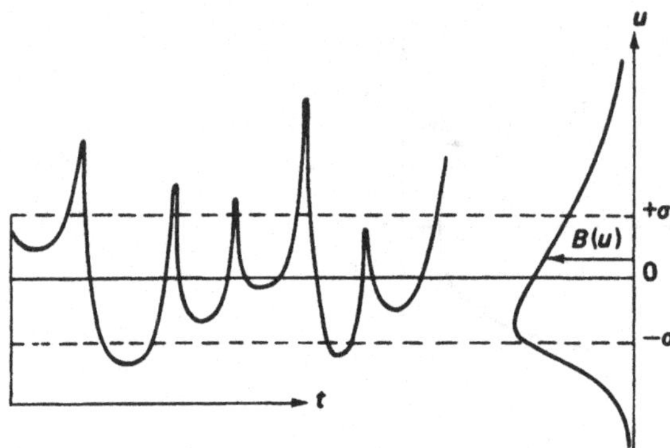

Figure 6.3. A function with positive skewness.

The square root of the variance, σ, is the familiar *standard deviation* (rms amplitude). The standard deviation is the most convenient measure of the width of $B(u)$.

The value of σ^2 is not affected by any lack of symmetry in $B(u)$ about the origin; if $B(u)$ is written as the sum of symmetric and antisymmetric parts, the latter does not contribute to σ^2. The *third moment,* however, defined by

$$\overline{u^3} \equiv \int_{-\infty}^{\infty} u^3 B(u)\, du, \tag{6.1.7}$$

depends only on the lack of symmetry in $B(u)$. If $B(u)$ is symmetric about the origin, $\overline{u^3} = 0$. It is customary to nondimensionalize $\overline{u^3}$ by σ^3, which gives a dimensionless measure of the asymmetry. This is called the *skewness* (S):

$$S \equiv \overline{u^3}/\sigma^3. \tag{6.1.8}$$

Figure 6.3 pictures a function with a positive value of S. The skewness is positive because large negative values of u^3 are not as frequent as large positive values of u^3.

The fourth moment, nondimensionalized by σ^4, is called *kurtosis* or *flatness factor*; it is represented by the symbol K:

$$K \equiv \frac{\overline{u^4}}{\sigma^4} = \frac{1}{\sigma^4}\int_{-\infty}^{\infty} u^4 B(u)\, du. \tag{6.1.9}$$

Two functions, one with a relatively small and the other with a relatively large kurtosis, are sketched in Figure 6.4. The value of the kurtosis is large if the values $B(u)$ in the tails of the probability density are relatively large. The

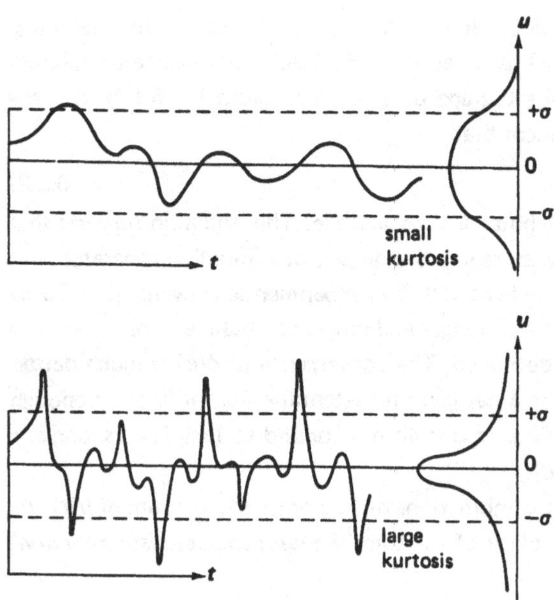

Figure 6.4. Functions with small and large kurtosis.

peaky function in Figure 6.4 frequently takes on values far away from the axis, so that its kurtosis is large. Because the fourth moment is nondimensionalized with σ^4, K contains no information on the width of the density $B(u)$.

6.2
Fourier transforms and characteristic functions

Although it is easy to see the physical significance of the probability density, it is often more convenient to work with another quantity, the *characteristic function*. This function is defined as the Fourier transform of $B(u)$. This means that we have to discuss Fourier transforms.

A Fourier-transform pair is defined by

$$\phi(k) \equiv \int_{-\infty}^{\infty} e^{iku} B(u)\, du, \quad B(u) \equiv \frac{1}{2\pi} \int_{-\infty}^{\infty} e^{-iku} \phi(k)\, dk. \tag{6.2.1}$$

We have used the probability density $B(u)$ and the corresponding characteristic function $\phi(k)$ as examples; we use other Fourier-transform pairs later. The conditions on the existence of $\phi(k)$ and on its ability to produce $B(u)$ upon integration are straightforward and need not concern us here.

In order to gain an appreciation for the usefulness of Fourier transforms, the behavior of $\phi(k)$ as reflected in $B(u)$ and conversely are explored. From the definition of the average of a function given in (6.1.4), and the definition of $\phi(k)$, it is evident that

$$\phi(k) = \overline{\exp\,[iku(t)]}. \tag{6.2.2}$$

As always, the overbar denotes a time average. This equation suggests that $\phi(k)$ can be measured by averaging the output of a function generator that converts $u(t)$ into $\sin u(t)$ and $\cos u(t)$. The experimental convergence of $B(u)$ is poor, because one must wait longer and longer to obtain a stable average as the window width Δu is decreased. The convergence of $\phi(k)$ is much better. Of course, there cannot be a net gain; to determine $\phi(k)$ accurately enough to obtain $B(u)$ from the Fourier transform is bound to take just as long as a direct measurement of $B(u)$.

If we have to deal with combinations of functions, say the sum of $u(t)$ and $v(t)$, the characteristic function of the sum (the *joint characteristic function*) is simply expressed by

$$\phi(k,\ell) = \overline{\exp\,[iku(t) + i\ell v(t)]}. \tag{6.2.3}$$

The corresponding probability density, which we encounter shortly, has no such simple form. This simplicity is one reason for the introduction of the characteristic function. We further discuss joint characteristic functions in Section 6.3.

The moments of $u(t)$ are related to $\phi(k)$ in a simple way. Differentiating the first of (6.2.1) with respect to k, we find that the moments are related to derivatives of $\phi(k)$ at the origin:

$$\left.\frac{d^n\phi(k)}{dk^n}\right|_{k=0} = i^n\,\overline{u^n}. \tag{6.2.4}$$

Because $\overline{u} = 0$, the slope of ϕ at the origin is zero. Because of (6.2.4), the characteristic function can be written as a Taylor series of the moments:

$$\phi(k) = \sum_{n=0}^{\infty} \frac{(ik)^n}{n!}\,\overline{u^n}. \tag{6.2.5}$$

Because no densities obtained in a laboratory have moments that are unbounded, the corresponding characteristic functions in principle have all derivatives. We say "in principle," because the larger the order of a moment is, the longer it takes to obtain a stable value. High-order moments are very strongly affected by large excursions from the mean, which seldom occur. Therefore, moments higher than the fourth are seldom measured, so that we never have more than the first few derivatives of $\phi(k)$.

If $B(u)$ is symmetric, $\phi(k)$ is real. This can be seen if the first of (6.2.1) is written in terms of sin ku and cos ku. This yields

$$\phi(k) = \int_{-\infty}^{\infty} \cos ku\, B(u)\, du + i \int_{-\infty}^{\infty} \sin ku\, B(u)\, du. \tag{6.2.6}$$

Only the antisymmetric part of $B(u)$ can contribute to the second integral. From (6.2.6) we also conclude that the real part of $\phi(k)$ is even in k, while the imaginary part is odd.

The modulus of $\phi(k)$ is given by

$$|\phi(k)| = \left| \int_{-\infty}^{\infty} e^{iku} B(u)\, du \right| \leqslant \int_{-\infty}^{\infty} |e^{iku}|\, B(u)\, du = \int_{-\infty}^{\infty} B(u)\, du = 1, \tag{6.2.7}$$

because $B(u) \geqslant 0$ and because the modulus of the exponential is unity. The last integral in (6.2.7) is equal to $\phi(0)$, so that we can write

$$|\phi(k)| \leqslant 1 = \phi(0). \tag{6.2.8}$$

The widths of $\phi(k)$ and of $B(u)$ are inversely related. Let us nondimensionalize the fluctuations u by σ, so that we have $u/\sigma = \eta$. Let us define a new probability density B' by

$$B'(\eta) \equiv \sigma B(u) = \sigma B(\sigma\eta). \tag{6.2.9}$$

Defined this way, the integral of B', according to (6.1.2), is equal to one. The characteristic function then becomes

$$\phi(k) = \int_{-\infty}^{\infty} e^{ik\sigma\eta} B'(\eta)\, d\eta. \tag{6.2.10}$$

A measure for the width of $\phi(k)$ can be defined as the value of k where the right-hand side of (6.2.10) is equal to $\frac{1}{2}$. This value is clearly proportional to $1/\sigma$, because $B'(\eta)$ has unit width. The effective width of $\phi(k)$ thus increases if σ decreases. If $\phi(k)$ is narrow, $B(u)$ is broad, and vice versa.

The effects of spikes and discontinuities Suppose $B(u)$ has a very high, narrow spike at some value of u, which we denote by s. This is pictured in Figure 6.5. The flat spots in the function $u(t)$ might be caused by a "dwell"

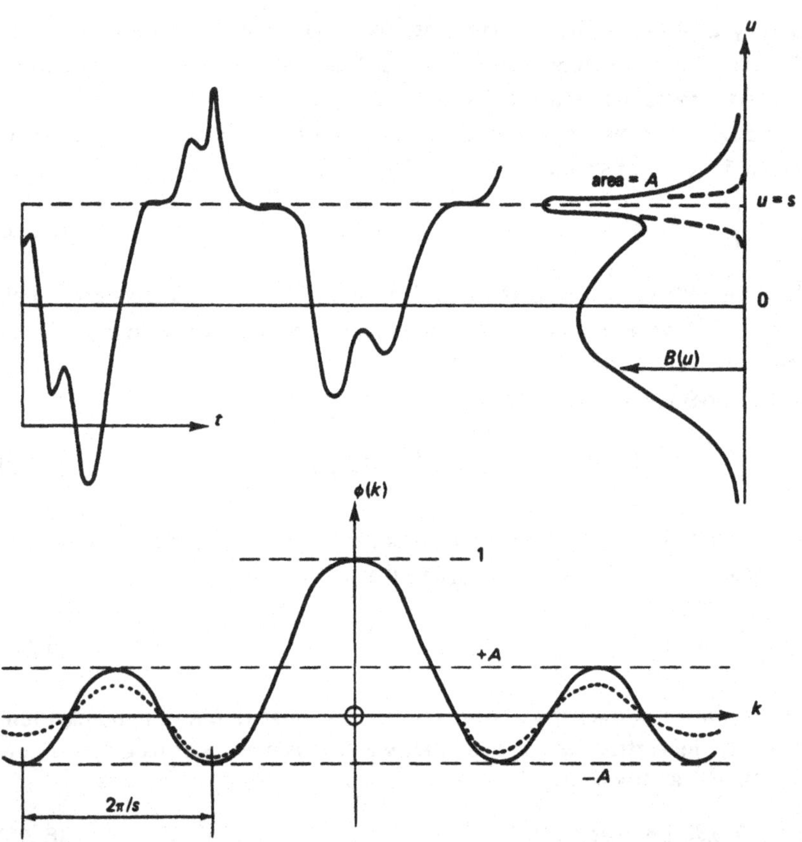

Figure 6.5. The characteristic corresponding to a probability density with a spike. The dotted line indicates the attenuation of $\phi(k)$ due to the finite width of the spike.

circuit of some kind. We assume that the area under the spike in $B(u)$ is A; the rest of the area enclosed by $B(u)$ is then $1 - A$. The spike in $B(u)$ produces a component of the characteristic function which behaves as $A \exp iks$. This component does not decay at infinity (Figure 6.5). In reality, of course, the spike is never infinitely high and narrow. If B_s is the spike component of B and ϕ_s is the spike component of ϕ, the latter can be written as

$$\phi_s(k) = \int_{-\infty}^{\infty} e^{iku} B_s(u)\, du = e^{iks} \int_{-\infty}^{\infty} e^{ik(u-s)}\, B_s(u)\, du$$

$$= e^{iks} \int_{-\infty}^{\infty} e^{ikx}\, B_s(s+x)\, dx = e^{iks}\, \phi_s'(k). \tag{6.2.11}$$

Here, $\phi'_s(k)$ is the transform of B_s, but shifted to the origin, so that it does not oscillate. Therefore, $\phi'_s(k)$ is a characteristic function with a width inversely proportional to the width of the spike. If the spike is infinitely narrow, $\phi'_s(k)$ is constant. If the spike has a finite width σ_s, $\phi'_s(k)$ decreases as k/σ_s, thus reducing the amplitude of $\exp iks$ (Figure 6.5).

If $B(u)$ has a discontinuity, so that its derivative has a spike, similar oscillations of $\phi(k)$ are generated. Integrating the first of (6.2.1) by parts once, we obtain

$$\phi(k) = -\int_{-\infty}^{\infty} \frac{e^{iku}}{k} \frac{dB(u)}{du} du. \tag{6.2.12}$$

If the spike in dB/du is infinitely narrow, we conclude that $\phi(k)$ behaves as $(1/k) \exp iks$ at large values of k. If the spike has finite width, $\phi(k)$ decreases somewhat faster. In general, if $B(u)$ and its first n derivatives are continuous, with a discontinuity in the $(n + 1)$st, $\phi(k)$ is proportional to $k^{-(n+2)} \exp iks$ asymptotically.

Three pairs of Fourier transforms are sketched in Figure 6.6. In the first example, $B(u)$ itself has a discontinuity, so that ϕ decays as k^{-1}. In the second example, B has a discontinuity of slope, so that ϕ decays as k^{-2}. The third example is the probability density of a sine wave; here B has a spike, but it is not infinitely narrow, so that ϕ does decay, though rather slowly.

Parseval's relation Consider two functions, f and g, with Fourier transforms F and G:

$$F(k) = \int_{-\infty}^{\infty} e^{ikx} f(x)\, dx, \quad G(k) = \int_{-\infty}^{\infty} e^{ikx} g(x)\, dx. \tag{6.2.13}$$

With a little algebra it can be shown that

$$\int_{-\infty}^{\infty} F(k) G^*(k)\, dk = 2\pi \int_{-\infty}^{\infty} f(x) g^*(x)\, dx, \tag{6.2.14}$$

where asterisks denote the complex conjugates. This is known as Parseval's relation; it can be used to see how an operation carried out on a function affects its Fourier transform. For example suppose that $f(x)$ is being averaged over an interval $-X \leqslant x \leqslant X$. This amounts to evaluating the integral on the right-hand side of (6.2.14) with the use of a function $g^*(x)$ that looks like the "top-hat" function at the top left of Figure 6.6:

$$\begin{aligned} g^*(x) &= \tfrac{1}{2} X^{-1} \quad \text{for } -X \leqslant x \leqslant X, \\ g^*(x) &= 0 \qquad\quad \text{otherwise.} \end{aligned} \tag{6.2.15}$$

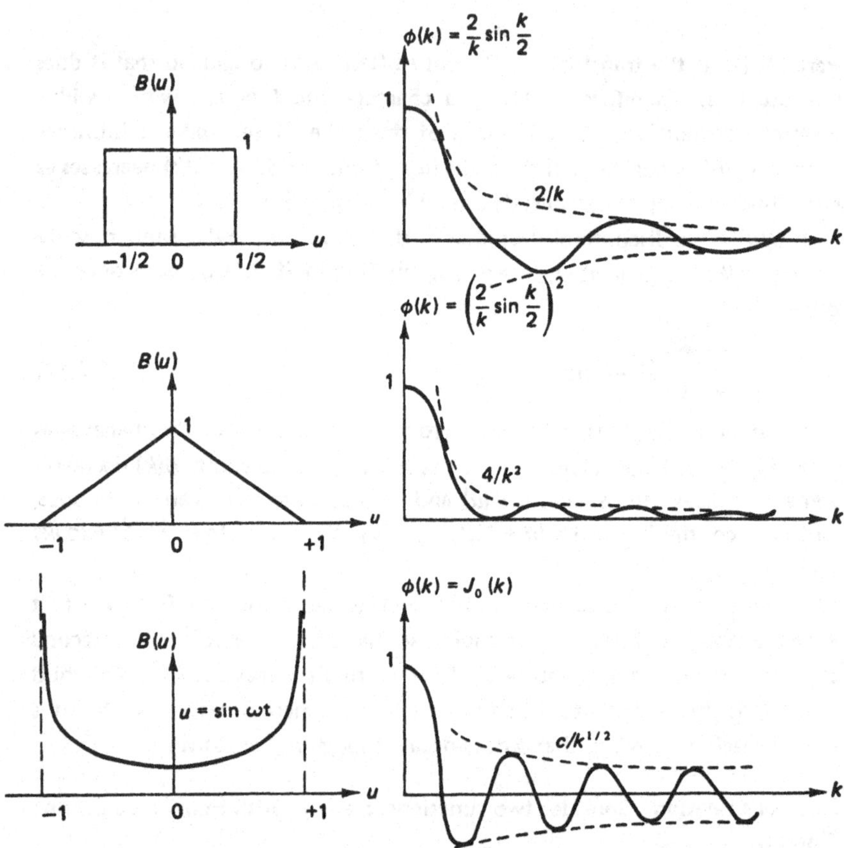

Figure 6.6. Some Fourier-transform pairs. Note that $\phi(k) = \phi(-k)$ because $B(u)$ is real.

The integrand on the left-hand side of (6.2.14) consists of the product of $F(k)$ and $G^*(k)$. The latter looks like the function on the top right of Figure 6.6. Now, as $g^*(x)$ becomes wider, $G^*(k)$ becomes narrower, as we saw earlier. If the averaging interval is quite long so that $G^*(k)$ is quite narrow, the integral on the left-hand side of (6.2.14) may be approximated by $F(0)$ times the integral of $G^*(k)$. Apparently, averaging a function is equivalent to selecting the value of its Fourier transform at the origin. If the physical variable is time, the transform variable is frequency; the origin in transform space corresponds to zero frequency. If we average something, the only thing left is the component at zero frequency; all other components become zero.

Similar problems arise when random variables are measured with sensors of finite dimensions. For example, a hot wire of finite length spatially averages the velocity fluctuations that are measured. The effects of this averaging on the output of the hot-wire instrument can be described in terms of the first two Fourier-transform pairs in Figure 6.6 (Uberoi and Kovasznay, 1953).

6.3
Joint statistics and statistical independence

Let us consider the probability density for two variables simultaneously. A simple way to visualize this is to imagine that one variable $u(t)$ is displayed on the x axis of an oscilloscope, while the other variable $v(t)$ is displayed on the y axis (Figure 6.7). We assume that u and v are variables with zero mean, for simplicity. The *joint probability density* $B(u,v)$ is proportional to the fraction of time that the moving spot in Figure 6.7 spends in a small window between u and $u + \Delta u$, v and $v + \Delta v$. If we took a time exposure of the screen, the intensity at a point would be proportional to the joint probability density. As before, the sum of all the amounts of time spent at all locations must be equal to the total time, and the time fractions cannot be negative. Thus,

$$B(u, v) \geqslant 0, \quad \iint_{-\infty}^{\infty} B(u, v)\, du\, dv = 1. \tag{6.3.1}$$

Also, if all of the values of v at a given value of u are combined, we should get the density of $u(t)$, which we call $B_u(u)$. On the oscilloscope, this amounts to turning the gain to zero on the y axis, so that the figure collapses to a horizontal line. A similar statement can be made about $B_v(v)$, so that we can write

$$\int_{-\infty}^{\infty} B(u, v)\, dv = B_u(u), \quad \int_{-\infty}^{\infty} B(u, v)\, du = B_v(v). \tag{6.3.2}$$

The moments of $u(t)$ and $v(t)$ can be obtained separately, or with (6.3.2). The most important *joint moment* is $\overline{uv}$, which is defined as

$$\overline{uv} \equiv \iint_{-\infty}^{\infty} uvB(u, v)\, du\, dv. \tag{6.3.3}$$

This is called the *covariance* or *correlation* between u and v. Students of mechanics will recognize that the covariance is equivalent to the product of inertia of a distribution of mass. The correlation is thus a measure of the asymmetry of $B(u, v)$. If the value of $B(-u, v)$ is the same as that of $B(u, v)$, then $\overline{uv} = 0$. A few examples are given in Figure 6.8.

As we discussed in Section 2.1, if $\overline{uv} = 0$, $u(t)$ and $v(t)$ are said to be

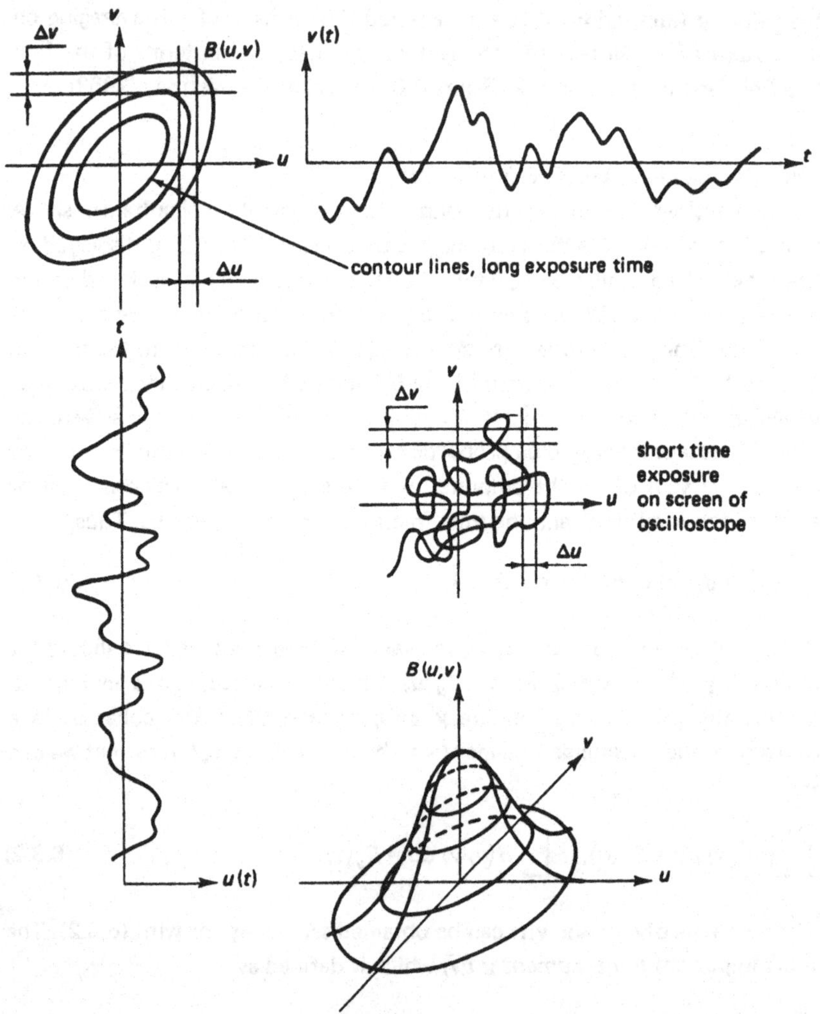

Figure 6.7. The joint probability density.

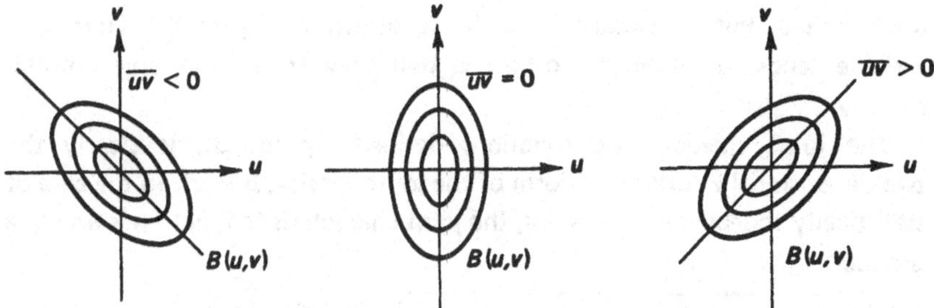

Figure 6.8. Examples of joint densities with various correlations.

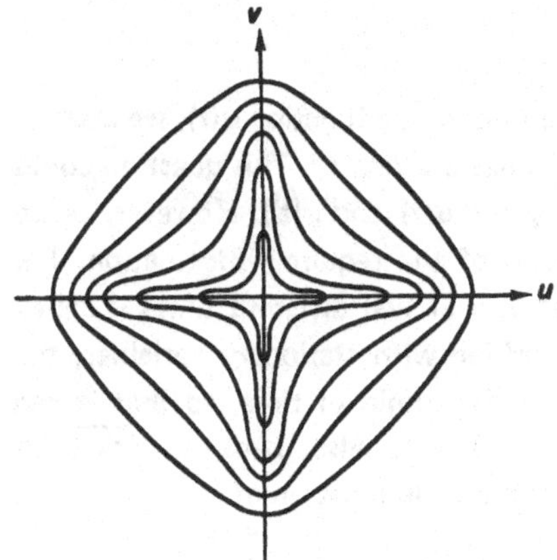

Figure 6.9. Probability density for two uncorrelated variables that tend to inhibit each other.

uncorrelated. Uncorrelated variables, however, are not necessarily independent of each other. The correlations in Figure 6.8 can be made zero by rotating the figures on the left and right until they are symmetric about one of the axes. In other words, it is possible to select two linear combinations of $u(t)$ and $v(t)$ and to create two new variables $u'(t)$ and $v'(t)$ which are uncorrelated. Clearly, the absence of correlation is no clue for the presence or absence of a dependence between the variables.

Two variables are *statistically independent* if

$$B(u, v) = B_u(u)\, B_v(v). \qquad (6.3.4)$$

The probability density of one variable is then not affected by the other variable, and vice versa. For variables that depend on each other, the joint density cannot be written as a product. An example of the joint density of uncorrelated, but dependent, variables is shown in Figure 6.9. Here, one variable tends to inhibit the other, so that they are seldom large simultaneously.

The joint characteristic function, defined by (6.2.3), is clearly the two-dimensional Fourier transform of the joint density, $B(u, v)$. In the case of statistically independent variables, the joint characteristic function is a simple product:

$$\phi(k,\ell) = \overline{\exp\,[iku + i\ell v]} = \overline{\exp\,[iku]}\ \overline{\exp\,[i\ell v]} = \phi_u(k)\,\phi_v(\ell). \tag{6.3.5}$$

6.4 Correlation functions and spectra

If we want to describe the evolution of a fluctuating function $u(t)$, we need to know how the values of u at different times are related. This question could be answered by forming a joint density for $u(t)$ and $u(t')$. However, as we have seen, the correlation provides much of the required information. The correlation $\overline{u(t)u(t')}$ between the values of u at two different times is called the *autocorrelation*. Because we are working with stationary variables, the autocorrelation gives no information on the origin of time, so that it can depend only on the time difference $\tau = t' - t$. Also, because $\overline{u(t)u(t')} = \overline{u(t')u(t)}$, the autocorrelation must be a symmetric function of τ.

Schwartz's inequality states that

$$|\,\overline{u(t)u(t')}\,| \leqslant [\overline{u^2(t)} \cdot \overline{u^2(t')}]^{1/2}. \tag{6.4.1}$$

For stationary variables, $\overline{u^2(t)} = \overline{u^2(t')} = \text{const}$, so that it is convenient to define an *autocorrelation coefficient* $\rho(\tau)$ by

$$\frac{\overline{u(t)u(t')}}{\overline{u^2}} \equiv \rho(\tau) = \rho(-\tau). \tag{6.4.2}$$

With (6.4.1) and (6.4.2), we obtain

$$|\rho| \leqslant 1 = \rho(0). \tag{6.4.3}$$

An autocorrelation coefficient similar to $\rho(\tau)$ was used in Section 2.3. The *integral scale* $\mathscr{T}$ is defined by

$$\mathscr{T} \equiv \int_0^\infty \rho(\tau)\,d\tau. \tag{6.4.4}$$

In turbulence, it is always assumed that the integral scale is finite. The value

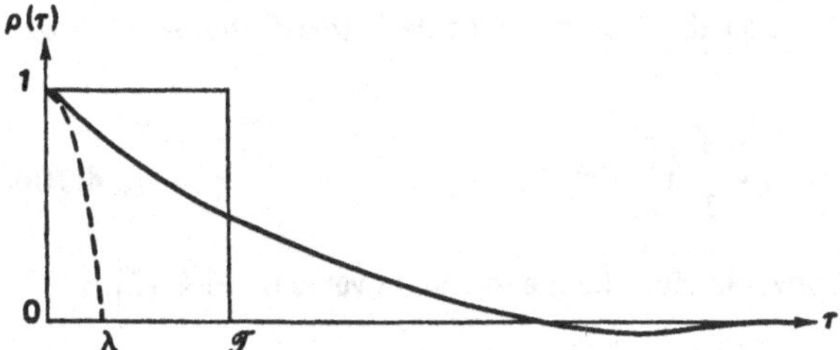

Figure 6.10. Sketch of an autocorrelation coefficient.

of $\mathscr{T}$ is a rough measure of the interval over which $u(t)$ is correlated with itself. A sketch of $\rho(\tau)$ is given in Figure 6.10.

Also shown in Figure 6.10 is the *microscale* λ, which is defined by the curvature of the autocorrelation coefficient at the origin:

$$d^2\rho/d\tau^2\Big|_{\tau=0} \equiv -2/\lambda^2. \tag{6.4.5}$$

Expanding ρ in a Taylor series about the origin, we can write, for small τ,

$$\rho(\tau) \cong 1 - \tau^2/\lambda^2. \tag{6.4.6}$$

The microscale is thus the intercept of the parabola that matches $\rho(\tau)$ at the origin (Figure 6.10). Because $u(t)$ is stationary, we can write

$$0 = \frac{d^2}{dt^2}(\overline{u^2}) = 2\,\overline{u\frac{d^2u}{dt^2}} + 2\,\overline{\left(\frac{du}{dt}\right)^2}. \tag{6.4.7}$$

From (6.4.5) and (6.4.7) we obtain

$$\overline{\left(\frac{du}{dt}\right)^2} = \frac{2\overline{u^2}}{\lambda^2}. \tag{6.4.8}$$

In Chapter 3, the Taylor microscale, defined in a similar way from the spatial velocity autocorrelation, was extensively used.

The convergence of averages Suppose we want to obtain the average value of a function $\tilde{u}(t)$ in the laboratory. Of course, we cannot integrate over an infinitely long time interval, so that we have to consider the error due to finite integration time. The average is

$$U_T = \frac{1}{T}\int_{t_0}^{t_0+T} \tilde{u}(t)\,dt. \tag{6.4.9}$$

The difference between U_T and the true mean value U (recall that $\tilde{u} = U + u$) is given by

$$U_T - U = \frac{1}{T}\int_0^T [\tilde{u}(t) - U]\, dt = \frac{1}{T}\int_0^T u(t)\, dt. \qquad (6.4.10)$$

Here we took $t_0 = 0$ for convenience. The mean-square value of (6.4.10) is

$$\overline{(U_T - U)^2} = \frac{\overline{u^2}}{T^2}\iint_0^T \rho(t' - t)\, dt\, dt' = \frac{2\overline{u^2}}{T}\int_0^T \left(1 - \frac{\tau}{T}\right)\rho(\tau)\, d\tau. \qquad (6.4.11)$$

If the integrating time T is much longer than the integral scale $\mathcal{T}$, $\tau/T \sim 0$ in the range of values of τ where $\rho(\tau) \neq 0$, so that, by virtue of (6.4.4), the mean-square error may be approximated by

$$\overline{(U_T - U)^2} \cong 2\,\overline{u^2}\;\mathcal{T}/T. \qquad (6.4.12)$$

It is clear that the average value can be determined to any accuracy desired if the integral scale is finite.

Ergodicity The requirement that a time average should converge to a mean value, that is, that the error should become smaller as the integration time increases and that the mean value found this way should always be the same, is called *ergodicity*. A variable is called *ergodic* if averages of all possible quantities formed from it converge. An ergodic variable not only becomes uncorrelated with itself at large time differences ($\tau \to \infty$), but it also becomes statistically independent of itself. A variable is ergodic if all integral scales that can be formed from it exist. Actually, this condition is not quite necessary; more general statements could be made. Let us consider a laboratory average of $\exp iku(t)$, which should differ little from the characteristic function $\phi(k) = \overline{\exp iku(t)}$ defined by (6.2.2). If the integral scale of $\exp iku(t)$ exists, the autocorrelation between $\exp iku(t)$ and $\exp iku(t')$ should vanish for large $t' - t$. Thus, for large $t' - t$,

$$\overline{[\exp iku(t) - \phi(k)]\,[\exp iku(t') - \phi(k)]} \to 0, \qquad (6.4.13)$$

so that

$$\overline{\exp\,[iku(t) + iku(t')]} \to \phi(k)\phi(k). \qquad (6.4.14)$$

From the definition (6.2.3) of a joint characteristic function, and the form (6.3.5) which it takes for statistically independent variables, it is clear that the left-hand side of (6.4.14) would not approach a simple product unless the joint density itself were a simple product. Thus, by virtue of (6.3.4), $u(t)$ and $u(t')$ are statistically independent at large time differences.

It is reasonable to expect that all the integral scales associated with $u(t)$ are about the same, because they are determined by the scale of the physical process that produces $u(t)$. The integral scale $\mathscr{T}$ of $u(t)$ itself is thus not only a measure of the time over which $u(t)$ is correlated with itself but also a measure of the time over which $u(t)$ is dependent on itself. For time intervals large compared to $\mathscr{T}$, $u(t)$ becomes statistically independent of itself, so that $\mathscr{T}$ is a measure for the time interval over which $u(t)$ "remembers" its past history.

Another look at this concept is obtained if the output of the discriminator circuit in Figure 6.1 is considered. Let us call this function $I(t)$; it is equal to one if $u(t)$ appears in the window between u and $u + du$, and zero otherwise. The mean value of $I(t)$ is the value of $B(u)\,\Delta u$ we wish to determine:

$$\overline{I(t)} = B(u)\,\Delta u. \tag{6.4.15}$$

The mean-square error in the measurement of $B(u)\,\Delta u$ is obtained as follows. The variance σ^2 of $I(t)$ is given by

$$\begin{aligned}\sigma^2 &= \overline{[I(t) - B\,\Delta u]^2} = \overline{I^2(t)} - 2\overline{I(t)}\,B\,\Delta u + (B\,\Delta u)^2 \\ &= \overline{I^2(t)} - (B\,\Delta u)^2 = B\,\Delta u - (B\,\Delta u)^2.\end{aligned} \tag{6.4.16}$$

The last step in (6.4.16) could be taken because $I(t)$ and $I^2(t)$ always have the same value (either one or zero). Applying the error estimate (6.4.12) to the laboratory average I_T (obtained by integrating $I(t)$ over a time T), we find, if T is large and $B\,\Delta u$ is small,

$$\overline{(I_T - B\,\Delta u)^2} = 2\,\mathscr{T}B\,\Delta u/T. \tag{6.4.17}$$

The mean-square relative error is then given by

$$\overline{(I_T/B\,\Delta u - 1)^2} = 2\mathscr{T}/(TB\,\Delta u). \tag{6.4.18}$$

Now $TB\,\Delta u$ is the amount of time spent by $u(t)$ between u and $u + \Delta u$ if the averaging time is T. Hence, (6.4.18) shows that the error is small if the averaging time is so long that the amount of time spent in the window Δu is large compared to the integral scale $\mathscr{T}$.

Another way to obtain $B\Delta u$ is to sample $I(t)$ at time intervals large enough to make the samples statistically independent of each other. With this procedure, the mean-square relative error is

$$\frac{1}{(B\,\Delta u)^2}\overline{\frac{1}{N}\sum_{n=1}^{N}[I(t_n)-B\,\Delta u]^2}\cong\frac{1}{NB\,\Delta u}. \tag{6.4.19}$$

Here, N is the total number of independent samples taken. If we compare (6.4.19) with (6.4.18), we see that $T/2\mathscr{T}$ may be regarded as the number of independent samples in a record of length T. Therefore, sampling once every two integral scales is adequate. We conclude that averages converge and integral scales exist if $u(t)$ may be regarded as consisting of a series of records of length $2\mathscr{T}$ (say, pieces of an analog tape), which are approximately statistically independent of each other.

The Fourier transform of $\rho(\tau)$ The autocorrelation coefficient $\rho(\tau)$ is a function that is equal to unity at the origin and is majorized by that value, that is real and symmetric, and that goes to zero faster than $1/\tau$, so that its integral scale exists. Referring back to Section 6.2, we conclude that $\rho(\tau)$ must be the Fourier transform of a continuous, symmetric, positive, real function $S(\omega)$ whose integral is unity. The transform of $\rho(\tau)$ must be continuous because ρ goes to zero faster than $1/\tau$; it must be symmetric because ρ is real; it must be real because ρ is symmetric; it must have a unit integral because $\rho = 1$ at the origin; it must be positive because ρ is majorized by its value at the origin.

The Fourier transform $S(\omega)$ of $\rho(\tau)$ is known as the *power spectral density*, or simply *spectrum;* it is defined by

$$\rho(\tau)=\int_{-\infty}^{\infty}e^{i\tau\omega}S(\omega)\,d\omega,\quad S(\omega)=\frac{1}{2\pi}\int_{-\infty}^{\infty}e^{-i\tau\omega}\rho(\tau)\,d\tau. \tag{6.4.20}$$

An appreciation for the relevance of $S(\omega)$ can be obtained by attempting to formulate a Fourier transform of $u(t)$ itself. Let us define

$$a_T(\omega,t)\equiv\frac{1}{T}\int_t^{t+T}e^{i\omega t'}u(t')\,dt'. \tag{6.4.21}$$

Let us recall the discussion on Parseval's relation at the end of Section 6.2. In this case, the function multiplying $u(t')$ is $g^*(t')$, which is given by

$$\begin{aligned} g^*(t') &= (1/T)\exp i\omega t' \quad \text{for } t\leqslant t'\leqslant t+T,\\ g^*(t') &= 0 \quad \text{otherwise.}\end{aligned} \tag{6.4.22}$$

The transform of $g^*(t')$ is

$$G^*(\omega') = \frac{\sin(\omega' - \omega)T/2}{(\omega' - \omega)T/2} \exp[-i(\omega' - \omega)(t + T/2)]. \tag{6.4.23}$$

The exponential has an absolute value of unity; it is present only because the midpoint of the integration interval T is a running variable. The first factor on the right-hand side of (6.4.23) is exactly the same as the function on the top right of Figure 6.6 but displaced to the center frequency ω. The average in (6.4.21) thus selects the value of the Fourier transform of $u(t)$ at the frequency ω rather than at the origin, if the time interval T is large enough.

Apparently, (6.4.21) is obtained by passing $u(t)$ through a filter that admits only frequences near ω. The width of the filter is about $1/T$. If we think of $u(t)$ as being synthesized from contributions at many frequencies, only the contributions close to ω form a square with $\exp i\omega t'$, so that only for those contributions does the integrand in (6.4.21) not oscillate. The contributions to $u(t)$ from all other frequencies cause the integrand to oscillate, so that they do not contribute to $a_T(\omega,t)$ if the integration time T is large (that is, if the bandwidth $1/T$ is small).

With a little algebra, it can be shown that the mean-square value of $a_T(\omega,t)$ is related to the spectrum $S(\omega)$ by

$$\lim_{T\to\infty} T\,\overline{|a_T(\omega, t)|^2} = \overline{u^2}\, S(\omega). \tag{6.4.24}$$

For a similar calculation, see Hinze (1959), Section 1-12. The spectrum thus represents the mean-square amplitude of the filtered signal or the mean-square amplitude of the Fourier coefficient of $u(t)$ at ω; it may be thought of as the energy in $u(t)$ at that frequency.

From (6.4.20) we conclude that the value of $S(\omega)$ at the origin is given by $S(0) = \mathcal{T}/\pi$. Also, if $\rho(\tau) \geqslant 0$ everywhere, $S(\omega)$ is maximized by its value at the origin. Conversely, if $S(\omega)$ has a peak away from the origin, then $\rho(\tau)$ must have negative regions. However, this does not imply that $S(\omega)$ must have a peak away from the origin if $\rho(\tau)$ is negative somewhere, as the Fourier transform pairs in Figure 6.6 demonstrate.

The spectrum of the derivative of a function is related to the spectrum of the function in a simple way. The autocorrelation of du/dt is given by

$$\overline{\frac{du(t)}{dt}\frac{du(t')}{dt'}} = \overline{u^2}\frac{d^2}{dt\,dt'}\rho(t' - t) = -\overline{u^2}\frac{d^2\rho}{d\tau^2}. \tag{6.4.25}$$

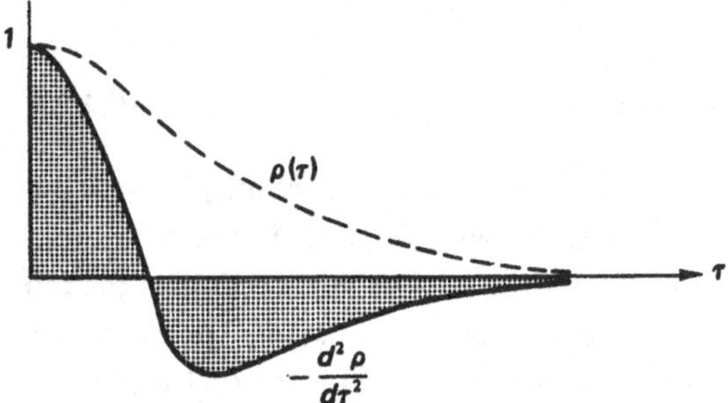

Figure 6.11. The autocorrelation coefficient of the first derivative of $u(t)$.

Differentiating the first of (6.4.20) twice, we obtain

$$-\frac{d^2\rho}{d\tau^2} = \int_{-\infty}^{\infty} e^{j\omega\tau}\omega^2 S(\omega)\, d\omega. \tag{6.4.26}$$

From (6.4.25, 6.4.26) we conclude that the spectrum of the first derivative is proportional to $\omega^2 S(\omega)$. This relation, of course, also can be applied to higher derivatives; for example, the spectrum of the second derivative is proportional to $\omega^4 S(\omega)$. Because $S(0)$ is finite (it is equal to $\mathcal{T}/\pi$, as we have seen), the spectra of derivatives vanish at the origin. This means that the integral scales of derivatives are zero. An example is given in Figure 6.11; the area under the curve is zero:

$$-\int_0^{\infty} \frac{d^2\rho}{d\tau^2}\, d\tau = \frac{d\rho}{d\tau}\bigg|_0^{\infty} = 0. \tag{6.4.27}$$

6.5
The central limit theorem

In the analysis of turbulence, many quantities can be written as averages of stationary variables. In Chapter 7 we find that such quantities frequently arise in the discussion of turbulent transport (diffusion, mixing). The question arises, do averages of stationary variables have a probability density that is independent of the nature of the variable that is being averaged? In other words, we wonder if the very process of averaging introduces its own characteristic pattern, which masks the characteristics of the variable that is averaged. Subject to some simple conditions, the answer to this question is yes; the probability density of averages of stationary variables always tends to have the same shape.

Let us consider N statistically independent quantities $x_n(t)$. We assume that all $x_n(t)$ have identical probability densities and that their mean values are zero. It is convenient to work with the characteristic function $\phi(k)$, which is defined by

$$\phi(k) \equiv \overline{\exp ikx_n(t)}. \tag{6.5.1}$$

Because the densities of all $x_n(t)$ are the same, (6.5.1) holds for all n between 1 and N. Let us define the sum $z(t)$ of all $x_n(t)$ by

$$z(t) \equiv \sum_{n=1}^{N} x_n(t). \tag{6.5.2}$$

The variance of z is given by

$$\overline{z^2} = \sum_{n=1}^{N} \sum_{m=1}^{N} \overline{x_n x_m} = \sum_{n=1}^{N} \overline{x_n^2} = N\sigma^2. \tag{6.5.3}$$

Here, σ^2 is the variance of x_n, which is the same for each x_n because they have identical densities. The double sum becomes a single sum because x_n and x_m are statistically independent and have zero mean, so that they are uncorrelated. The variance of z increases as N increases, so that it is more convenient to define a new quantity $w(t)$ by

$$w(t) \equiv N^{-1/2} z(t). \tag{6.5.4}$$

The variance of $w(t)$ is equal to σ^2, no matter how large N becomes. Can we predict the probability density of $w(t)$? First it is convenient to compute the characteristic function $\phi_w(k)$ of $w(t)$. We obtain

$$\phi_w(k) \equiv \overline{\exp ikw(t)} = \overline{\exp\left(\frac{ik}{N^{1/2}} \sum_{n=1}^{N} x_n\right)} = [\phi(kN^{-1/2})]^N. \tag{6.5.5}$$

The last step in (6.5.5) could be taken because the x_n are statistically independent, so that the mean of the product of all $\exp(ikx_nN^{-1/2})$ is equal to the product of all ϕ. If the first few moments of the probability density of x_n exist, $\phi(kN^{-1/2})$ may be expanded in a Taylor series:

$$\phi(kN^{-1/2}) = 1 - k^2\sigma^2/2N + \mathcal{O}(k^3 N^{-3/2}). \tag{6.5.6}$$

This expansion is based on (6.2.5); the last term in (6.5.6) indicates that the remainder is of order $k^3 N^{-3/2}$, so that it can be made as small as desired by selecting a sufficiently large value of N. Substituting (6.5.6) into (6.5.5), we obtain, for very large N,

$$\phi_w(k) = \lim_{N\to\infty} (1 - k^2\sigma^2/2N)^N = \exp(-k^2\sigma^2/2). \tag{6.5.7}$$

This is called the *central limit theorem;* the characteristic function $\phi_w(k)$ is called a *Gaussian characteristic function.* The probability density $B(w)$ corresponding to $\phi_w(k)$ can be computed from the definition (6.2.1) of the Fourier transform pair and the shape (6.5.7) of $\phi_w(k)$; the result is

$$B(w) = \frac{\exp(-w^2/2\sigma^2)}{(2\pi\sigma^2)^{1/2}}. \tag{6.5.8}$$

This is called a *Gaussian probability density*. The function $\exp -k^2$ is the only one that preserves its shape under a Fourier transformation. We conclude that asymptotically (as $N \to \infty$), the sum of a large number of identically distributed independent variables has a Gaussian probability density, regardless of the shape of the density of the variables themselves.

The statistics of integrals Let us now consider an integral of $u(t)$ over a time interval T. Because $u(t)$ is a stationary random variable, the value of the integral is also a stationary random variable which depends on the origin of the time interval. If the integration is performed in the laboratory, the probability distribution of the integral could be obtained by repeating the experiment many times.

An integral is like a sum, so that the central limit theorem may govern its probability distribution under suitable conditions. If the integration time T is large compared to the integral scale $\mathcal{T}$, the integral may be broken up into sections of length larger than $2\mathcal{T}$, so that the sections are approximately independent (recall the discussion of (6.4.18) and (6.4.19)):

$$\int_0^T u(t)\,dt = \int_0^{n\mathcal{T}} u(t)\,dt + \int_{n\mathcal{T}}^{2n\mathcal{T}} u(t)\,dt + \ldots. \tag{6.5.9}$$

As n increases, the sections of integral become more nearly independent, because adjacent sections depend on each other only near the ends. If the length of each section is $n\mathcal{T}$ and the total integration time is T, the number of sections is $T/n\mathcal{T}$. It is easy to arrange this in such a way that both $n\mathcal{T}$ and $T/n\mathcal{T}$ go to infinity as $T \to \infty$. We then have more and more sections, and they become less and less dependent, so that the probability distribution of the integral on the left-hand side of (6.5.9) becomes Gaussian under favorable conditions.

The primary question is whether the sections of the integral become independent fast enough. It is possible to show (although we cannot do it here) that, as long as all integral scales exist and are nonzero, the sections become independent fast enough for the central limit theorem to apply. For a full discussion, see Lumley, 1972a.

The condition on the behavior of the correlation at large separations may be translated into a condition on the behavior of the spectrum near the origin, as we recall from the discussion in Section 6.2. From the definition (6.4.20) of $S(\omega)$ we conclude, in analogy with (6.2.4), that the derivatives of the spectrum near the origin are the moments of the correlation coefficient $\rho(\tau)$; if the moments exist, the derivatives do too, and vice versa. The condition that the correlation should be integrable to a value $\neq 0$ then becomes the condition that the spectrum near the origin be finite and nonzero.

A secondary question, which is not apparent in terms of correlations, becomes clear when stated in terms of the spectrum. We know from the discussion following Parseval's relation (6.2.14) that the average of $u(t)$ is equivalent to an operation on the Fourier transform of $u(t)$. In fact, the top-hat function at the top left of Figure 6.6 corresponds to an average. Evidently, averaging $u(t)$ is equivalent to multiplying the Fourier transform of $u(t)$ by the "filter" function at the top right of Figure 6.6. As the top-hat function representing the average becomes wider, the filter function on the right becomes narrower. The requirement that the spectrum be nonzero at the origin guarantees that the product of the Fourier transform and the filter function gets narrower as the integration time increases.

It is easy to find a violation of this condition. Consider du/dt; near the origin, its spectrum is proportional to ω^2, because $S(\omega)$ is approximately constant at small ω. The Fourier transform of du/dt must then be proportional to ω near the origin. However, the filter function on the top right of Figure 6.6 behaves as ω^{-1}. Hence, the product remains of constant width; it does not become narrower as the integration time increases. Therefore, we do not expect that the integral of du/dt will become Gaussian. This is obvious, because the integral of du/dt is $u(t)$ itself, which certainly does not need to be Gaussian.

A generalization of the theorem On the basis of the preceding discussion the central limit theorem can be simplified and generalized. Any variable having finite integral scales takes on a Gaussian distribution if it is filtered with a filter that is narrow enough; it becomes more Gaussian as the filter becomes progressively narrower. Clearly, we are not limited to simple averages. A variable $u(t)$ may be multiplied by any function before it is integrated; the only condition is that the Fourier transform of that function be a filter that, multiplied by the Fourier transform of $u(t)$, makes the product progressively narrower.

For example, a second integral may be written as

$$\int_{-T}^{T} dt \int_{-t}^{t} u(t')\, dt' = 2T \int_{-T}^{T} \left(1 - \frac{|t|}{T}\right) u(t)\, dt. \tag{6.5.10}$$

The factor $2T$ in front of the integral on the right-hand side need not concern us here. It is merely a normalizing factor that affects the variance of the double integral but not the applicability of the central limit theorem. The multiplying function in (6.5.10) has the same shape as the triangular function at the center left of Figure 6.6. Hence, the corresponding filter function decreases as ω^{-2}. If the Fourier transform of $u(t)$ rises more slowly than ω^2, the integral (6.5.10) becomes asymptotically Gaussian. This implies that a double integral of the first derivative of a stationary function $u(t)$ becomes Gaussian, even though a single integral of du/dt does not.

More statistics of integrals In the derivation (6.5.1–6.5.8) of the central limit theorem, the sum of the variables was normalized, so that the variance of $\omega(t)$ remained finite. That was a matter of convenience only; if the sum were not normalized, it would still have a Gaussian distribution, but with a variance that would increase with N.

Let us define an integral $X(T)$ of a stationary variable $u(t)$ by

$$X(T) \equiv \int_0^T u(t)\, dt. \tag{6.5.11}$$

The variance of $X(T)$ becomes (see (6.4.11))

$$\overline{X^2} = \overline{u^2} \iint_0^T \rho(t' - t)\, dt\, dt' = 2T\, \overline{u^2} \int_0^T \left(1 - \frac{\tau}{T}\right) \rho(\tau)\, d\tau \cong 2T\, \overline{u^2}\, \mathscr{T}. \tag{6.5.12}$$

The characteristic function $\phi_X(k)$ of $X(T)$ is Gaussian:

$$\phi_X(k) = \overline{\exp ikX(T)} \cong \exp(-k^2 \overline{u^2}\, T\mathscr{T}). \qquad (6.5.13)$$

The probability density $B(X)$ corresponding to (6.5.13) is

$$B(X) = \frac{\exp(-X^2/4\overline{u^2}\,T\mathscr{T})}{(4\pi\, \overline{u^2}\,T\mathscr{T})^{1/2}}. \qquad (6.5.14)$$

If a double integral $W(T)$ is defined by

$$W(T) \equiv \int_0^T dt \int_0^t u(t')\, dt', \qquad (6.5.15)$$

it can be shown that the variance of $W(T)$ is given by

$$\overline{W^2} = \tfrac{2}{3}\, \overline{u^2}\, T^3 \int_0^T \left(1 - \frac{3\tau}{2T} + \frac{\tau^3}{T^3}\right) \rho(\tau)\, d\tau \cong \tfrac{2}{3}\, \overline{u^2}\, T^3 \mathscr{T}. \qquad (6.5.16)$$

The characteristic function of $W(T)$ is Gaussian:

$$\overline{\exp ik\, W(T)} = \exp\,(-k^2 \overline{u^2}\, T^3\, \mathscr{T}/3). \qquad (6.5.17)$$

We use these relations in Chapter 7.

Problems

6.1 Fluctuating velocity derivatives are associated with vorticity and strain-rate fluctuations. Will the skewness of a velocity-derivative signal ever be zero? Experiments have shown that the kurtosis of velocity derivatives is large if the Reynolds number is large. Use the simple model of Problem 3.2 to make estimates of the skewness and kurtosis.

6.2 Consider a stationary random variable with zero mean and a Gaussian probability density. Derive an approximate expression for the probability of exceeding amplitudes much larger than the standard deviation σ. What is the probability of exceeding 3σ? What is the probability of exceeding 10σ?

6.3 Compute the autocorrelation curve of a sine wave. What is the corresponding Fourier transform? What is the value of the integral scale?

6.4 In turbulent flow at large Reynolds numbers, the Taylor microscale λ is very small compared to the integral scale $\mathscr{T}$, and some investigators find it convenient to approximate the autocorrelation coefficient by $\rho(\tau) =$

$\exp(-|\tau|/\mathscr{T})$. What is the shape of the spectrum corresponding to this approximation? Also, is the spectrum of the derivative well behaved? Compare your results with the spectra given in Chapter 8.

6.5 Estimate the form of the spectrum of ocean waves in the range of frequencies where the Fourier coefficients of the wave amplitudes are determined by the frequency and the acceleration of gravity only.

6.6 Consider a sum of two statistically independent Gaussian variables, one of much lower frequency content than the other, both having zero mean. What do the autocorrelation and the spectrum look like? Suppose there is a gap between the spectra of the two, and the averaging time is long enough to average the fast one but not the slow; what do the correlation and spectrum look like in this case? What is the integral scale?

6.7 Consider one Gaussian variable modulated by another. The variables are statistically independent of each other; the second has a lower frequency content than the first. Both variables have zero mean. The product of the two variables appears to be "intermittent," that is, the low-frequency modulation appears to turn the high-frequency signal on and off. What is the kurtosis? What is the spectrum? Also consider a product of three independent variables, or of four. What is the kurtosis? If there are gaps between the spectra of the individual spectra, how does the measured kurtosis depend on the averaging time? Try to construct a continuous model, in which the logarithm of the signal is represented as the integral of a stationary process. Use the central limit theorem.

7

TURBULENT TRANSPORT

As a turbulent flow moves, it carries fluid from place to place. A tiny parcel of fluid (small, say, compared to the Kolmogorov microscale, but large compared to molecular scales) gradually wanders away from its initial location. This is the mechanism that is responsible for the large transfer rates observed in turbulent flows. In the preceding chapters, the transport capability of turbulence was represented by such quantities as the momentum flux $-\rho\overline{uv}$ and the heat flux $-\rho c_p\overline{\theta v}$; estimates for these were obtained by similarity arguments and dimensional reasoning. Here, we study the details of the process of transport. We first analyze how turbulent motion transports fluid points; then, in the second half of this chapter, we deal with the transport (dispersion, mixing) of contaminants.

7.1 Transport in stationary, homogeneous turbulence

We would like to be able to predict transport in real flows, which generally are inhomogeneous and nonstationary. This is the heart of the turbulence problem; unfortunately, it is impossible to describe the details of transport in other than very simple cases. Let us first discuss the motion of a single fluid "point" in stationary, homogeneous turbulence without mean velocity. This is an idealized situation, because turbulence without a mean velocity gradient has no source of energy, so that it decays and cannot be stationary. More important, this idealized case may not even be relevant to transport in real decaying flows, because (as we later see) the "memory time" of a fluid point is usually of the order of the decay time, so that a real decaying flow never appears even approximately stationary to a wandering point. Consequently, we have to be careful in generalizing the conclusions we obtain for this idealized turbulence; we should not be surprised if the conclusions have qualitative significance only.

Stationarity Before we start the analysis, let us ask when we may expect the velocity of a wandering point to be a stationary (statistically steady) function of time. This question, of course, bears on the applicability of the central limit theorem (Section 6.5). Clearly, it is necessary that the flow be stationary itself. If the flow is also homogeneous, we are assured that the velocity of the wandering point is stationary. This case is discussed first. If the flow is not homogeneous and unbounded in the direction of inhomogeneity, the moving point wanders into regions of progressively different characteristics. For example, in a boundary layer the flow is distinctly inhomogeneous in the cross-stream direction. As time proceeds, the boundary layer grows and a wandering point moves progressively farther away from the wall into regions where the turbulence properties are different. In such a case, the velocity of a wandering point is not stationary. In a pipe flow, on the other hand, the flow is homogeneous in the streamwise direction and inhomogeneous, but also bounded, in the cross-stream direction. A wandering point may then move toward one wall, but it eventually returns and moves toward the other. Hence, we expect its velocity to be stationary. We conclude that the velocity of a wandering point is stationary if the flow is stationary and bounded in all directions of inhomogeneity.

Stationary, homogeneous turbulence without mean velocity Let us analyze the motion of a fluid point in stationary, homogeneous turbulence without mean velocity (Figure 7.1). The velocity at time t of a moving point which was at the point $x_i = a_i$ at $t = 0$ will be called $v_i(\mathbf{a}, t)$. The use of vector notation (denoted by boldface letters) in the argument of v_i prevents confusion of indices. As we discussed above, $v_i(\mathbf{a}, t)$ is a stationary (statistically steady) function; it is called a *Lagrangian velocity.*

The position of the wandering point is the integral of its velocity:

$$X_i(\mathbf{a}, t) = a_i + \int_0^t v_i(\mathbf{a}, t')\, dt', \tag{7.1.1}$$

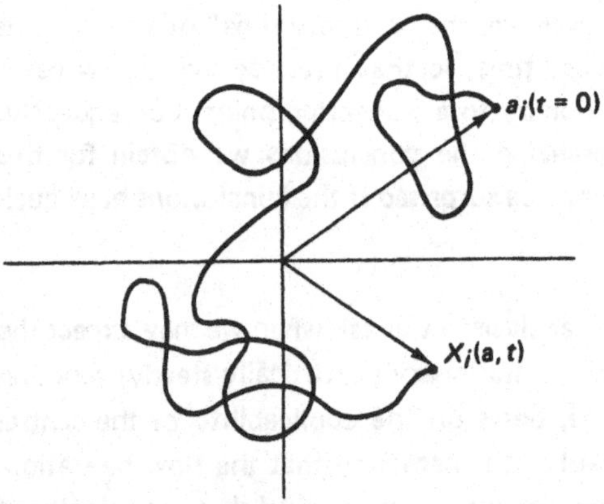

$a_i(t = 0)$

$X_i(\mathbf{a}, t)$

Figure 7.1. The motion of a wandering point

where $X_i(\mathbf{a}, 0) = a_i$. The Lagrangian position is X_i, and the Eulerian position is x_i. The velocity of the moving point is equal to the velocity of the fluid at the point where it happens to be. The velocity $u_i(\mathbf{x}, t)$ measured at the location x_i at time t is called the *Eulerian velocity*; it is related to v_i by

$$v_i(\mathbf{a}, t) = u_i(\mathbf{X}(\mathbf{a}, t), t). \tag{7.1.2}$$

The study of transport is very difficult because of (7.1.2). Eulerian velocities (u_i) can be measured by putting a fixed probe in the fluid, but the measurement of Lagrangian velocities (v_i) requires that the motion of "tagged" fluid points be followed with photographic or radioactive tracer techniques. Often, only Eulerian measurements are made; however, the statistics of u_i are not related to those of v_i in a simple way. The problem is that one needs to know v_i in order to find X_i in order to find u_i. The problem is similar to that of the passage of light through air with turbulent fluctuations in the index of refraction n. The path of a light ray depends on the fluctuations in the n it sees. The path tends to curve around regions with high n and tends to veer away from regions with low n, so that the statistics of n experienced by the light ray are different from those seen on a straight line through the turbulent air.

However, because v_i is a stationary function presumably having nonzero integral scales the central limit theorem (Section 6.5) can be applied to the integral (7.1.1). Consider one component of $X_i - a_i$, and call this $X_\alpha - a_\alpha$. Here, α may be equal to 1, 2, or 3, but we stipulate that the index summation convention does not apply to the index α. Because v_i is stationary, $X_\alpha - a_\alpha$ asymptotically has a Gaussian probability density; its variance is given by (Taylor, 1921)

$$\overline{(X_\alpha - a_\alpha)^2} = 2\,\overline{v_\alpha^2}\,t \int_0^t \left(1 - \frac{\tau}{t}\right) \rho_{\alpha\alpha}(\tau)\, d\tau \cong 2\,\overline{v_\alpha^2}\,t\,\mathcal{T}_{\alpha\alpha}. \tag{7.1.3}$$

The *Lagrangian autocorrelation coefficient* $\rho_{\alpha\alpha}(\tau)$ is defined by

$$\overline{v_\alpha^2}\,\rho_{\alpha\alpha}(\tau) \equiv \overline{v_\alpha(\mathbf{a}, t)\, v_\alpha(\mathbf{a}, t + \tau)}. \tag{7.1.4}$$

The integral scale of $\rho_{\alpha\alpha}$ is $\mathcal{T}_{\alpha\alpha}$; it is called the *Lagrangian integral scale.* The shape of $\rho_{\alpha\alpha}$ looks approximately like the curve in Figure 6.10.

A great deal of effort has been spent in attempts to predict $\mathcal{T}_{\alpha\alpha}$ from Eulerian data, with very little success. A relatively simple prediction is made shortly. We also have to consider the problem of determining $\overline{v_\alpha^2}$.

The set of equations (7.1.1–7.1.4) is also applicable to molecular diffusion

(X_i would be the position of a molecule and v_i would be its velocity). A Lagrangian time integral scale for molecular motion in gases is of the order of a few collision times ξ/a (ξ is the mean free path, a is the speed of sound; see Section 2.2). At ordinary temperatures and pressures, the time scales of interest in diffusion problems are much larger than ξ/a, so that the asymptotic form of (7.1.3) applies. The dispersion $\overline{(X_\alpha - a_\alpha)^2}$ is then proportional to t and the coefficient $a^2(\xi/a) \sim a\xi$ is the molecular diffusivity, which is of the same order as the kinematic viscosity ν (Section 2.2). In turbulence, however, the time span before (7.1.3) reaches its asymptotic form is not too short to be of interest. In fact, by the time the integral reaches the "diffusion limit" the wandering point has usually left the (approximately homogeneous) part of the flow field where it started. Still, the asymptotic form of (7.1.3) is a useful, though rather crude, approximation in many cases of practical interest. Note that the asymptotic form of (7.1.3) is equivalent to assuming that the eddy diffusivity $\overline{v_\alpha^2}\,\mathscr{T}_{\alpha\alpha}$ is constant.

The probability density of the Lagrangian velocity In order to make use of (7.1.3), we need to know $\overline{v_\alpha^2}$. The easiest way to predict this is to exploit the fact that an incompressible fluid moving in a box always fills the box. This simple-looking statement has surprising consequences. If we want to integrate a quantity over all the moving fluid points in the box, we can integrate either over their present locations (an Eulerian integral) or over their initial locations (a Lagrangian integral). Because the fluid continues to fill the box as it moves around, either way each point is counted only once, so that it is immaterial which integral we take. Suppose $F(\mathbf{x}, t)$ is the function we wish to integrate over the volume V of the box; the integral statement then reads

$$\iiint_V F(\mathbf{X}(\mathbf{a}, t), t)\, da_1\, da_2\, da_3 = \iiint_V F(\mathbf{x}, t)\, dx_1\, dx_2\, dx_3. \tag{7.1.5}$$

If an incompressible flow is not confined to a box, a similar statement can be made. The only problem is that the integration volume on the left-hand side is not the same as that on the right-hand side. Points that were initially on the boundaries of the volume V move, so that the new boundaries gradually wander away from the original ones. However, if the velocities involved are of order u, the boundaries move a distance of order ut in a time t, so that the volume difference between the new and the old boundaries is of order utL^2 ($L = V^{1/3}$ is the length scale of the integration volume). The volume fraction involved is of order ut/L, which, at any fixed time interval t, can be made as small as desired by making L large enough. Hence, for an unbounded flow the equivalent of (7.1.5) reads

$$\lim_{V\to\infty} \frac{1}{V} \iiint F(\mathbf{X}(\mathbf{a}, t), t)\, da_1\, da_2\, da_3 = \lim_{V\to\infty} \frac{1}{V} \iiint F(\mathbf{x}, t)\, dx_1\, dx_2\, dx_3. \tag{7.1.6}$$

Now, let $F(\mathbf{x}, t) = \exp\,[i\mathbf{k}\cdot\mathbf{u}(\mathbf{x}, t)]$. The average value of this gives the characteristic function of the Eulerian velocity field (note that we use vectors **k** and **u** here; all three components of u_i are treated simultaneously). On the other hand, after averaging, $F(\mathbf{X}(\mathbf{a}, t), t)$ gives the characteristic function of the Lagrangian velocity field. Substituting the Eulerian and Lagrangian characteristic functions into (7.1.6) and taking averages, we obtain

$$\begin{aligned}
&\overline{\lim_{V\to\infty} \frac{1}{V} \iiint \exp[i\mathbf{k}\cdot\mathbf{v}(\mathbf{a},t)]\, da_1\, da_2\, da_3} \\
&\quad = \lim_{V\to\infty} \frac{1}{V} \iiint \overline{\exp\,[i\mathbf{k}\cdot\mathbf{v}(\mathbf{a},t)]}\, da_1\, da_2\, da_3 \\
&\quad = \overline{\exp[i\mathbf{k}\cdot\mathbf{v}(\mathbf{a}, t)]} \\
&\quad = \overline{\lim_{V\to\infty} \frac{1}{V} \iiint \exp[i\mathbf{k}\cdot\mathbf{u}(\mathbf{x}, t)]\, dx_1\, dx_2\, dx_3} \\
&\quad = \lim_{V\to\infty} \frac{1}{V} \iiint \overline{\exp[i\mathbf{k}\cdot\mathbf{u}(\mathbf{x}, t)]}\, dx_1\, dx_2\, dx_3 \\
&\quad = \overline{\exp\,[i\mathbf{k}\cdot\mathbf{u}(\mathbf{x}, t)]}.
\end{aligned} \tag{7.1.7}$$

The characteristic functions can be removed from under the integrals because the turbulence is homogeneous, so that the characteristic functions are independent of position. We conclude that the characteristic functions, and therefore also the probability densities, of the Lagrangian and Eulerian velocity fields are identical in homogeneous turbulence in an incompressible fluid. This implies that in homogeneous, incompressible flow

$$\overline{v_\alpha^2} = \overline{u_\alpha^2}. \tag{7.1.8}$$

Therefore, we do not need to determine $\overline{v_\alpha^2}$ in (7.1.3) by direct methods; a relatively easy measurement of $\overline{u_\alpha^2}$ suffices.

The result (7.1.8) might have been expected, but the method used can also be applied to more complex problems. For example, consider fully developed turbulent pipe flow (Section 5.2). Let us take $F(\mathbf{x}, t) = \tilde{u}_1(\mathbf{x}, t)$, which is the instantaneous total axial velocity in the pipe (x_1 is the streamwise direction). Since pipe flow is homogeneous in the x_1 direction and bounded in the x_2, x_3 plane, application of (7.1.6) gives

$$\frac{1}{\pi r^2}\iint \overline{\tilde{v}_1(\mathbf{a}, t)}\, da_2\, da_3 = \frac{1}{\pi r^2}\iint \overline{\tilde{u}_1(\mathbf{x}, t)}\, dx_2\, dx_3. \tag{7.1.9}$$

We may expect that $\overline{\tilde{v}_1}(\mathbf{a}, t)$ will be homogeneous in a cross section of the pipe if t is large enough, because no matter where a moving point starts from, it eventually wanders all around the cross section. With the usual notation convention, $\overline{\tilde{v}}_1 = V_1$ and $\overline{\tilde{u}}_1 = U_1$, so that (7.1.9) becomes, for large t,

$$V_1 = \frac{1}{\pi r^2}\iint \overline{\tilde{u}_1(\mathbf{x}, t)}\, dx_2\, dx_3 = \frac{1}{\pi r^2}\iint U_1(\mathbf{x})\, dx_2\, dx_3 \equiv U_b. \tag{7.1.10}$$

The mean axial velocity of a moving fluid point in a pipe is thus equal to the *bulk velocity* U_b of the fluid.

The mean-square fluctuation in the axial Lagrangian velocity is obtained in the same way:

$$\frac{1}{\pi r^2}\iint \overline{(\tilde{v}_1(\mathbf{a}, t) - V_1)^2}\, da_2\, da_3 = \frac{1}{\pi r^2}\iint \overline{v_1^2}\, da_2\, da_3$$
$$= \frac{1}{\pi r^2}\iint \overline{[\tilde{u}_1(\mathbf{x}, t) - U_b]^2}\, dx_2\, dx_3. \tag{7.1.11}$$

Again, the left-hand-side integrand may be expected to be homogeneous. The right-hand-side integrand is not homogeneous; however, with $\tilde{u}_1 = U_1 + u_1$, we obtain

$$\overline{[\tilde{u}_1(\mathbf{x}, t) - U_b]^2} = \overline{u_1^2} + [U_1(\mathbf{x}) - U_b]^2. \tag{7.1.12}$$

Hence, (7.1.11) becomes

$$\overline{v_1^2} = \frac{1}{\pi r^2}\iint [\overline{u_1^2} + (U_1 - U_b)^2]\, dx_2\, dx_3. \tag{7.1.13}$$

The Lagrangian axial velocity variance thus receives contributions both from the Eulerian velocity variance and from the square of the difference between the Eulerian mean velocity and the bulk velocity. Clearly, as a moving point wanders around in the pipe, its axial velocity fluctuates not only because the Eulerian velocity fluctuates but also because it wanders from time to time into regions where the mean velocity is different from the bulk velocity. The results (7.1.10, 7.1.13) are used in Section 7.2.

It would be tempting to extend this approach to the determination of the Lagrangian correlation. However, no useful results would evolve, because the analysis would yield Lagrangian space-time correlations, which not only are beyond the scope of this book, but also are relatively poorly understood.

The Lagrangian integral scale The second problem associated with applications of the dispersion formula (7.1.3) is the determination of $\mathcal{T}_{\alpha\alpha}$. From simple dimensional reasoning, we know that the Lagrangian (time) integral scale must be proportional to ℓ/u in turbulence with a single length scale ℓ and a single velocity scale u. In Section 2.3, extremely crude mixing-length arguments were used to show that

$$\nu_T = \frac{1}{2}\frac{d}{dt}(\overline{X_2^2}) = \overline{u_2^2}\mathcal{T}_{22}. \tag{7.1.14}$$

In wakes, the eddy viscosity is given by (Table 4.1)

$$\nu_T \cong 2.8\, u_* \ell_*. \tag{7.1.15}$$

Here, u_* is defined on basis of the Reynolds stress and ℓ_* is based on the maximum slope of the mean velocity profile. If we take u'_2 (the rms value of u_2) to be equal to u_* and if we identify ℓ_* with the length ℓ defined by $\ell = (u'_2)^3/\epsilon$ (ϵ is the dissipation rate), we obtain from (7.1.14) and (7.1.15)

$$\mathcal{T}_{22} \cong \ell/2.8\, u'_2. \tag{7.1.16}$$

Now, wakes are the most nearly homogeneous flows we have examined, so that (7.1.15) may be approximately valid for homogeneous turbulence. However, (7.1.14) is known to be incorrect because by the time the "diffusion limit" is valid, wandering points have moved to regions of different properties, even in the nearly homogeneous turbulence of a wake. Therefore, an independent estimate of $\mathcal{T}_{22}$, which does not rely on (7.1.14), would be welcome.

Corrsin (1963a) derived an estimate of $\mathcal{T}_{22}$ from spectral similarity considerations. His analysis is discussed in Section 8.5; the result is

$$\mathcal{T}_{22} \cong \ell/3u'_2. \tag{7.1.17}$$

The good agreement between (7.1.16) and (7.1.17) should not be taken too seriously. If we are honest, all we really can state is that $u_2'\mathscr{T}_{22}/\ell \sim 1$, which we should interpret as "somewhere between $\frac{1}{3}$ and 3". Nevertheless, the estimates (7.1.16) and (7.1.17) are quite successful in practice; the coefficient 3 in (7.1.17) may be regarded as an experimentally determined constant (much like the von Kármán constant).

The diffusion equation In homogeneous turbulence, the Lagrangian velocity variance is given by (7.1.8) and the Lagrangian integral scale may be estimated with (7.1,17). The asymptotic form of the diffusion equation (7.1.3) then becomes

$$\overline{(X_\alpha - a_\alpha)^2} = \frac{2\overline{u_\alpha^2}\ell t}{3u_\alpha'} = \frac{2u_\alpha'\ell t}{3}. \tag{7.1.18}$$

The length ℓ is defined by $\ell = (u'_\alpha)^3/\epsilon$, as was stated before. It is often more convenient to use Eulerian integral scales instead of ℓ. The analysis in Section 8.5 shows that the relations between ℓ and the Eulerian integral scales L_{11} and L_{22} (downstream and cross-stream integral scales, respectively), may be estimated as

$$\mathscr{T}_{22} \cong \frac{1}{3}\frac{\ell}{u_2'} \cong \frac{2}{3}\frac{L_{11}}{u_2'} \cong \frac{4}{3}\frac{L_{22}}{u_2'}. \tag{7.1.19}$$

7.2 Transport in shear flows

The case of homogeneous, stationary turbulence discussed in Section 7.1 is rather unrealistic, because turbulence cannot be maintained without mean shear. In this section, we discuss transport in a uniform shear flow and transport in pipes and channels.

Uniform shear flow Consider turbulent flow with uniform mean shear ($\partial U_1/\partial x_2$ = constant). The turbulence will be homogeneous in planes normal to the mean velocity U_1; however, Lagrangian velocities are not stationary, because the mean flow has no length scale, so that all length scales slowly grow in the streamwise direction, much as in grid turbulence (see Lumley, in Batchelor and Moffatt, 1970). Nevertheless, the rate of growth of the length scales is fairly slow; we may get a qualitative impression of the effects of mean shear by assuming that the Eulerian velocity field is homogeneous in all directions, so that the Lagrangian velocity field is stationary.

If the mean flow is defined by

$$U_1 = Sx_2, \quad U_2 = U_3 = 0, \tag{7.2.1}$$

the position of a moving point is given by

$$X_1(\mathbf{a}, t) = a_1 + \int_0^t [SX_2(\mathbf{a}, t') + v_1(\mathbf{a}, t')]\, dt', \tag{7.2.2}$$

$$X_2(\mathbf{a}, t) = a_2 + \int_0^t v_2(\mathbf{a}, t')\, dt', \tag{7.2.3}$$

$$X_3(\mathbf{a}, t) = a_3 + \int_0^t v_3(\mathbf{a}, t')\, dt'. \tag{7.2.4}$$

Because the turbulence is stationary and homogeneous, the fluctuating Lagrangian velocities v_1, v_2, and v_3 are stationary. From the central limit theorem we conclude directly that X_2 and X_3 asymptotically have Gaussian distributions, whose variance is given by (7.1.18). However, the downstream transport has to be determined separately because of the presence of the mean shear S. As a wandering point moves in the x_2 direction, it moves into a region with a different mean velocity, so that it tends to move faster (or slower, as the case may be) than in a flow without shear.

If the mean value of (7.2.3) is combined with the mean value of (7.2.2), there results

$$\overline{X_1} = a_1 + Sa_2 t. \tag{7.2.5}$$

This states that the mean position moves with the mean velocity of the initial location ($X_2(0) = a_2$). Subtracting (7.2.5) from (7.2.2), we obtain after differentiation

$$\frac{d}{dt}(X_1 - \overline{X}_1) = S(X_2 - a_2) + v_1. \tag{7.2.6}$$

The variance of $X_2 - a_2$ grows linearly at large times, but the variance of v_1 is constant. Hence, for large times the first term of (7.2.6) dominates and the second term may be neglected. Differentiating (7.2.6) once more, we obtain

$$\frac{d^2}{dt^2}(X_1 - \overline{X}_1) = S\frac{dX_2}{dt} = Sv_2. \tag{7.2.7}$$

This shows that $X_1-\overline{X}_1$ is a double integral of a stationary function. According to (6.5.16), it asymptotically has a Gaussian distribution, whose variance is given by (Corrsin, 1953)

$$\overline{(X_1-\overline{X}_1)^2} = \tfrac{2}{3} S^2\, \overline{u_2^2}\, t^3 \mathscr{T}_{22}. \tag{7.2.8}$$

The dispersion in the x_1 direction thus increases much faster than the dispersion in the x_2 and x_3 directions. The latter are given by

$$\overline{(X_2-a_2)^2} = 2\,\overline{u_2^2}\, t\, \mathscr{T}_{22}, \quad \overline{(X_3-a_3)^2} = 2\,\overline{u_3^2}\, t\, \mathscr{T}_{33}. \tag{7.2.9}$$

In (7.2.8) and (7.2.9) the Lagrangian variance $\overline{v_\alpha^2}$ has been replaced by $\overline{u_\alpha^2}$ because the turbulence is homogeneous (7.1.8).

Joint statistics If we want to predict the average shape of a patch of pollutant (smoke particles, say) released in a shear flow, the joint statistics of $X_1-\overline{X}_1$ and X_2-a_2 have to be analyzed. With a considerable amount of algebra, it can be shown that $X_1-\overline{X}_1$ and X_2-a_2 are jointly Gaussian at large times and that their covariance is given by

$$\overline{(X_1-\overline{X}_1)(X_2-a_2)} = \overline{u_2^2} S t^2 \int_0^t \left(1-\frac{\tau}{t}\right) \rho(\tau)\, d\tau = \overline{u_2^2} S t^2 \mathscr{T}_{22}. \tag{7.2.10}$$

At large times, the correlation coefficient between $X_1-\overline{X}_1$ and X_2-a_2 is $\frac{1}{2}\sqrt{3}$; contours of constant probability density are given by

$$\frac{x^2}{\sigma_1^2} - \sqrt{3}\,\frac{xy}{\sigma_1\sigma_2} + \frac{y^2}{\sigma_2^2} = \text{const.} \tag{7.2.11}$$

Here, $x = X_1-\overline{X}_1$, $y = X_2-a_2$; the variance σ_1^2 of x is given by (7.2.8) and the variance σ_2^2 of y is given by (7.2.9). The contours defined by (7.2.11) are ellipses; normalized with the standard deviation, as in (7.2.11), the ellipses have a constant aspect ratio, with a major axis of length $(1+\frac{1}{2}\sqrt{3})^{1/2} \cong 1.37$ and a minor axis of length $(1-\frac{1}{2}\sqrt{3})^{1/2} \cong 0.36$. The angle α between the major axis and the x_1 direction is given by

$$\tan\alpha = (\sigma_2/\sigma_1)^{1/2} = \sqrt{3}/St. \tag{7.2.12}$$

As the patch moves downstream, the major axis rotates towards the horizontal (Figure 7.2). At large times, the patch becomes quite elongated.

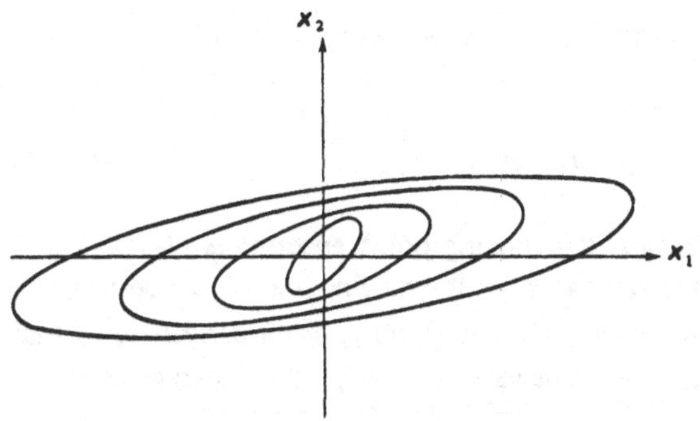

Figure 7.2. Dispersion in uniform shear flow. Equal increments of time are shown; all times are large compared to $\mathcal{T}_{22}$ (adapted from Corrsin, 1953).

Longitudinal dispersion in channel flow Let us now consider dispersion in a channel flow. The mean flow is in the x_1, x_2 plane, $U_2 = U_3 = 0$, the height of the channel is $2h$. The position of a moving point is given by

$$X_1(\mathbf{a}, t) - a_1 = \int_0^t [U_1(X_2(\mathbf{a}, t')) + u_1(\mathbf{X}(\mathbf{a}, t'), t')] \, dt', \tag{7.2.13}$$

$$X_2(\mathbf{a}, t) - a_2 = \int_0^t v_2(\mathbf{a}, t') \, dt', \tag{7.2.14}$$

$$X_3(\mathbf{a}, t) - a_3 = \int_0^t v_3(\mathbf{a}, t') \, dt'. \tag{7.2.15}$$

Here, $u_1(\mathbf{X}(\mathbf{a}, t'))$ is the Eulerian velocity fluctuation at the position of the moving point. For the reasons discussed in Section 7.1, v_2 and v_3 are stationary. Because $X_3 - a_3$ is the integral of a stationary function which itself is not a derivative of a stationary function, $X_3 - a_3$ asymptotically has a Gaussian distribution. Although $X_2 - a_2$ is also the integral of a stationary function, it does not have a Gaussian distribution because of the constraints imposed on v_2. Clearly, $X_2 - a_2$ itself is a stationary function because a moving point has to stay inside the channel. Therefore, v_2 is the derivative of a stationary function; its integral scale must be zero, and its spectrum behaves as ω^2 near the origin, so that the central limit theorem does not apply (see Section 6.5).

The mean axial velocity of a moving point is given by (7.1.10); integrating this, we obtain

$$\overline{X}_1 = a_1 + U_b t. \tag{7.2.16}$$

Here, as in (7.1.10), U_b is the bulk velocity. Substitution of (7.2.16) into (7.2.13) yields

$$X_1 - \overline{X}_1 = \int_0^t [U_1(X_2) - U_b + u_1]\, dt'. \tag{7.2.17}$$

Because $X_2 - a_2$ is stationary, the moving point encounters $U_1\ (X_2)$ and $u_1\ (\mathbf{X}(\mathbf{a},t'),t')$ in a stationary way: as far as the moving point is concerned, the integrand in (7.2.17) is stationary. The integrand is not a derivative, so that $X_1 - X_1$ is Gaussian at large times. The variance of $X_1 - \overline{X}_1$ is then given by

$$\overline{(X_1 - \overline{X}_1)^2} = 2\,\overline{v_1^2}\, t\mathscr{T}. \tag{7.2.18}$$

The Lagrangian velocity variance $\overline{v_1^2}$ was computed for pipe flow in (7.1.13); transposing this to channel flow, we have

$$\overline{v_1^2} = \frac{1}{2h}\int_{-h}^{h} [\overline{u_1^2} + (U_1 - U_b)^2]\, dx_2. \tag{7.2.19}$$

In the core region of channel flow at large Reynolds numbers, we may write (Section 5.2)

$$(U_1 - U_b)/u_* = F(\eta), \quad \overline{u_1^2}/u_*^2 = g(\eta). \tag{7.2.20}$$

Here, $\eta = x_2/h$. Substituting (7.2.20) into (7.2.19), we obtain

$$\overline{v_1^2} = u_*^2 \left(\int_0^1 F^2\, d\eta + \int_0^1 g\, d\eta \right) = A u_*^2. \tag{7.2.21}$$

The constant A is approximately equal to 5. In the wall layer, (7.2.20) is not valid; however, the wall layer is so thin that it makes a negligible contribution to the integrals. The integral scale $\mathscr{T}$ in (7.2.18) should be of order h/u_*, because u_* and h are the velocity and length scales of the core region of channel flow (Section 5.2). With this estimate and (7.2.21), (7.2.18) becomes

$$\overline{(X_1 - \overline{X}_1)^2} = C u_* h t, \tag{7.2.22}$$

where C should be approximately equal to 10.

Channel flow is difficult to set up in a laboratory; however, experimental values for C in pipe flow indeed range around 10 if the pipe radius instead of the channel half-width h is used in the formula for the variance (Monin and Yaglom, 1971). Of course, (7.2.22) is valid only for $t >> h/u_*$.

Bulk velocity measurements in pipes Equations (7.2.16) and (7.2.22) may be applied to the problem of determining U_b in pipes with tracer methods. If the pipe radius $D/2$ is used instead of the channel half-width h in (7.2.22), the relative error in the measurement of bulk velocity is

$$\frac{\overline{[(X_1 - \overline{X}_1)^2]}^{1/2}}{\overline{X}_1 - a_1} = \left(\frac{Cu_* D}{2U_b^2 t}\right)^{1/2} = \left(\frac{u_*}{U_b}\right)^{1/2} \left(\frac{CD}{2U_b t}\right)^{1/2}. \tag{7.2.23}$$

The measurement is performed by releasing a patch of tracer material at $X_1 = a_1$ at time $t = 0$; the time interval t between release and the passage of the patch at some downstream location is measured. The factor $U_b t/D$ is the streamwise distance in diameters; clearly, the accuracy of the measurement improves as this distance increases. The ratio u_*/U_b is the square root of the friction coefficient; at typical Reynolds numbers, its value is about 0.04, so that $(u_*/U_b)^{1/2} \cong 0.2$. If $C = 10$, a streamwise separation of 100 diameters gives about a 4% standard deviation in the measurement of U_b. The accuracy can be improved considerably if the streamwise concentration distribution of the patch of tracer material is measured at the downstream location.

7.3
Dispersion of contaminants

So far, we have discussed only the dispersion of moving points and assumed that it would be possible to mark or tag a Lagrangian "point" in such a way that it would keep its identity. In the two examples given in Section 7.2 we assumed without discussion that the motion of a minute tracer particle is identical to the motion of the Lagrangian point of fluid that would occupy the position of the particle if it were not there. Now, we have to consider more realistic dispersion problems. Two questions arise. First, contaminants are commonly released with some initial concentration distribution, so that the concentration distribution at later times has to be predicted. Second, contaminants are also dispersed by molecular transport, which may interact with the turbulent transport. We will discuss these problems separately.

The concentration distribution Let us consider contaminants which are not dispersed by molecular motion. This is an idealization; however, in liquids the molecular transport of contaminants (salinity, heat) is poor and in air the molecular transport of minute tracer particles (smoke, say) is poor, so that the assumption of zero diffusivity should be fairly realistic in those cases.

The transport of a contaminant with zero diffusivity is governed by

$$\frac{\partial \tilde{c}}{\partial t} + \tilde{u}_i \frac{\partial \tilde{c}}{\partial x_i} = 0. \tag{7.3.1}$$

Here, $\tilde{c}$ is the instantaneous concentration at a point x_i, t, and $\tilde{u}_i$ is the instantaneous fluid velocity at that point. The solution of (7.3.1) is

$$\tilde{c}(\mathbf{X}(\mathbf{a}, t), t) = \tilde{c}(\mathbf{a}, 0). \tag{7.3.2}$$

This states that the concentration at each moving point remains equal to its value at the time of release. Because there is no molecular diffusion, this result is obvious. If we want to predict the mean concentration $C(\mathbf{x}, t)$, (7.3.2) has to be inverted. This is a backward dispersion problem: instead of asking where a point that started from a_i at time $t = 0$ will go to, we are asking where a point that arrives at x_i at time t came from. In other words, we need a Lagrangian displacement integral like (7.1.1), but with time running backwards.

If the Lagrangian velocity field is stationary, the backward and forward dispersion problems are the same. If $B(\mathbf{X}, \mathbf{a}, t)$ is the probability density of $X_i(\mathbf{a}, t)$ for points that started at a_i at time $t = 0$, then $B(\mathbf{a}, \mathbf{x}, t)$ is the probability density of the original positions $a_i(\mathbf{x}, t)$ of points that arrive at x_i at time t. If $C(\mathbf{x}, t)$ is the mean concentration, we can write

$$C(\mathbf{x}, t) = \iiint\limits_{-\infty}^{\infty} \tilde{c}(\mathbf{a}, 0) B(\mathbf{a}, \mathbf{x}, t)\, da_1\, da_2\, da_3. \tag{7.3.3}$$

This states that the mean concentration at a point is the concentration carried by a particle times the probability of the particle being there, integrated over all particles that could be there.

If the initial concentration is all at one point (a_i^0, say), we have

$$\tilde{c}(\mathbf{a}, 0) = 0 \text{ for all } a_i \neq a_i^0, \tag{7.3.4}$$

$$\iiint\limits_{-\infty}^{\infty} \tilde{c}(\mathbf{a}, 0)\, da_1\, da_2\, da_3 = 1. \tag{7.3.5}$$

The integral (7.3.5) has been normalized for convenience. Equations (7.3.4) and (7.3.5) define a Dirac delta function $\delta(\mathbf{a} - \mathbf{a}^0)$; the integral (7.3.3) reduces to

$$C(\mathbf{x}, t) = \iiint\limits_{-\infty}^{\infty} \delta(\mathbf{a} - \mathbf{a}^0)\, B(\mathbf{a}, \mathbf{x}, t)\, da_1\, da_2\, da_3 = B(\mathbf{a}^0, \mathbf{x}, t). \tag{7.3.6}$$

The mean concentration is then equal to the probability density of the position of a moving point leaving from a_i^0. This suggests that $B(\mathbf{a}^0, \mathbf{x})$ can be measured by introducing a small point source of contamination at a_i^0 and measuring the mean concentration throughout the field. The omission of the argument t in B is intentional: in practice, continuous point sources with constant flux are used, so that C and B are independent of time.

It is clear from (7.3.6) that minute tracer particles and nondiffusing contaminants indeed may be used to mark Lagrangian points. The conclusions obtained in Sections 7.1 and 7.2 thus apply to the concentration distribution as well as to the probability density; however, it must be kept in mind that the identification can be made only if the Lagrangian velocity field is stationary.

The effects of molecular transport If the contaminant has a finite molecular diffusivity γ, the conservation equation for $\tilde{c}$ becomes

$$\frac{\partial \tilde{c}}{\partial t} + \tilde{u}_i \frac{\partial \tilde{c}}{\partial x_i} = \gamma \frac{\partial^2 \tilde{c}}{\partial x_i \partial x_i}. \tag{7.3.7}$$

The presence of molecular diffusion makes it impossible to write (7.3.2), so that we have to proceed in a different way. The general problem raised by (7.3.7) is intractable; we consider the special case of a small spot of contaminant, centered around a moving point. Let us change to coordinates moving with the wandering point. If ξ_i is the difference between the Eulerian position x_i and the position of the Lagrangian point X_i, (7.3.7) becomes

$$\frac{\partial \tilde{c}}{\partial t} + \frac{\partial}{\partial \xi_i}[\tilde{c}(\tilde{u}_i(\xi) - \tilde{u}_i(0))] = \gamma \frac{\partial^2 \tilde{c}}{\partial \xi_i \partial \xi_i}. \tag{7.3.8}$$

Here, the continuity equation has been used to bring $\tilde{u}_i$ inside the derivative; of course, seen from a coordinate system moving with a Lagrangian point, the Eulerian velocity is not $\tilde{u}_i(\mathbf{x})$ but $\tilde{u}_i(\xi) - \tilde{u}_i(0)$. Equation (7.3.8) describes dispersion relative to a moving point. If the patch of contaminant is smaller than the Kolmogorov microscale, the velocity distribution in the neighborhood of the moving point is approximately linear:

$$\tilde{u}_i(\xi) - \tilde{u}_i(0) = \xi_j \frac{\partial \tilde{u}_i}{\partial \xi_j}(0). \tag{7.3.9}$$

The velocity field around the moving point is then a combination of a solid-body rotation (corresponding to the skew-symmetric part of $\partial \tilde{u}_i / \partial \xi_j$) and a pure strain (corresponding to the symmetric part of $\partial \tilde{u}_i/\partial \xi_j$). The value of $\partial \tilde{u}_i/\partial \xi_j$ at $\xi_i = 0$, of course, generally varies in time.

Substitution of (7.3.9) into (7.3.8) yields

$$\frac{\partial \tilde{c}}{\partial t} + \frac{\partial \tilde{u}_i}{\partial \xi_j} \xi_j \frac{\partial \tilde{c}}{\partial \xi_i} = \gamma \frac{\partial^2 \tilde{c}}{\partial \xi_i \partial \xi_i}. \tag{7.3.10}$$

It is easy to see that the total amount of contaminant in the spot must be conserved:

$$\iiint_{-\infty}^{\infty} \tilde{c}\, d\xi_1\, d\xi_2\, d\xi_3 = 1. \tag{7.3.11}$$

The integral has been normalized for convenience. The shape and size of the spot can be measured by I_{pq}, which is defined by

$$\iiint_{-\infty}^{\infty} \xi_p \xi_q \tilde{c}\, d\xi_1\, d\xi_2\, d\xi_3 = I_{pq} \ . \tag{7.3.12}$$

The sum of the diagonal components of I_{pq} is I_{pp}; this is proportional to the square of the average spot radius. The equation for I_{pq} can be obtained from (7.3.10); it reads

$$\frac{d\, I_{pq}}{dt} - I_{pj} \frac{\partial \tilde{u}_q}{\partial \xi_j} - I_{qj} \frac{\partial \tilde{u}_p}{\partial \xi_j} = 2\gamma \delta_{pq}. \tag{7.3.13}$$

If $\partial \tilde{u}_i/\partial \xi_j$ is equal to zero, the solution of (7.3.13) is straightforward:

$$I_{pq} = 2\, \gamma\, t\, \delta_{pq}. \tag{7.3.14}$$

This states that, in the absence of relative motion near a point, the spot of contaminant is round ($I_{pq} = 0$ if p and q are different) and that it spreads by molecular diffusion in all directions. The radius of the spot is proportional to $I_{pp}^{1/2}$; clearly, the radius increases as $(\gamma t)^{1/2}$, as in all diffusion problems.

The effect of pure, steady strain Equation (7.3.13) cannot easily be solved for a general velocity field. However, the solution of a special case is instructive. Let us restrict the analysis to the effects of pure strain. Take a two-dimensional strain-rate field in which $\partial u_1/\partial \xi_1 = s$, $\partial u_2/\partial \xi_2 = -s$, $\partial u_3/\partial \xi_3 = 0$, and in which all off-diagonal components of $\partial \tilde{u}_i/\partial \xi_j$ are zero. This represents pure, plane strain with stretching in the ξ_1 direction and compression in the ξ_2 direction. The approximation (7.3.9) implies that s is uniform; we also assume that it does not vary in time. The choice of the symbol s is not an arbitrary one. We found in Chapter 3 that the strain-rate fluctuations s_{ij} in turbulence are quite large. We recall that $s_{ij} \sim u/\lambda$ (λ is the Taylor microscale); these large strain rates are associated with the small-scale motion. For the small spot we are considering here, we may thus consider s to be of order u/λ.

For steady, plane strain, (7.3.13) becomes

$$dl_{11}/dt - 2sl_{11} = 2\gamma, \tag{7.3.15}$$

$$dl_{22}/dt + 2sl_{22} = 2\gamma, \tag{7.3.16}$$

$$dl_{12}/dt = 0. \tag{7.3.17}$$

The solution of (7.3.15–7.3.17) is

$$l_{11} = 2\gamma \frac{\exp(2st) - 1}{2s}, \quad l_{22} = 2\gamma \frac{1 - \exp(-2st)}{2s}, \tag{7.3.18}$$

$$l_{11} + l_{22} = 4\gamma \frac{\sinh(2st)}{2s}, \tag{7.3.19}$$

$$l_{12} = 0. \tag{7.3.20}$$

For very small total strain st, $\sinh(2st) \cong 2st$, so that $l_{11} + l_{22} \cong 4\gamma t$, which agrees with (7.3.14). However, as the strain st increases, (7.3.19) increases much faster than t, so that the spot spreads much faster than it would as a result of molecular transport alone. The straining motion thus accelerates molecular diffusion of small spots. In turbulence this effect is quite pronounced, because the fluctuating strain rates are so large.

The cause of the accelerated diffusion is easy to understand. As a spot of contaminant is drawn out in the ξ_1 direction (Figure 7.3), the concentration gradients in that direction are reduced. Because the diffusion of contaminant is proportional to the concentration gradient, the rate of spread in the ξ_1 direction is reduced. In the ξ_2 direction, however, the spot is being compressed, so that the gradients and the molecular diffusion in the ξ_2 direction increase. At small values of st, the increase in the ξ_2 gradient is about equal to the decrease in the ξ_1 gradient, but at large values of st the increase of diffusion in the ξ_2 direction is much larger than the decrease of diffusion in the ξ_1 direction, so that the net rate of diffusion increases as indicated by (7.3.19).

The interaction of turbulent and molecular transport thus results in much

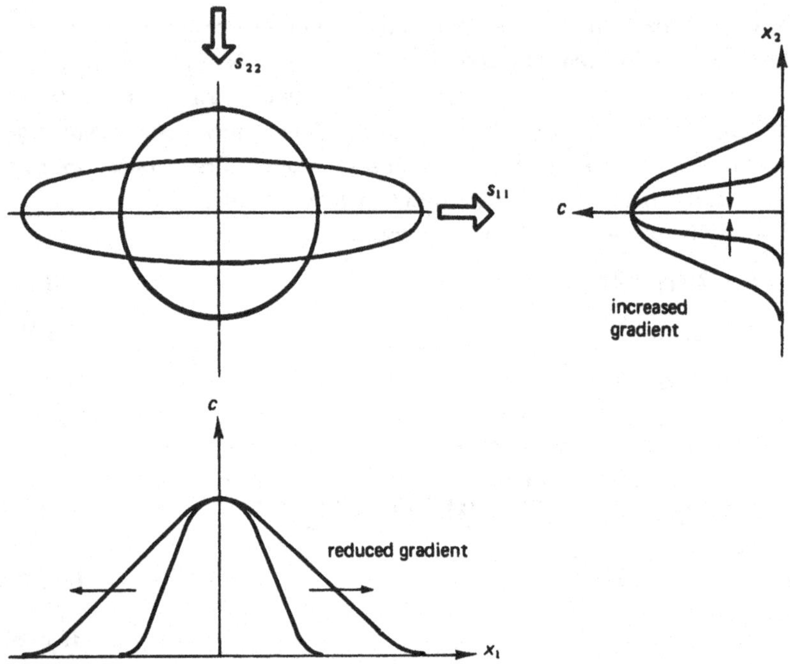

Figure 7.3. Effect of strain on concentration gradients.

faster spreading of the spot. This is one of the reasons why turbulent mixing is so effective. If there were no molecular transport, turbulent mixing would carry thin sheets and filaments of contaminant to every part of the flow. However, there would still be large inhomogeneities at small scales, because the filaments would be separated by regions of uncontaminated fluid, which would have to be filled by unaccelerated molecular diffusion.

The expressions (7.3.18) provide support for the calculations of the minimum scale in cases with $\gamma/\nu < 1$, given in Chapter 3 (see (3.3.68), (3.4.7), and Figure 3.6). Examining the expression for l_{22}, we see that it never gets smaller than γ/s, no matter how large the total strain becomes. The minimum scale then is $(\gamma/s)^{1/2}$. On substitution of s by $u/\lambda \sim (\epsilon/\nu)^{1/2}$, the contaminant microscale becomes $(\gamma/\nu)^{1/2}\eta$.

The assumption that the strain-rate field is steady is not unrealistic. As we saw in Chapter 3, time derivatives of the vorticity and strain-rate fields are of order $R_\ell^{-1/2}$ relative to $1/s$. In other words, the straining goes on for many times $1/s$. Of course, the strain rate eventually changes sign, so that the rate of spreading is controlled not by st, but by $s(t\mathscr{T})^{1/2}$, where $\mathscr{T}$ is the Lagrangian integral scale of the strain-rate field. The assumption that the strain-rate field is infinite compared to the spot size clearly corresponds to $\gamma/\nu \ll 1$. The assumption that the vorticity is zero is not vital to the argument; a more exact calculation that includes the vorticity does not change the conclusions obtained here (Lumley, 1972 b).

Transport at large scales The effects of turbulence-accelerated molecular transport are mainly confined to small scales because the strain-rate fluctuations are most intense at small scales. As we have seen above, molecular diffusion rapidly removes the small-scale concentration inhomogeneities created by the straining motion. This interaction tends to make the concentration distribution approximately homogeneous at small scales. The time needed for homogenizing may be large compared to $(\nu/\epsilon)^{1/2}$, but $(\nu/\epsilon)^{1/2} \sim R_\ell^{-1/2}\ \ell/u$ (1.5.15), so that this time scale is likely to be small compared to the large-eddy time scale ℓ/u.

If the instantaneous concentration $\tilde{c}$ is decomposed into a mean concentration C and concentration fluctuations c, the conservation equation for C becomes (in the absence of mean flow)

$$\frac{\partial C}{\partial t} + \frac{\partial}{\partial x_i}\overline{cu_i} = \gamma \frac{\partial^2 C}{\partial x_i \partial x_i}. \tag{7.3.21}$$

The transport term on the left-hand side of (7.3.21) is of order Cu/ℓ_c (ℓ_c is a length scale characteristic of mean concentration gradients). The molecular diffusion term is of order $\gamma C/\ell_c^2$. The ratio of these is $u\ell_c/\gamma$. Because turbulence-accelerated diffusion increases ℓ_c rapidly, $u\ell_c/\gamma$ (which is comparable to the Reynolds number if $\gamma/\nu \cong 1$, as in gases) tends to become large, so that the effects of molecular diffusion on the mean concentration distribution can often be neglected. This conclusion, of course, is identical to the one obtained for the transport of mean momentum (Section 2.1).

7.4 Turbulent transport in evolving flows

In the preceding sections we have discussed only cases in which the Lagrangian velocities were stationary. The problem becomes much more difficult if they are not. Nonstationary Lagrangian velocities arise if the Eulerian flow field is nonstationary or inhomogeneous (or both); in this section, we discuss the dispersion of contaminants in self-preserving, inhomogeneous, statistically steady flows.

Thermal wake in grid turbulence Consider transport of heat released from a line source in grid turbulence. The turbulence is produced in a wind tunnel. The mean velocity U is in the x direction; it is assumed to be uniform. The turbulence is homogeneous in the y, z plane, but it decays downstream. The (Eulerian) integral scale increases as $x^{1/2}$ downstream, while the turbulent energy $\frac{1}{2}u^2$ decreases as x^{-1}, to a first approximation (Section 3.2).

The line source could be a heated wire stretched across the wind tunnel; we assume that the heat supply is steady. The wire produces a small temperature rise in all the material points that happen to pass through its boundary layer. The heated wake of the wire is slowly broadened by the turbulence-accelerated molecular transport, but it is also carried from side to side by larger eddies (Figure 7.4). If the mean temperature difference between any point within the thermal wake and the unheated fluid is called Θ and if the temperature fluctuations are designated by θ, the equation for Θ reads

$$U\frac{\partial\Theta}{\partial x}+\frac{\partial}{\partial x}(\overline{u\theta})+\frac{\partial}{\partial y}(\overline{v\theta})=\gamma\frac{\partial^2\Theta}{\partial x^2}+\gamma\frac{\partial^2\Theta}{\partial y^2}. \tag{7.4.1}$$

The second and fourth terms of (7.4.1) are small, as can easily be demonstrated by repeating the order-of-magnitude analysis for plane wakes (Section 4.1). The last term of (7.4.1) is also small, but we will retain it to see what effect molecular transport has on the distribution of Θ. Consequently, (7.4.1) is approximated by

$$U\frac{\partial\Theta}{\partial x}+\frac{\partial}{\partial y}(\overline{v\theta})=\gamma\frac{\partial^2\Theta}{\partial y^2}. \tag{7.4.2}$$

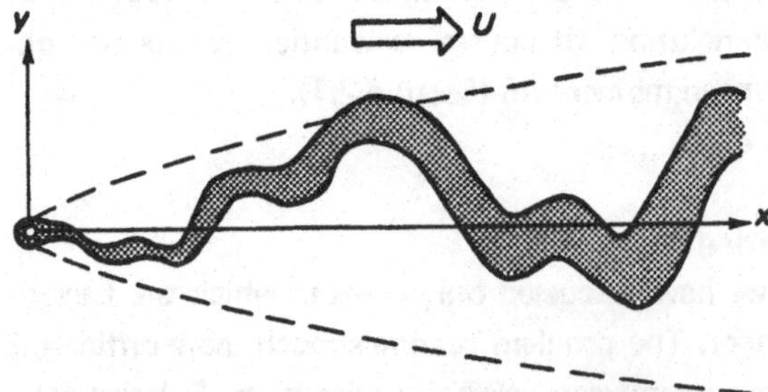

Figure 7.4. Definition sketch for plane thermal wake.

Integrating (7.4.2), we obtain

$$U \int_{-\infty}^{\infty} \Theta \, dy = \frac{H}{\rho c_p}. \tag{7.4.3}$$

The total flux of heat past any downstream location is thus constant. This relation is similar to the momentum integral in ordinary wakes.

Self-preservation We are looking for a self-preserving solution to (7.4.2, 7.4.3). Immediately, a problem arises. The turbulence has a length scale ℓ, whose growth is fixed; if the temperature distribution has another length scale, which might increase at a different rate, self-preservation cannot exist. If the virtual origin of the thermal wake is the same as the virtual origin of the turbulence, this problem would not arise. This could be arranged by putting the heated wire very close to the grid or, even better, by heating one of the bars of the grid. If the heated wire is at some distance from the grid, however, self-preservation does not seem possible. If the mean temperature difference at the center line of the wake is called Θ_0 and if the length scale of the thermal wake is ℓ_θ, the turbulent transport term in (7.4.2) is of order

$$-\frac{\partial}{\partial y}(\overline{v\theta}) = \mathcal{O}(u\Theta_0/\ell_\theta). \tag{7.4.4}$$

If the thermal wake is self-preserving, because the heated wire is located near the grid, the transport term is

$$-\frac{\partial}{\partial y}(\overline{v\theta}) = \mathcal{O}(u\Theta_0/\ell). \tag{7.4.5}$$

The values of Θ_0 in (7.4.4) and (7.4.5) are not the same; r ne heat flux is the same in both cases, the value of Θ_0 at some given downstream distance x from the grid is larger for the wake of the wire that is closest o x. Also, close behind that wire $\ell_\theta \ll \ell$, so that (7.4.4) produces abnormally large turbulent transport in the y direction. This causes rapid broadening of the temperature distribution, so that we may expect ℓ_θ to catch up with ℓ (Figure 7.5).

Another way to understand this effect is to take account of the fact that the width of the distribution Θ increases roughly proportionally to the square root of the time since release for all but very small times. At a given mean velocity U, the width thus increases as the square root of the distance from the wire; if the distance from the wire is much smaller than the distance from

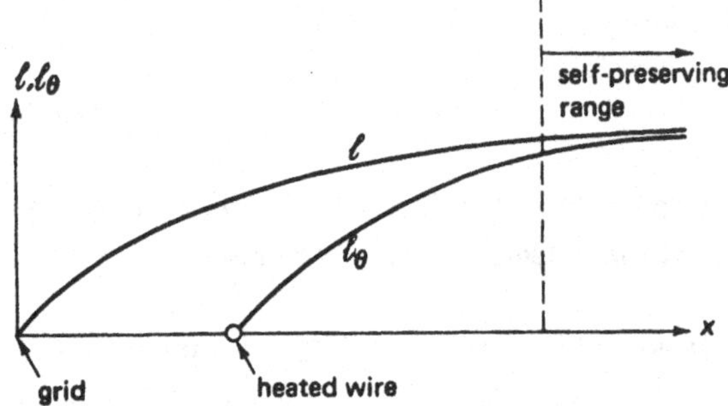

Figure 7.5. The growth of ℓ_θ for a source not located near the virtual origin of the turbulence.

the turbulence-producing grid, ℓ_θ increases faster than ℓ. Therefore, if we allow some distance for ℓ_θ to become comparable to ℓ, a self-preserving solution should be feasible.

The assumption of self-preservation consists of

$$\Theta = \Theta_0 f(y/\ell), \quad -\overline{\theta v} = \Theta_0 u g(y/\ell). \tag{7.4.6}$$

Here, $\Theta_0 = \Theta_0(x)$ and $\ell = \ell(x)$. We have assumed that $\ell_\theta = \ell$; of course, the self-preserving decay of the turbulence prescribes $u \propto x^{-1/2}$, $\ell \propto x^{1/2}$. Substitution of (7.4.6) into (7.4.2) yields

$$\frac{U}{u}\frac{\ell}{\Theta_0}\frac{d\Theta_0}{dx}f - \frac{U}{u}\frac{d\ell}{dx}f'\eta = g' + \frac{\gamma}{u\ell}f''. \tag{7.4.7}$$

Here, primes denote differentiation with respect to $\eta(=y/\ell)$. Self-preservation can be obtained only if the coefficients in (7.4.7) are constant:

$$\frac{U}{u}\frac{\ell}{\Theta_0}\frac{d\Theta_0}{dx} = A, \quad \frac{U}{u}\frac{d\ell}{dx} = B, \quad \frac{\gamma}{u\ell} = \frac{1}{P}. \tag{7.4.8}$$

Because $u \propto x^{-1/2}$ and $\ell \propto x^{1/2}$, the second and third of (7.4.8) are satisfied (P is a Péclet number). The first of (7.4.8) can be satisfied by any power law $\Theta_0 \propto x^n$, but the heat flux integral (7.4.3) requires that $\Theta_0 \ell$ be constant, so that Θ_0 varies as $x^{-1/2}$. This is not surprising, because Θ_0 is similar to the center-line velocity difference U_s in momentum wakes (Section 4.1). It is convenient that the molecular transport term is also self-preserving; it will be retained. With these results, (7.4.7) becomes

$$-B(f+\eta f') = g' + P^{-1} f''. \tag{7.4.9}$$

Integration of (7.4.9) yields

$$-B\eta f = g + P^{-1} f'. \tag{7.4.10}$$

Let us assume that the eddy diffusivity is constant. This is a much better assumption than in the wakes studied in Section 4.1, because this flow has no edges, is not intermittent, and is homogeneous in the cross-stream direction. If the eddy diffusivity is γ_T and the nondimensional group $u\ell/\gamma_T$ is called the turbulent Péclet number P_T, (7.4.10) becomes

$$-B\eta f = (P_T^{-1} + P^{-1}) f'. \tag{7.4.11}$$

It is convenient to define ℓ by

$$B = \frac{U}{u}\frac{d\ell}{dx} \equiv P_T^{-1} + P^{-1}. \tag{7.4.12}$$

The solution of (7.4.11) then becomes

$$f = \exp\left(-\tfrac{1}{2}\eta^2\right). \tag{7.4.13}$$

The mean temperature difference Θ thus has a Gaussian distribution, just like the momentum deficit in wakes (4.2.15).

It is clear from (7.4.11, 7.4.12) that, to the degree of approximation used here, the effect of molecular transport on the mean temperature distribution is additive. If P_T is of the same order as R_T in plane wakes ($R_T = 12.5$, see Table 4.1) and if P is at all large, the additional spreading due to molecular transport is negligible.

Dispersion relative to the decaying turbulence It has been assumed that the width of the temperature wake scales with the length of the decaying turbulence, which increases as $x^{1/2}$. This implies that the dispersion, nondimensionalized with the local length scale, does not increase as soon as self-preservation has been attained. Clearly, wandering points are not being dispersed in the sense used earlier in this chapter. This peculiar behavior arises because the grid turbulence "disperses" its own length scales at a rate consistent with the dispersion of contaminants; it is characteristic of dispersion in evolving flows such as jets, wakes, and boundary layers.

If the heated wire is not located close to the grid, self-preservation is unlikely to be observed experimentally. The time scale ℓ/u of the turbulence is of the same order as $t = x/U$ (3.2.32). It takes the length scale of the temperature wake several ℓ/u to catch up with the length scale of the turbulence. However, ℓ/u is also the time scale of decay, so that after several ℓ/u the turbulence is no longer self-preserving, but has entered the final period of decay, in which ℓ and u change downstream in a different way.

The Gaussian distribution The result that the distribution of Θ is Gaussian (within the assumption of constant eddy diffusivity) is in good agreement with experimental data, even at small distances from the point of release. This should not be construed as support for a constant eddy diffusivity, because the probability density of the velocity fluctuations is also observed experimentally to be approximately Gaussian at all times, so that the Gaussian distribution of Θ would seem to be an unavoidable result. In fact, the position of a wandering point, nondimensionalized with the local length scale ℓ, itself becomes a stationary variable at a large distance from the grid; there is no reason why it should have a Gaussian distribution, except for the dynamics of turbulence which happen to make it so.

Dispersion in shear flows The analysis presented in this section may also be applied to dispersion by other self-preserving flows, such as jets, wakes, plumes, and boundary layers. Some time after release, the plume of contaminant will have spread throughout the turbulent part of the flow; beyond that, dispersion of momentum and dispersion of contaminant go hand in hand, just as in the thermal plume discussed in Section 4.6. Because the contaminant cannot spread beyond the edges of the flow, the length scale of the contaminant distribution remains the same as the length scale of the flow.

If the point of release of contaminant does not coincide with the virtual origin of the flow, we cannot expect self-preservation near the point of release. Because shear flows exhibit no cross-stream homogeneity, the initial dispersion problem is extremely complicated. Sometimes, approximate solutions are obtained by assuming that the turbulence is homogeneous and that the mean velocity U is approximately constant in the neighborhood of the point of release; the initial dispersion can then be described with the analysis of Section 7.1, where the time t since release is replaced by x/U. The effect of mean shear is sometimes accounted for by assuming that the mean velocity gradient is approximately constant; the results obtained in Section 7.2 may then lead to qualitatively correct conclusions.

Problems

7.1 A chemical is added at the center line of a fully developed turbulent pipe flow. The reaction rate is large, so the total reaction time is determined by turbulent transport. How many diameters are required for the reaction to be completed?

7.2 A kilogram of a half-and-half mixture of two fluids is being homogenized by a 25-watt mixer. The two fluids have about the same viscosity and density (about 10^{-5} m^2/sec and 1 kg/m^3, respectively); the diffusivity of one fluid into the other is about 3×10^{-6} times the viscosity. This situation occurs if one of the fluids is a dilute solution of high molecular weight polymers. Make a conservative estimate of the mixing time required for homogeneity of the mixture. Suppose that in the mixing process it is necessary to use only strain rates small compared to 5×10^2 sec^{-1}, because larger strain rates tend to tear the polymer molecules apart. If you limit yourself to strain rates one-tenth of this value, what mixing time is required? What is the power of the mixer in this case? If the mixer paddle is 5 cm in diameter, is the flow turbulent?

7.3 A smokestack located in the lower part of the atmospheric boundary layer releases a steady stream of neutrally buoyant smoke. Estimate the downstream position of the point of maximum pollutant concentration at the surface. What is the effect of the stack height on the maximum surface concentration?

8

SPECTRAL DYNAMICS

In Chapter 6, the energy spectrum, defined as the Fourier transform of the autocorrelation, was introduced. There would be relatively little value in working with the spectrum, however, if it did not have its own physical interpretation. We shall find that spectral analysis allows us to draw conclusions that are almost unattainable in any other way. Spectra are decompositions of the measured function into waves of different periods or wavelengths. The value of the spectrum at a given frequency or wavelength is the mean energy in that wave, as we found in Section 6.4. Spectra thus give us an opportunity to think about the way in which waves, or eddies, of different sizes exchange energy with each other. This is the central issue in this chapter, because turbulence commonly receives its energy at large scales, while the viscous dissipation of energy occurs at very small scales. We shall find that there often exists a range of eddy sizes which are not directly affected by the energy maintenance and dissipation mechanisms; this range is called the inertial subrange.

8.1
One- and three-dimensional spectra

A turbulent flow varies randomly in all three space directions and in time. Experimental measurements, say of velocity, may be made along a straight line at a fixed time, at a fixed point as a function of time, or following a moving fluid point as a function of time. A measurement of this kind generates a random function of position or time. If the function is stationary or homogeneous, an autocorrelation can be formed and a spectrum can be computed. If the autocorrelation is a function of a time interval, the transform variable is a frequency; if the autocorrelation is a function of a spatial separation, the transform variable is a *wave number* (with dimensions length^{-1}). Spectra obtained in this way are called *one-dimensional spectra* because the measurements producing them were taken in one dimension.

Aliasing in one-dimensional spectra One-dimensional spectra do not seem very appropriate for the description of turbulence, because it is three dimensional. In a way, one-dimensional spectra give misleading information about three-dimensional fields. Suppose that we are making measurements along a straight line and that we are looking for components of wave number κ.

Because we are measuring along a line, we cannot distinguish between disturbances of wave number κ whose wave-number vector is aligned with the direction of measurement and disturbances of wave numbers larger than κ whose wave-number vector is oblique to the line of measurement (Figure 8.1). Thus, a one-dimensional spectrum obtained in a three-dimensional field contains at wave number κ contributions from components of all wave numbers larger than κ. This is called *aliasing*. Measured one-dimensional spectra ordinarily have a finite value at the origin (proportional to the integral scale). This does not mean that there is finite energy at zero wave number; the energy merely has been aliased from higher wave numbers to zero.

The problem of aliasing is not serious at high wave numbers, however. This is because small eddies tend to have about the same size in all directions, so that there is little chance that the situation depicted in Figure 8.1 occurs at small scales.

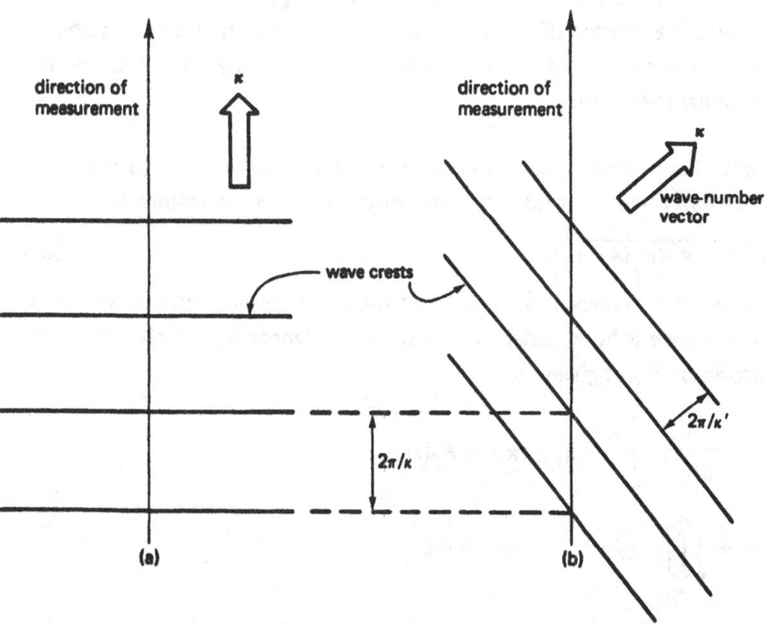

Figure 8.1. Aliasing in a one-dimensional spectrum: (a) a wave of true wave number κ, aligned with the line of measurement, (b) a wave of wave number $\kappa' > \kappa$, with wave-number vector oblique to the line of measurement.

The three-dimensional spectrum In order to avoid the aliasing problem, we can take measurements not just along a line but in all possible directions. This produces a correlation that is a function of the separation vector. The three-dimensional Fourier transform of such a correlation produces a spectrum that is a function of the wave-number vector κ_i. Unfortunately, this gives much more information than we can handle. The addition of the directional information eliminates the aliasing problem in exchange for a complexity that makes physical reasoning difficult. In order to remove the directional information, the spectrum is usually integrated over spherical shells around the origin of wave-number space. In this way, we obtain a spectrum that is a function of the scalar wave-number magnitude κ and whose value represents the total energy at that wave-number magnitude without aliasing. This is called a *three-dimensional spectrum.*

One additional problem remains. Often, the velocity components u_1, u_2, and u_3 are measured separately. However, for spectral analysis we need a spectrum that represents all of the kinetic energy at a given wave number. Therefore, the spectra of u_1, u_2, and u_3 are commonly added together; it is the spectrum of the total energy which is always referred to as the three-dimensional spectrum.

The correlation tensor and its Fourier transform Let us now formalize what we have described in words. The *correlation tensor* R_{ij} is defined by

$$R_{ij}(\mathbf{r}) \equiv \overline{u_i(\mathbf{x}, t)u_j(\mathbf{x} + \mathbf{r}, t)}. \tag{8.1.1}$$

The correlation tensor is a function of the vector separation $\mathbf{r}$ only, providing the turbulence is homogeneous. The spectrum tensor ϕ_{ij}, which is the Fourier transform of R_{ij}, is given by

$$\begin{aligned} \phi_{ij}(\boldsymbol{\kappa}) &= \frac{1}{(2\pi)^3} \iiint_{-\infty}^{\infty} \exp(-i\boldsymbol{\kappa} \cdot \mathbf{r})\, R_{ij}(\mathbf{r})\, d\mathbf{r}, \\ R_{ij}(\mathbf{r}) &= \iiint_{-\infty}^{\infty} \exp(i\boldsymbol{\kappa} \cdot \mathbf{r})\, \phi_{ij}(\boldsymbol{\kappa})\, d\boldsymbol{\kappa}. \end{aligned} \tag{8.1.2}$$

Unlike the definition of the spectrum used in Section 6.4, the correlation here has not been normalized; the form (8.1.2) is customary in the literature. Of primary interest is the sum of the diagonal components of ϕ_{ij}, which is

$\phi_{ii} = \phi_{11} + \phi_{22} + \phi_{33}$, because it represents the kinetic energy at a given wave-number vector. This becomes clear by considering $R_{ii}(0)$, which is

$$R_{ii}(0) = \overline{u_i u_i} = 3u^2 = \iiint_{-\infty}^{\infty} \phi_{ii}(\boldsymbol{\kappa})\, d\boldsymbol{\kappa}. \tag{8.1.3}$$

The directional information in $\phi_{ii}(\boldsymbol{\kappa})$ is removed by integration over a spherical shell of radius κ (κ is the modulus of the vector $\boldsymbol{\kappa}$; that is, $\kappa^2 = \boldsymbol{\kappa}\cdot\boldsymbol{\kappa} = \kappa_i \kappa_i$). If we call the surface element of the shell $d\sigma$, we can write

$$E(\kappa) = \frac{1}{2} \oiint \phi_{ii}(\boldsymbol{\kappa})\, d\sigma. \tag{8.1.4}$$

The purpose of the factor $\frac{1}{2}$ is to make the integral of the three-dimensional spectrum $E(\kappa)$ equal to the kinetic energy per unit mass:

$$\int_0^\infty E(\kappa)\, d\kappa = \frac{1}{2} \int_0^\infty \left[\oiint \phi_{ii}(\boldsymbol{\kappa}) d\sigma \right] d\kappa = \frac{1}{2} \iiint_{-\infty}^{\infty} \phi_{ii}(\boldsymbol{\kappa})\, d\boldsymbol{\kappa}$$

$$= \tfrac{1}{2} \overline{u_i u_i} = \tfrac{3}{2} u^2. \tag{8.1.5}$$

Two common one-dimensional spectra The one-dimensional spectra that are most often measured are the one-dimensional Fourier transforms of $R_{11}(r,0,0)$ and $R_{22}(r,0,0)$. The geometry involved in measuring R_{11} and R_{22} is sketched in Figure 8.2; $R_{11}(r,0,0)$ is called a *longitudinal correlation*, $R_{22}(r,0,0)$ is called a *transverse correlation*. Correspondingly, F_{11} is called a *longitudinal spectrum* and F_{22} is called a *transverse spectrum*.

The one-dimensional spectra $F_{11}(\kappa_1)$ and $F_{22}(\kappa_1)$ are defined by

$$R_{11}(r, 0, 0) \equiv \int_{-\infty}^{\infty} \exp(i\kappa_1 r)\, F_{11}(\kappa_1)\, d\kappa_1, \tag{8.1.6}$$

$$R_{22}(r, 0, 0) \equiv \int_{-\infty}^{\infty} \exp(i\kappa_1 r)\, F_{22}(\kappa_1)\, d\kappa_1. \tag{8.1.7}$$

The relations between F_{11}, F_{22}, and E are quite complicated. This can be seen by considering the relation between R_{11} and ϕ_{ij}, which is

$$R_{11}(r, 0, 0) = \int_{-\infty}^{\infty} \exp(i\kappa_1 r) \left(\iint_{-\infty}^{\infty} \phi_{11}(\boldsymbol{\kappa})\, d\kappa_2\, d\kappa_3 \right) d\kappa_1. \tag{8.1.8}$$

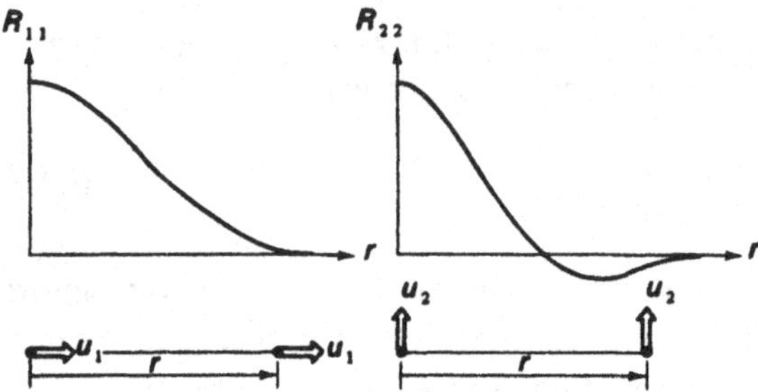

Figure 8.2. The longitudinal and transverse correlations.

Comparing (8.1.8) with (8.1.6), we find

$$F_{11}(\kappa_1) = \iint_{-\infty}^{\infty} \phi_{11}(\boldsymbol{\kappa})\, d\kappa_2\, d\kappa_3. \tag{8.1.9}$$

This demonstrates the aliasing problem. The integration is over a slice of wave-number space at a given value of κ_1, so that energy from high wave numbers which are not located near the κ_1 axis is aliased to κ_1.

The shapes of F_{11} and F_{22} are somewhat different. Measured values of R_{11} do not ordinarily go negative (though there is no reason why they should not); this means that F_{11} has a maximum at the origin. Because F_{11} is majorized by its value at the origin, it curves downward parabolically away from $\kappa_1 = 0$ (note that F_{11} is symmetric because R_{11} is real).

The transverse correlation R_{22}, however, does become negative for some values of r (Figure 8.2). It is of interest to see why this occurs. Consider a plane perpendicular to the x_2 direction. Across this plane there should be no net mass flux, because the mean value of u_2 is zero. Therefore, the integral of u_2 over the entire x_1, x_3 plane should be zero:

$$\iint_{-\infty}^{\infty} u_2(x_1, 0, x_3)\, dx_1\, dx_3 = 0. \tag{8.1.10}$$

If the integral is multiplied by u_2 at some given point, there results, after averaging,

$$\iint_{-\infty}^{\infty} R_{22}(r_1, 0, r_3)\, dr_1\, dr_3 = 0. \tag{8.1.11}$$

This means that R_{22} must go negative somewhere in the x_1, x_3 plane. This merely states that backflow is necessary somewhere in the plane in order to keep the net mass flux zero. If the turbulence is isotropic (meaning, as we saw in Chapter 3, that its statistics are invariant under reflections or rotations of the coordinate system), $R_{22}(r_1, 0, r_3)$ can be a function only of the distance $r = (r_1^2 + r_3^2)^{1/2}$. In this case, (8.1.11) becomes

$$\int_{-\infty}^{\infty} r R_{22}(r, 0, 0)\, dr = 0. \tag{8.1.12}$$

The transverse correlation thus must be negative somewhere. In Chapter 6 we found that the corresponding spectrum, F_{22}, is likely to have a peak away from the origin.

Experimental one-dimensional spectra are commonly obtained by moving a probe so rapidly through the turbulence that the velocity field does not change appreciably during the time of measurement. The probe sees a fluctuating velocity, which is a function of time; if the traversing speed U of the probe is large enough, the velocity signal $u(t)$ may be identified with $u(x/U)$. This approximation is known as *Taylor's hypothesis*; it is also referred to as the frozen-turbulence approximation. The substitution $t = x/U$ is a good approximation only if $u/U \ll 1$ (Hinze, 1959, Sec. 1.8; Lumley, 1965). This is an important constraint in the design of turbulence experiments.

Isotropic relations In general, the relations between F_{11}, F_{22}, and E are quite complicated. This is unfortunate; it seems natural to base physical reasoning on $E(\kappa)$, but most measurements give one-dimensional spectra like F_{11} and F_{22}. If the turbulence is isotropic, however, the relations between F_{11}, F_{22}, and E are fairly simple. The derivation of these isotropic relations is beyond the scope of this book. Two of the most useful relations are (Batchelor, 1953; Hinze, 1959)

$$E(\kappa) = \kappa^3 \frac{d}{d\kappa}\left(\frac{1}{\kappa}\frac{dF_{11}}{d\kappa}\right), \tag{8.1.13}$$

$$\frac{d}{d\kappa_1} F_{22}(\kappa_1) = -\frac{\kappa_1}{2}\frac{d^2}{d\kappa_1^2} F_{11}(\kappa_1). \tag{8.1.14}$$

The first of these is often used to obtain E from measured values of F_{11} at high wave numbers. This procedure is legitimate because turbulence is very nearly isotropic at high wave numbers.

According to (8.1.13) and (8.1.14), $F_{11} \propto \kappa_1^n$ and $F_{22} \propto \kappa_1^n$ if $E \propto \kappa^n$. The exponent in the power law is the same, but the coefficients are different. We shall find shortly that in a major part of the spectrum $E \propto \kappa^{-5/3}$; it is encouraging to know that F_{11} and F_{22} exhibit the same power law. If all spectra are proportional to $\kappa^{-5/3}$, (8.1.14) gives $F_{22} = \frac{4}{3}F_{11}$. This relation is often used to examine turbulence for evidence of isotropy.

The isotropic relations also give some indication of the shapes of F_{11}, F_{22}, and E near the origin. Because R_{11} is real and positive (no experiments are known in which $R_{11} < 0$ anywhere), F_{11} is symmetric and majorized by its value at the origin:

$$F_{11}(\kappa) = A_1 - B\kappa_1^2 + C\kappa_1^4 + \ldots. \tag{8.1.15}$$

Substituting this into (8.1.13, 8.1.14), we obtain

$$E(\kappa) = 8C\kappa^4 + \ldots, \tag{8.1.16}$$

$$F_{22}(\kappa_1) = A_2 + \tfrac{1}{2}B\kappa_1^2 + \ldots. \tag{8.1.17}$$

Thus, E begins from zero, quartically upward, while F_{22} curves upward parabolically, so that it has a peak away from the origin. The quartic behavior of $E(\kappa)$ deserves special attention. Physically, the point is that there is no energy at zero wave number, so that $\phi_{ii}(0) = 0$. Because ϕ_{ii} is symmetric, it begins parabolically ($\propto \kappa^2$). Now, $E(\kappa)$ is an integral of ϕ_{ii} over a spherical shell whose area is proportional to κ^2, so that $E(\kappa)$ must be proportional to κ^4 near the origin. A much more careful analysis is needed to show that this result is not restricted to isotropic turbulence. Also, the coefficient in the parabolic form of ϕ_{ii} near the origin is not the same in all directions, but that has no effect on the behavior of E (Lumley, 1970). It should be kept in mind, however, that the large-scale structure of turbulence is unlikely to be isotropic, so that (8.1.13) and (8.1.14) should not be used to obtain quantitative results at small wave numbers.

Spectra of isotropic simple waves We may get an impression of the shapes of one- and three-dimensional spectra by examining a rather artificial case. Consider an isotropic field of waves that all have the same wavelength $2\pi/\kappa_*$, but whose wave-number vectors have random directions. For this isotropic field, ϕ_{ii} is zero, except on a shell of radius κ_*, where it has a uniform distribution. Therefore, $E(\kappa)$ is zero everywhere, except for a spike at $\kappa = \kappa_*$. The shape of F_{11} can be computed from (8.1.9). If the plane of integration is beyond κ_*, F_{11} is zero; if $\kappa_1 \leq \kappa_*$, the integration yields

$$F_{11}(\kappa_1) = \frac{A}{2\kappa_*^3}(\kappa_*^2 - \kappa_1^2). \qquad (8.1.18)$$

Here, A is an arbitrary constant, related to the area under the spike in $E(\kappa)$. Substitution of (8.1.18) into (8.1.14) gives an expression for F_{22}; again, $F_{22} = 0$ for $\kappa_1 > \kappa_*$, while for $\kappa_1 \leqslant \kappa_*$

$$F_{22}(\kappa_1) = \frac{A}{4\kappa_*^2}(\kappa_*^2 + \kappa_1^2). \qquad (8.1.19)$$

These spectra are shown in Figure 8.3a. It should be noted that F_{11} curves parabolically downward, while F_{22} curves upward.

In a general isotropic field, $E(\kappa)$ can be thought of as being made up from spikes of different amplitudes at different wave numbers, so that F_{11} and

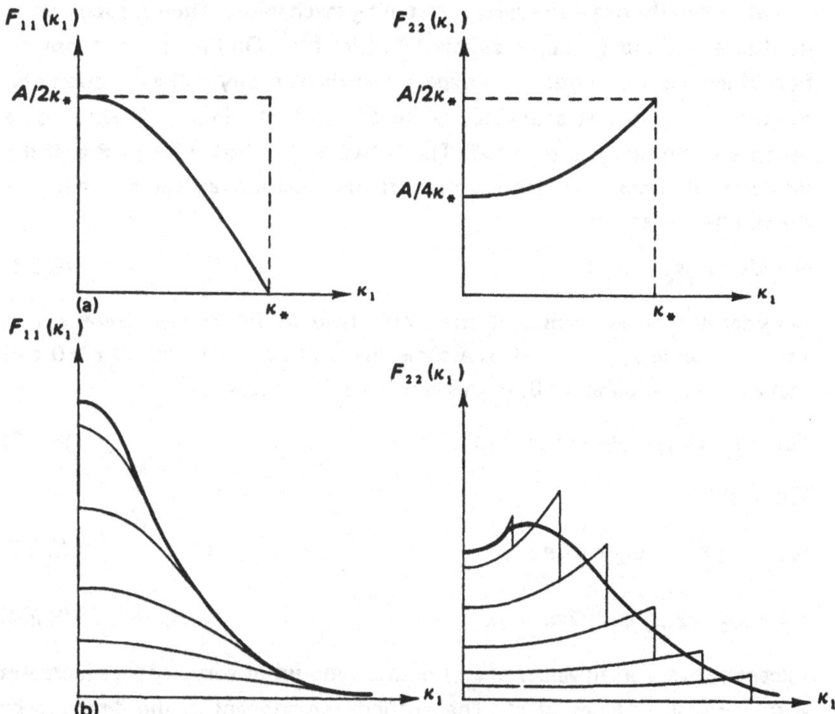

Figure 8.3. Longitudinal and transverse spectra of fields of isotropic simple waves: (a) spectra for a field of simple waves of wave number κ_*, (b) composite spectra for a field of waves with different wave numbers (adapted from Corrsin, 1959).

F_{22} can be constructed by adding many spectra of the type sketched in Figure 8.3a. It is evident that the longitudinal spectrum is likely to be a monotone decreasing function, while the transverse spectrum is likely to rise at first before it decreases (Figure 8.3b). Of course, both spectra should go to zero as $\kappa_1 \to \infty$, because the total area under the curves is proportional to the kinetic energy.

8.2 The energy cascade

The existence of energy transfer from large eddies to small eddies, driven by vortex stretching and leading to viscous dissipation of energy near the Kolmogorov microscale, was demonstrated in Chapter 3. Here, we discuss how the energy exchange takes place.

Let us briefly recall the vortex-stretching mechanism. When vorticity finds itself in a strain-rate field, it is subject to stretching. On the basis of conservation of angular momentum, we expect that the vorticity in the direction of a positive strain rate is amplified, while the vorticity in the direction of a negative strain rate is attenuated. This effect is sketched in Figure 8.4. If the influence of viscosity is ignored, the vorticity equation reads (recall that s_{ij} is the strain-rate tensor)

$$d\omega_i/dt = \omega_j s_{ij}. \tag{8.2.1}$$

Consider the two-dimensional strain-rate field in Figure 8.4. Here, $s_{11} = -s_{22} = s$, while $s_{12} = 0$. Let us assume that s is a constant for all $t > 0$ and that $\omega_1 = \omega_2 = \omega_0$ at $t = 0$. In this case, (8.2.1) reduces to

$$d\omega_1/dt = s\omega_1, \quad d\omega_2/dt = -s\omega_2. \tag{8.2.2}$$

This yields

$$\omega_1 = \omega_0 e^{st}, \quad \omega_2 = \omega_0 e^{-st}, \tag{8.2.3}$$

$$\omega_1^2 + \omega_2^2 = 2\omega_0^2 \cosh 2st. \tag{8.2.4}$$

Except for very small values of st, the total amount of vorticity thus increases with increasing values of st. The vorticity component in the direction of stretching increases rapidly, while the vorticity component in the direction of compression (shrinking) decreases slowly at large st. This is similar to the rate of growth of a spot of contaminant (Section 7.3); of course, the same stretch-

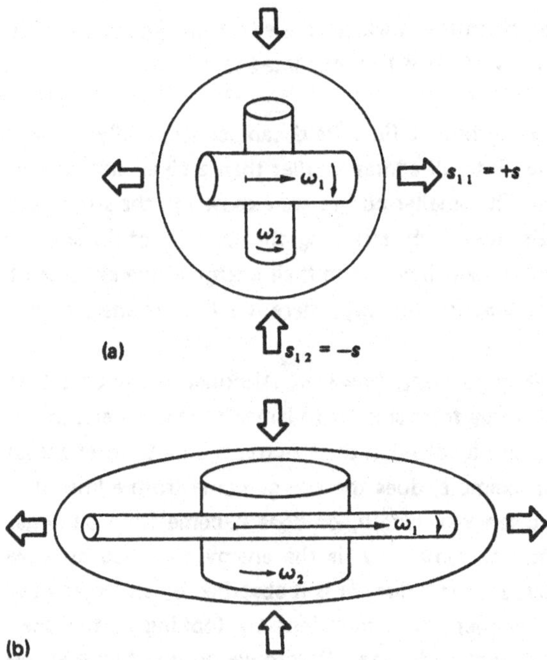

Figure 8.4. Vorticity stretching in a strain-rate field: (a) before stretching, (b) after stretching.

ing mechanism is involved. Note, however, that viscous effects are not accounted for in (8.2.1, 8.2.2).

Vortex stretching involves an exchange of energy, because the strain rate performs deformation work on the vortices that are being stretched. We learned in Chapter 3 that the amount of energy gained by a disturbance with velocity components u_i, u_j in a strain rate s_{ij} is equal to $-u_i u_j s_{ij}$ per unit mass and time. In the plane strain-rate field of Figure 8.4, the energy exchange rate is

$$T = s(u_2^2 - u_1^2). \tag{8.2.5}$$

Now, the vorticity component ω_1 is increased, which corresponds to an increase in u_2 and u_3; also, ω_2 is decreased, which corresponds to a decrease in u_1 and u_3. We thus expect that u_2^2 increases and u_1^2 decreases, while u_3^2 increases fairly slowly. Hence, the difference $u_2^2 - u_1^2$, although starting from zero at $t = 0$, becomes positive. This means that T also becomes positive, so that the strain rate indeed performs work on the eddies in Figure 8.4. The total amount of energy in the vortices is thus expected to increase.

Spectral energy transfer A turbulent flow field can (conceptually, at any rate) be imagined as divided into all eddies smaller than a given size and all eddies larger than that size. The smaller eddies are exposed to the strain-rate field of the larger eddies. Because of the straining, the vorticity of the smaller eddies increases, with a consequent increase in their energy at the expense of the energy of the larger eddies. In this way, there is a flux of energy from larger to smaller eddies.

The situation is not yet quite clear, however. Although we expect that there will be a net flux of energy from smaller to larger wave numbers, we do not know which eddy sizes are involved in the spectral energy transfer across a given wave number. For example, does the energy come from eddies that are slightly larger than a given wavelength, or does it come from all larger eddies indiscriminately? In the same way, is the energy absorbed at wave numbers slightly larger than a given value, or is it absorbed by all larger wave numbers? We attempt to answer these questions by looking at the characteristic strain rates of different eddy sizes. Before we do this, however, we need a better mental picture of the concept of an eddy.

A simple eddy Let us recall that an autocorrelation and the corresponding spectrum are a Fourier-transform pair. If the correlation is a function of spatial separation, the spectrum is a function of wave number. A certain eddy size, say ℓ, is thus associated with a certain wave number, say κ. An "eddy" of wave number κ may be thought of as some disturbance containing energy in the vicinity of κ. It would be tempting to think of an eddy as a disturbance contributing a narrow spike to the spectrum at κ. However, a narrow spike in the spectrum creates slowly damped oscillations (of wavelength $2\pi/\kappa$) in the correlation, as we discovered in Section 6.2. Such a correlation is characteristic of wavelike disturbances, but not of eddies; we expect eddies to lose their identity because of interactions with others within one or two periods or wavelengths. Therefore, the contribution of an eddy to the spectrum should be a fairly broad spike, wide enough to avoid oscillatory behavior ("ringing") in the correlation.

It is convenient to define an eddy of wave number κ as a disturbance containing energy between, say 0.62κ and 1.62κ. This choice centers the energy around κ on a logarithmic scale, because $\ln(1.62) = \ln(1/0.62) \cong \frac{1}{2}$; it also makes the width of the contribution to the spectrum equal to κ (Figure 8.5). We recall from Section 6.2 that the transform of a narrow band around κ is a wave of wavelength $2\pi/\kappa$, with an envelope whose width is the inverse of the bandwidth. Now, because the bandwidth selected is κ, the width of the envelope of the eddy is of order $1/\kappa$. This is sketched in Figure 8.5; we see that an eddy defined this way is indeed the relatively compact disturbance we want it to be. The eddy size ℓ is roughly equal to $2\pi/\kappa$.

The schematic eddy presented in Figure 8.5 suffices for the development of energy-cascade concepts. This model, however, cannot deal with all of the problems associated with the distinction between waves and eddies. The Fourier transform of a velocity field is a decomposition into waves of different wavelengths; each wave is associated with a single Fourier coefficient. An eddy, however, is associated with many Fourier coefficients and the phase relations among them. Fourier transforms are used because they are convenient (spectra can be measured easily); more sophisticated transforms are needed if one wants to decompose a velocity field into eddies instead of waves (Lumley, 1970).

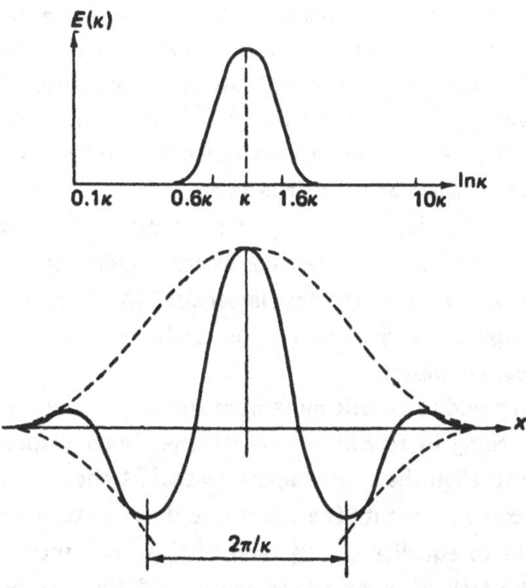

Figure 8.5. An eddy of wave number κ and wavelength $2\pi/\kappa$.

The energy cascade Let us return to the role of the strain rates of different eddy sizes in the spectral energy exchange mechanism. The energy of all eddies of size $2\pi/\kappa$ is roughly proportional to $E(\kappa)$ times the width of the eddy spectrum, which is κ. Hence, a characteristic velocity is given by $[\kappa E(\kappa)]^{1/2}$. The size of the eddy is about $2\pi/\kappa$, so that the characteristic strain rate (and the characteristic vorticity) of an eddy of wave number κ is given by

$$s(\kappa) = \frac{(\kappa E)^{1/2}}{2\pi/\kappa} = \frac{(\kappa^3 E)^{1/2}}{2\pi}. \tag{8.2.6}$$

We recall from Section 3.2 that the strain rate of large eddies, which contain most of the energy, is of order u/ℓ (ℓ is an integral scale), while the small-scale strain-rate fluctuations are of order u/λ (λ is a Taylor microscale). Therefore, we should expect that the strain rate $s(\kappa)$ increases with wave number. We find in the next section that $E(\kappa) \propto \kappa^{-5/3}$ in the central part of the spectrum; this gives $s(\kappa) \propto \kappa^{2/3}$, so that $s(\kappa)$ indeed increases with κ. We shall use this result for convenience.

The energy spectrum is continuous; for the purposes of this discussion, however, we may think of the spectrum as being made up from eddies of discrete sizes. The strain rate imposed on eddies of wave number κ due to the eddies of the next larger size (which extends from 0.24κ to 0.62κ, centered around 0.38κ) is $s(0.38\kappa)$, which is about $\frac{1}{2}s(\kappa)$ if $s(\kappa) \propto \kappa^{2/3}$. The strain rate due to eddies two sizes larger than $2\pi/\kappa$ (whose energy extends from 0.09κ to 0.24κ, centered around 0.15κ) is again about half as large. Adding all of the strain rates of eddies larger than $2\pi/\kappa$, we conclude that of the total strain rate felt by an eddy of wave number κ, one-half comes from eddies of the next larger size, and another quarter from the next larger size. Therefore, we expect that most of the energy crossing a given wave number comes from eddies with slightly smaller wave numbers.

The question now is which eddies benefit most from the energy transfer across wave number κ. According to (8.2.5), energy transfer depends upon the ability of the strain rate to align the smaller eddies so that u_2^2 and u_1^2 (of the eddies in Figure 8.4) become different. The strain rate thus has to overcome the tendency of eddies to equalize u_1^2, u_2^2, and u_3^2. This tendency is called *return to isotropy*; the lack of isotropy (or anisotropy) that can be generated by the strain rate depends on the time scale for return to isotropy relative to the time scale of the straining motion. Because the strain rate has dimensions of time^{-1}, the time scale of return to isotropy is roughly $1/s(\kappa)$ for eddies of wave number κ. This means that those eddies would return to isotropy in a time of order $1/s(\kappa)$ if the strain rate were removed. Because smaller eddies have larger strain rates, small eddies return to isotropy rapidly.

If $\mathscr{S}$ is the combined strain rate of all eddies with wave numbers below κ, the time scale of the applied strain rate is of order $1/\mathscr{S}$. If $\mathscr{S}$ is large compared to $s(\kappa)$, the anisotropy is large; if $\mathscr{S}$ is small compared to $s(\kappa)$, the relatively rapid return to isotropy prevents the creation of a large anisotropy. It appears reasonable to assume that the degree of anisotropy is proportional to $\mathscr{S}/s(\kappa)$. The energy transferred from all larger eddies to an eddy of wave number κ is then approximately $\mathscr{S}^2 \kappa E(\kappa)/s(\kappa)$, by virtue of (8.2.5). The energy absorbed by eddies of the next smaller size (with energy between 1.6κ and 4.2κ, centered around 2.6κ) is about $\frac{1}{4}\mathscr{S}^2 \kappa E(\kappa)/s(\kappa)$, because $s(2.6\kappa) \cong 2s(\kappa)$ and $2.6\kappa E(2.6\kappa) \cong \frac{1}{2}\kappa E(\kappa)$ if $E(\kappa) \propto \kappa^{-5/3}$. Eddies of wave number κ thus receive about two-thirds of the total energy transfer, those of the next smaller size receive about one-sixth, and all smaller eddies combined also receive about one-sixth.

A crude picture is beginning to emerge. Most of the energy that is exchanged across a given wave number apparently comes from the next larger eddies and goes to the next smaller eddies. It seems fair to describe the energy transfer as a cascade, much like a series of waterfalls, each one filling a pool that overflows into the next one below. This concept proves to be exceptionally useful, because the largest eddies and the smallest eddies clearly have no direct effect on the energy transfer at intermediate wave numbers. However, we should not expect too much from the cascade model. After all, it is a very leaky cascade if half the water crossing a given level comes directly from all other pools uphill.

In the development of the cascade model, a number of crude assumptions have been made, some of which are not likely to be valid throughout the spectrum. One major assumption is clearly not valid at very small scales. The time scale of an eddy has been estimated as $1/s(\kappa)$; however, there is a viscous lower limit on time scales, as we saw in Chapters 1 and 3. The smallest time scale is $(\nu/\epsilon)^{1/2}$ and the strain rate of very small eddies is of order $(\epsilon/\nu)^{1/2}$, so that the model developed here is not valid if $s(\kappa)$ and $(\epsilon/\nu)^{1/2}$ become of the same order of magnitude. The cascade model is inviscid; it should be applied only to eddy sizes whose Reynolds number $s(\kappa)/\kappa^2\nu$ is large.

8.3
The spectrum of turbulence

We have found that the anisotropy of eddies depends on the time-scale ratio $\mathscr{S}/s(\kappa)$. The strain rate of the large, or "energy-containing" eddies is comparable to the strain rate of the mean flow (recall that $\partial U/\partial y \sim u/\ell$). Therefore, large eddies have a steady anisotropy due to the strain rate of the mean flow, which maintains a steady orientation. On the other hand, the strain rate of small eddies is large compared to that of the mean flow and of the large eddies (recall that $s_{ij} \sim u/\lambda$), so that no permanent anisotropy can be induced at small scales. This does not mean that small eddies are isotropic, because energy transfer is possible only if eddies are aligned with a strain rate. However, the anisotropy discussed in the preceding section is temporary; eddies of a given size are stretched mainly by somewhat larger eddies, whose strain-rate field is constantly shifting in magnitude and direction. As the eddy size becomes smaller,the permanent anisotropy decreases,so that at small scales the strain-rate field itself may be expected to be isotropic in the mean. In other words, turbulence is increasingly "scrambled" at small scales, and any permanent sense of direction is lost. This concept is called *local isotropy*; it was proposed by A. N. Kolmogorov in 1941 (see Friedlander and Topper, 1962). The adjective "local" refers to small scales (large wave numbers).

Local isotropy does not exist if the Reynolds number is not large enough. The strain rate of the mean flow is of order u/ℓ, the strain rate of the smallest eddies is of order $u/\lambda \sim (\epsilon/\nu)^{1/2}$. We probably need $u/\lambda \geqslant 10u/\ell$ in order to have local isotropy at the smallest scales. Consequently, $\ell/\lambda \sim R_\ell^{1/2}$ (3.2.17) needs to be at least 10, giving a Reynolds number of at least 100.

In the part of the spectrum in which local isotropy prevails, time scales are short compared to those of the mean flow. This means that small eddies respond quickly to changing conditions in the mean flow. Therefore, small eddies always are in approximate equilibrium with local conditions in the mean flow, even though the latter may be evolving. For this reason, the range of wave numbers exhibiting local isotropy is called the *equilibrium range*. It begins at a wave number where $s(\kappa)$ first becomes large compared to the mean strain rate, and it includes all higher wave numbers.

The spectrum in the equilibrium range In the equilibrium range, time scales are so short that the details of the energy transfer between the mean flow and the turbulence (which occurs mainly at large scales) cannot be important.

However, the amount of energy cascading down the spectrum should be a major parameter. Because all energy is finally dissipated by viscosity, the total amount of energy transfer must be equal to the dissipation rate ϵ, and the second major parameter should be the viscosity itself. If no other parameters are involved, we have $E = E(\kappa, \epsilon, \nu)$, which can have only one nondimensional form:

$$\frac{E(\kappa)}{\nu^{5/4}\epsilon^{1/4}} = \frac{E(\kappa)}{v^2\eta} = f(\kappa\eta). \tag{8.3.1}$$

This scaling law was derived by Kolmogorov; as before, $\eta = (\nu^3/\epsilon)^{1/4}$ is the Kolmogorov microscale and $v = (\nu\epsilon)^{1/4}$ is the Kolmogorov velocity. The Kolmogorov spectrum (8.3.1) is supported by a large amount of experimental data. Because the turbulence in the equilibrium range is isotropic, the isotropic relations (8.1.13, 8.1.14) may be used to compute E and F_{22} from measured F_{11}.

The similarity between (8.3.1) and the law of the wall in turbulent boundary layers (Chapter 5) is striking. Close to a rigid wall, the momentum flux is ρu_*^2; if u_*^2 and ν are the only parameters, $U = U(y, u_*^2, \nu)$, so that $U/u_* = f(yu_*/\nu)$. In boundary layers, the spatial momentum flux is involved; in the spectrum, it is the spectral energy flux.

Most of the viscous dissipation of energy occurs near the Kolmogorov microscale η, as we discussed in Chapter 3. The equilibrium range thus includes the *dissipation range* of wave numbers, much like the wall layer of a boundary layer includes the viscous sublayer (Section 5.2). It can be shown that the spectrum of the dissipation, $D(\kappa)$, is given by (Batchelor, 1953; Hinze, 1959)

$$D(\kappa) = 2\nu\kappa^2 E(\kappa). \tag{8.3.2}$$

The dissipation is proportional to the square of velocity gradients; the factor κ^2 in (8.3.2) arises because differentiation corresponds to multiplication by wave number. The dissipation rate ϵ is given by

$$\epsilon = 2\nu\overline{s_{ij}s_{ij}} = \int_0^\infty D(\kappa)\, d\kappa = 2\nu \int_0^\infty \kappa^2 E\, d\kappa. \tag{8.3.3}$$

If most of the dissipation occurs within the equilibrium range, we obtain, with (8.3.1),

$$\int_0^\infty (\kappa\eta)^2 f(\kappa\eta)\, d(\kappa\eta) = \tfrac{1}{2}. \tag{8.3.4}$$

The value of ϵ often is determined by integrating (8.3.3) with a measured energy spectrum.

The large-scale spectrum For small wave numbers, the spectrum scales in a different way. If the spectral Reynolds number $s(\kappa)/\kappa^2\nu$ is large, we do not expect viscosity to be relevant. The principal parameters are those that describe the energy transfer from the mean flow to the turbulence and the energy transfer from large to small scales. The turbulence receives its energy from the mean strain rate S and transfers energy to small scales at a rate ϵ, so that the scaling of the large-scale part of the spectrum should be based on S and ϵ. If these are the only relevant parameters, we must have $E = E(\kappa,\epsilon,S)$. For convenience, we define S and ℓ by the relations $S \equiv u/\ell$ and $\epsilon \equiv u^3/\ell$. The spectrum then becomes

$$\frac{E(\kappa)}{\epsilon^{3/2}S^{-5/2}} = \frac{E(\kappa)}{u^2\ell} = F(\kappa\ell). \tag{8.3.5}$$

This relation, of course, is not universal, but differs in flows with different geometries. In a family of flows with the same geometry, however, we expect the large-scale part of the spectrum to scale like (8.3.5).

The inertial subrange The Kolmogorov spectrum (8.3.1) is related to a limit process in which $s(\kappa)/S \to \infty$. Evaluating $s(\kappa)$ with (8.2.6) and (8.3.1), we find that this limit corresponds to

$$\frac{s(\kappa)}{S} = \frac{R_\ell^{1/2}}{2\pi}\,[(\kappa\eta)^3\, f(\kappa\eta)]^{1/2} \to \infty\,. \tag{8.3.6}$$

Here we used $S = u/\ell$ and $R_\ell = u\ell/\nu$. It is clear that the Kolmogorov spectrum is valid for $\kappa\eta = \mathcal{O}(1)$ in the limit as $R_\ell \to \infty$. On the other hand, the large-scale spectrum (8.3.5) applies to wave numbers for which $s(\kappa)/\kappa^2\nu \to \infty$. With (8.2.6) and (8.3.5), this limit becomes

$$\frac{s(\kappa)}{\kappa^2\nu} = \frac{R_\ell}{2\pi}\,[(\kappa\ell)^{-1}\, F(\kappa\ell)]^{1/2} \to \infty. \tag{8.3.7}$$

This implies that (8.3.5) is valid for $\kappa\ell = \mathcal{O}(1)$ and $R_\ell \to \infty$. We thus have viscous scaling at high wave numbers and inertial scaling at low wave numbers, both valid in the limit as $R_\ell \to \infty$. This is similar to the scaling laws for channel flow (Section 5.2), where we used an inviscid description for

$y/h = \mathcal{O}(1)$ and a viscous description for $yu_*/\nu = \mathcal{O}(1)$, both in the limit as $R_* \to \infty$. We found that those scaling laws had a common region of validity; perhaps we can do the same here.

The existence of a region of "overlap" depends on the possibility of taking the limits $\kappa\eta \to 0$ and $\kappa\ell \to \infty$ simultaneously. In other words, we should be able to go to the small-scale end of the large-scale spectrum and to the large-scale end of the Kolmogorov spectrum simultaneously, without violating the condition $R_\ell \to \infty$ required by (8.3.6, 8.3.7). Take $\kappa\ell = R_\ell^n$ $(n > 0)$ and recall that $\ell/\eta \sim R_\ell^{3/4}$ (1.5.14). We obtain

$$\kappa\eta = \kappa\ell(\eta/\ell) \sim \kappa\ell R_\ell^{-3/4} = R_\ell^{n-3/4}, \tag{8.3.8}$$

so that we need $0 < n < 3/4$ in order to obtain $\kappa\eta \to 0$. Because we do not know how $f(\kappa\eta)$ and $F(\kappa\ell)$ vary, we cannot tell if (8.3.6) and (8.3.7) will indeed be satisfied. We assume that they are, though, and verify the conditions after the matching has been performed.

With $0 < n < 3/4$, it is possible to have $\kappa\ell \to \infty$ and $\kappa\eta \to 0$ simultaneously, so that we expect that (8.3.1) and (8.3.5) can be matched. Equating the two and using $\kappa\ell = R_\ell^n$, $\kappa\eta \sim R_\ell^{n-3/4}$, we obtain

$$u^2\ell F(R_\ell^n) = v^2\eta f(R_\ell^{n-3/4}), \tag{8.3.9}$$

which becomes

$$R_\ell^{5/4} F(R_\ell^n) = f(R_\ell^{n-3/4}). \tag{8.3.10}$$

This has to be satisfied for any n in the interval between 0 and 3/4. The solution of (8.3.10) is

$$F(\kappa\ell) = \alpha(\kappa\ell)^{-5/3}, \quad f(\kappa\eta) = \alpha(\kappa\eta)^{-5/3}. \tag{8.3.11}$$

In the literature, this spectrum is often presented in nonnormalized form. Substitution of (8.3.11) into (8.3.1) or (8.3.5) gives

$$E(\kappa) = \alpha\epsilon^{2/3}\kappa^{-5/3}. \tag{8.3.12}$$

This expression is valid for $\kappa\ell \to \infty$, $\kappa\eta \to 0$, $R_\ell \to \infty$. Experimental data indicate that $\alpha = 1.5$ approximately. The range of wave numbers for which (8.3.12) is valid is called the *inertial subrange*; it is the spectral equivalent of the inertial sublayer in boundary layers. In the inertial subrange, the one-dimensional spectra F_{11} and F_{22} are also proportional to $\epsilon^{2/3}\kappa_1^{-5/3}$.

With (8.3.11), the conditions (8.3.6, 8.3.7) become

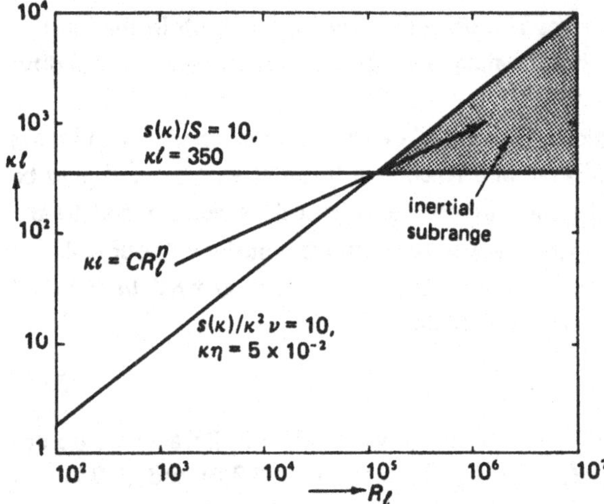

Figure 8.6. The inertial subrange.

$$\frac{s(\kappa)}{S} = \frac{n^{1/2}}{2\pi} R_\ell^{2n/3} \to \infty, \tag{8.3.13}$$

$$\frac{s(\kappa)}{\kappa^2 \nu} = \frac{n^{1/2}}{2\pi} R_\ell^{1-4n/3} \to \infty, \tag{8.3.14}$$

so that these are indeed satisfied if $R_\ell \to \infty$ and $0 < n < 3/4$.

Recall that the mean velocity profile in the inertial sublayer can be obtained simply by postulating that $\partial U/\partial y$ is a function of u_* and y only. As we found in Section 5.2, this gives $\partial U/\partial y = u_*/\kappa y$. In a similar way, (8.3.12) can be obtained simply by postulating that $E = E(\epsilon,\kappa)$ for $1/\ell \ll \kappa \ll 1/\eta$. The point of obtaining (8.3.12) with such care is to delineate the conditions of its validity. A graphical representation of these conditions is given in Figure 8.6. The horizontal line corresponds to $s(\kappa)/S = 10$, which makes the eddies of that size marginally independent of the mean strain rate S and therefore of the turbulence-production process ($\mathscr{P} = -\overline{u_i u_j} S_{ij}$). The line with a slope of $\frac{3}{4}$ corresponds to $s(\kappa)/\kappa^2\nu = 10$, which should make eddies of that size approximately independent of viscosity. It is clear that no inertial subrange exists unless the Reynolds number is quite large. With the conditions used previously, the Reynolds number needs to be at least 10^5; if the conditions are relaxed to $s(\kappa)/S > 3$, $s(\kappa)/\kappa^2\nu > 3$, the Reynolds number needs to be larger than 4×10^3. This is still a rather stringent condition. We conclude that it is unlikely that we would encounter an inertial subrange in laboratory flows;

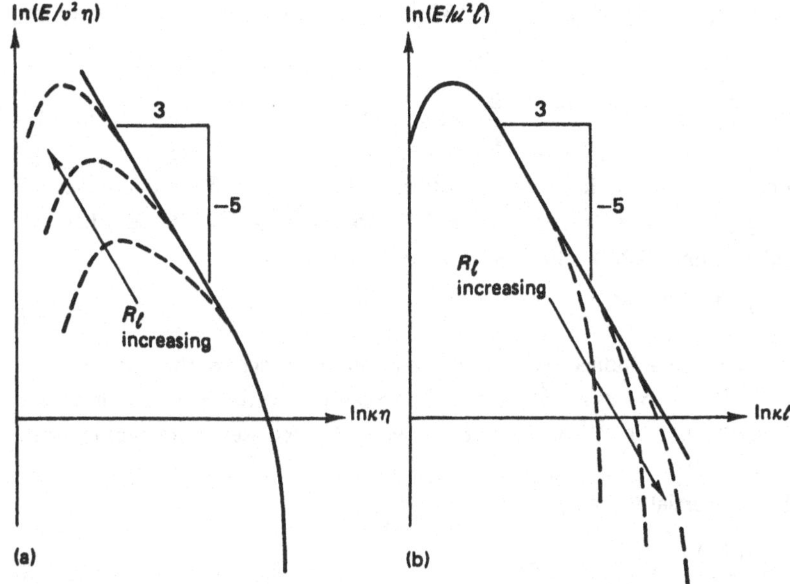

Figure 8.7. The spectrum of turbulence at different Reynolds numbers: (a) small-scale normalization, (b) large-scale normalization.

however, it is frequently observed in geophysical flows. The emergence of an inertial subrange in the spectrum of turbulence with increasing Reynolds numbers is sketched in Figure 8.7.

8.4
The effects of production and dissipation

In the inertial subrange, no energy is added by the mean flow and no energy is taken out by viscous dissipation, so that the energy flux T across each wave number is constant. In other words, the central part of the energy cascade is conservative, much like a cascade of waterfalls without any springs or drains. Because the total amount of energy dissipated per unit mass and time is ϵ, the spectral energy flux T in the inertial subrange is equal to ϵ.

In this section, we want to get a qualitative impression of how $E(\kappa)$ behaves near the ends of the inertial subrange. Recall that eddies of wave number κ get their energy, $\kappa E(\kappa)$, mainly from eddies of wave number 0.38κ, whose strain rate is $s(0.38\kappa) \cong \frac{1}{2}s(\kappa)$. The anisotropy produced by the larger eddies is proportional to $s(0.38\kappa)/s(\kappa)$, so that the energy flux T may be represented by

$$T(\kappa) = \frac{8\pi}{\alpha^{3/2}}\,\kappa\, E(\kappa)\,\frac{s^2(0.38\kappa)}{s(\kappa)} = \frac{2\pi}{\alpha^{3/2}}\,\kappa E(\kappa) s(\kappa). \tag{8.4.1}$$

The numerical factor has been chosen in such a way that $T = \epsilon$ in the inertial subrange, with $s(\kappa) = (\kappa^3 E)^{1/2}/2\pi$ (8.2.6) and $E(\kappa) = \alpha\epsilon^{2/3}\kappa^{-5/3}$ (8.3.12).

Substituting (8.2.6) into (8.4.1), we may write T as

$$T(\kappa) = \alpha^{-3/2} E^{3/2} \kappa^{5/2}. \tag{8.4.2}$$

This gives some indication of the variation of T across the spectrum. If $E \propto \kappa^{n-5/3}$, then $T \propto \kappa^{3n/2}$, so that T increases when the spectrum decreases less rapidly than $\kappa^{-5/3}$ and decreases when E decreases more rapidly than $\kappa^{-5/3}$.

In the inertial subrange, $s(\kappa)$ is given by

$$s(\kappa) = \frac{(\kappa^3 E)^{1/2}}{2\pi} = \frac{\alpha^{1/2}}{2\pi}\,\epsilon^{1/3}\kappa^{2/3}. \tag{8.4.3}$$

It turns out to be convenient to use the right-hand side of (8.4.3) not only inside the inertial subrange but also beyond its edges. Substituting the right-hand side of (8.4.3) into (8.4.1), we obtain

$$T(\kappa) = \alpha^{-1}\epsilon^{1/3}\kappa^{5/3} E(\kappa). \tag{8.4.4}$$

This estimate of $T(\kappa)$, however crude it may be, is attractive because it represents the spectral flux at some wave number as a local flux, determined only by the value of E and the inertial time scale $\epsilon^{-1/3}\kappa^{-2/3}$ at that wave number, and because it makes the relation between T and E a linear one. Of course, (8.4.2) is also a local estimate for $T(\kappa)$; however, it is not linear in E, so that it produces poorly behaved spectra.

Before we use (8.4.4) to calculate spectra outside the inertial subrange, let us take a close look at the assumptions underlying (8.4.2) and (8.4.4). In the inertial subrange, $T = \epsilon = \alpha^{-3/2} E^{3/2} \kappa^{5/2}$; comparing this with (8.4.2), we see that we are relaxing the condition $T = \epsilon$. Clearly, (8.4.2) assumes that T is proportional to $E^{3/2}\kappa^{5/2}$ even if $T \neq \epsilon$. In other words, we use inertial scaling for T even though we are outside the inertial subrange. This approximation can be justified only if the effects of viscosity and those of the mean strain rate are small, so that the difference between T and ϵ is small. Equation

(8.4.4) is even cruder, because it relaxes the condition $T = \epsilon$ but retains the inertial-subrange expression for the strain rate.

Both (8.4.2) and (8.4.4) may be thought of as spectral interpretations of mixing-length theory. If $-\overline{uv} \sim u\ell\, \partial U/\partial y$, the production of turbulent energy is proportional to $u\ell(\partial U/\partial y)^2$. Making the substitutions $u \to (\kappa E)^{1/2}$, $\ell \to 1/\kappa$, and $(\partial U/\partial y)^2 \to \kappa^2(\kappa E)$, we obtain (8.4.2); of course, turbulence production is now interpreted as transfer of energy from larger eddies to smaller eddies rather than transfer from the mean flow to all eddies. In a similar way, if we use $\epsilon \sim u^3/\ell$ to substitute for u in $u\ell(\partial U/\partial y)^2$, we obtain $\epsilon^{1/3}\ell^{4/3}(\partial U/\partial y)^2$. With $\ell \to 1/\kappa$ and $(\partial U/\partial y)^2 \to \kappa^2(\kappa E)$, this produces (8.4.4). Realizing how crude mixing-length theory is, we should not be too concerned about the relative merits of (8.4.2) and (8.4.4). It should be kept in mind that these spectral mixing-length models, like their spatial counterparts, can be used only in situations with a single length scale and a single time scale.

The effect of dissipation The viscous dissipation in a wave-number interval $d\kappa$ is equal to $2\nu\kappa^2 E(\kappa)\,d\kappa$, as we found in (8.3.2). This loss of energy is taken out of the energy flux $T(\kappa)$, so that we must have

$$dT/d\kappa = -2\nu\kappa^2 E. \tag{8.4.5}$$

If we substitute for $T(\kappa)$ with (8.4.4) and integrate the resulting equation, we obtain

$$E(\kappa) = \alpha\epsilon^{2/3}\kappa^{-5/3} \exp[-\tfrac{3}{2}\alpha(\kappa\eta)^{4/3}]. \tag{8.4.6}$$

This result, first given by Corrsin (1964) and later by Pao (1965), agrees very well with experimental data up to the largest values of $\kappa\eta$ that have been measured. Because virtually no data are available beyond $\kappa\eta = 1$, this is not a very severe test. In fact, the use of $s(\kappa) \sim \epsilon^{1/3}\kappa^{2/3}$ is unwarranted beyond $\kappa\eta = 1$ because viscosity limits the maximum strain rate to $(\epsilon/\nu)^{1/2}$. Also, of course, the use of (8.4.2) in a region in which viscous time scales are important is incorrect. The exponential decay of (8.4.6), which allows it to be integrated or differentiated without creating problems at large $\kappa\eta$, is thus merely a happy coincidence.

The dissipation spectrum corresponding to (8.4.6) is

$$D(\kappa) = 2\nu\kappa^2 E(\kappa) = 2\alpha\,\nu\epsilon^{2/3}\kappa^{1/3} \exp[-\tfrac{3}{2}\alpha(\kappa\eta)^{4/3}]. \tag{8.4.7}$$

In the inertial subrange ($\kappa\eta \ll 1$), the dissipation spectrum is proportional to

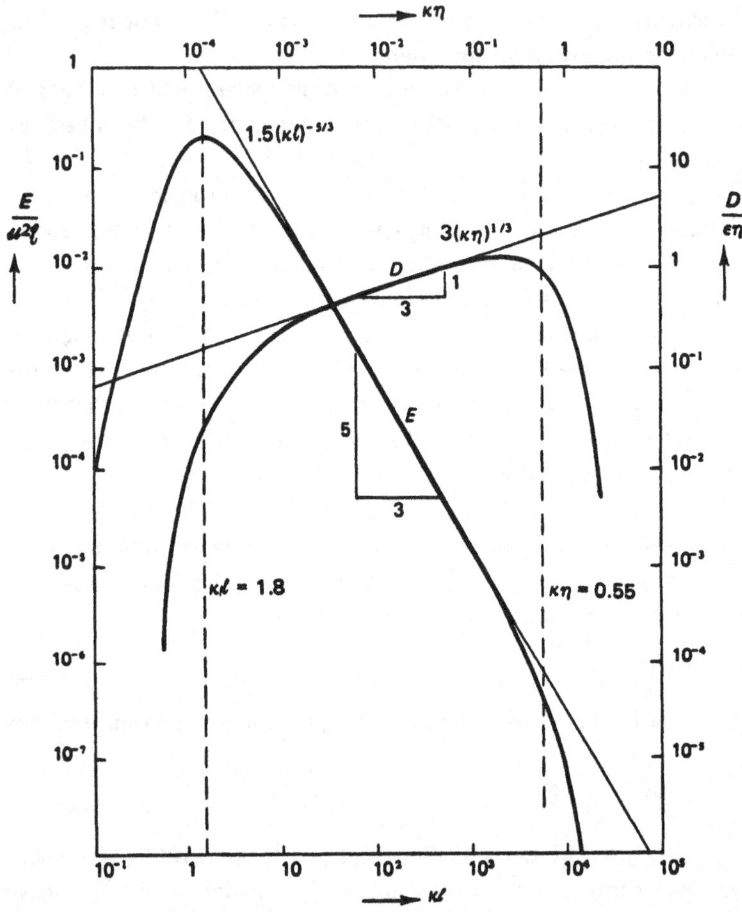

Figure 8.8. Normalized energy and dissipation spectra for $R_\ell = 2 \times 10^5$. The dashed lines indicate cutoffs for the approximate spectra to be described later.

$\kappa^{1/3}$. Figure 8.8 gives an impression of the shapes of $E(\kappa)$ and $D(\kappa)$. The curves also show how $E(\kappa)$ and $D(\kappa)$ trail off at low wave numbers; the analysis leading to this follows.

If $\alpha = 1.5$, the peak of the dissipation spectrum occurs at $\kappa\eta = 0.2$ and the value of $D(\kappa)$ at the peak is $D = 1.4\epsilon\eta$. These numbers agree well with most of the experimental data.

The effect of production Eddies near the lower edge of the inertial subrange, where $\kappa\ell$ is not necessarily very large, receive most of their energy from slightly larger eddies, but they also absorb some energy directly from the strain rate S of the mean flow. The anisotropy induced by the mean strain rate in eddies of wave number κ is proportional to $S/s(\kappa)$, so that the work done by the mean strain rate per unit wave number and per unit time is proportional to $ES^2/s(\kappa)$. Using (8.4.3) to substitute for $s(\kappa)$, we obtain for the production spectrum $P(\kappa)$,

$$P(\kappa) = \frac{2\pi\beta}{\alpha^{1/2}} \frac{S^2}{\epsilon^{1/3}} \kappa^{-2/3} E(\kappa). \tag{8.4.8}$$

The constant β is undetermined. In the inertial subrange, the production spectrum is proportional to $\kappa^{-7/3}$; this agrees fairly well with experimental evidence.

The spectral energy transfer $T(\kappa)$ increases wherever energy is being added. If the total amount of energy does not change and if viscous dissipation can be neglected, we have

$$dT/d\kappa = P(\kappa). \tag{8.4.9}$$

When (8.4.4) and (8.4.8) are substituted into (8.4.9), there results, after the equation is integrated,

$$E(\kappa) = \alpha\,\epsilon^{2/3}\kappa^{-5/3} \exp[-\tfrac{3}{2}\pi\beta\,\alpha^{1/2}(\kappa\ell)^{-4/3}]. \tag{8.4.10}$$

Here we have defined ℓ by $\ell = u^3/\epsilon$ and we have taken $S = u/\ell$ for convenience, as in (8.3.5). Although (8.4.10) is well behaved at all wave numbers, the assumptions $s(\kappa) \sim \epsilon^{1/3}\kappa^{2/3}$ and (8.4.8) on which it is based are not valid for small values of $\kappa\ell$. The value of β can be determined by requiring that the integral of (8.4.10) be equal to the total energy $\frac{1}{2}\overline{u_i u_i} = \frac{3}{2}u^2$; this yields $\beta = 0.3$. The maximum of (8.4.10) occurs at $\kappa\ell = 1.3$ approximately; its value is about $0.2 u^2 \ell$. Figure 8.8 gives a sketch of $E(\kappa)$ and $D(\kappa) = 2\nu\kappa^2 E(\kappa)$ as predicted by (8.4.10). These curves are in qualitative agreement with most experimental data. The Reynolds number in Figure 8.8 is $R_\ell = 2 \times 10^5$, which corresponds to $\ell/\eta = 10^4$. The graph suggests that there are only about two decades of inertial subrange at this Reynolds number.

Approximate spectra for large Reynolds numbers From the appearance of Figure 8.8 we are tempted to approximate $E(\kappa)$ by $1.5 u^2 \ell(\kappa\ell)^{-5/3}$ between a wave number somewhere near $\kappa\ell = 1$ and a wave number near $\kappa\eta = 1$, and to put it equal to zero outside that range. In fact, for many purposes such an approximation is quite satisfactory, provided that the Reynolds number is large enough.

Of course, the spectrum should have a correct integral. If the limits of integration of the truncated spectrum are κ_0 and κ_d, we have

$$\frac{3}{2} u^2 = \int_0^\infty E(\kappa)\, d\kappa \cong \int_{\kappa_0}^{\kappa_d} 1.5 u^2 \ell(\kappa\ell)^{-5/3} d\kappa \cong 1.5 u^2 \int_{\kappa_0 \ell}^\infty x^{-5/3} dx. \qquad (8.4.11)$$

We may set the upper limit of integration equal to ∞ if ℓ/η is large. This requires large Reynolds numbers. The integral condition (8.4.11) serves to determine $\kappa_0\ell$; the result is $\kappa_0\ell = (\frac{3}{2})^{3/2} \cong 1.8$. The other end of the range can be determined by requiring that the integral of the dissipation spectrum be correct. We can write

$$\epsilon = \int_0^\infty D(\kappa)\, d\kappa \cong \int_{\kappa_0}^{\kappa_d} 3\epsilon\eta(\kappa\eta)^{1/3} d\kappa \cong 3\epsilon \int_0^{\kappa_d \eta} x^{1/3} dx. \qquad (8.4.12)$$

The lower limit has been put at zero because the Reynolds number is presumed to be large. From (8.4.12), we obtain $\kappa_d\eta = 0.55$ approximately. The cutoffs $\kappa_0\ell = 1.8$ and $\kappa_d\eta = 0.55$ are indicated with dashed vertical lines in Figure 8.8.

The one-dimensional spectra (F_{11} and F_{22}) corresponding to this truncated three-dimensional spectrum can be computed fairly easily if we assume that the turbulence is isotropic. Because $E(\kappa)$ is equal to zero in the range $0 \leq \kappa\ell < 1.8$, F_{11} curves parabolically downward in that range (8.1.18) and F_{22} curves parabolically upward (8.1.19). In the range where $E(\kappa) \propto \kappa^{-5/3}$, F_{11} and F_{22} have the same slope (see Section 8.1); the coefficients involved can be computed with the isotropic relations (8.1.13, 8.1.14). For $0 \leq \kappa_1\ell < \kappa_0\ell$, there results

$$F_{11}(\kappa_1) = \frac{33}{110}\,\alpha u^2 \ell(\kappa_0\ell)^{-5/3}\,[1 - \tfrac{5}{11}\,(\kappa_1/\kappa_0)^2], \tag{8.4.13}$$

$$F_{22}(\kappa_1) = \frac{1}{2}\cdot\frac{33}{110}\,\alpha u^2 \ell(\kappa_0\ell)^{-5/3}\,[1 + \tfrac{5}{11}\,(\kappa_1/\kappa_0)^2]. \tag{8.4.14}$$

For $\kappa_1\ell > \kappa_0\ell$, the results are

$$F_{11}(\kappa_1) = \frac{9}{55}\,\alpha u^2 \ell(\kappa_1\ell)^{-5/3} \tag{8.4.15}$$

$$F_{22}(\kappa_1) = \frac{4}{3}\,F_{11}(\kappa_1) = \frac{4}{3}\cdot\frac{9}{55}\,\alpha u^2 \ell(\kappa_1\ell)^{-5/3}. \tag{8.4.16}$$

Of course, F_{11} and F_{22} are truncated at the same point as E; that is, $\kappa_1\eta$ = 0.55. The integrals of F_{11} and F_{22} over all κ_1 are equal to u^2 by virtue of (8.1.6, 8.1.7), so that the integrals of F_{11} and F_{22} over all positive κ_1 are equal to $\frac{1}{2}u^2$. In the literature, F_{11} and F_{22} are sometimes normalized in such a way that their integrals over all positive κ_1 are equal to u^2; in that case, the coefficient $\frac{9}{55}$ in (8.4.15, 8.4.16) becomes $\frac{18}{55}$, with corresponding changes in (8.4.13) and (8.4.14). Note that (8.4.13–8.4.16) describe F_{11} and F_{22} for $\kappa_1 > 0$. The spectra given by (8.4.13–8.4.16) are sketched in Figure 8.9. The parabolic part of F_{11} matches the $\kappa^{-5/3}$ part at κ_0 without a discontinuity of slope, but the slope of F_{22} changes sign at κ_0.

The values of F_{11} and F_{22} at the origin ($\kappa_1 = 0$) are of interest because they determine the longitudinal and transverse integral scales L_{11} and L_{22}:

$$F_{11}(0) = \frac{1}{2\pi}\int_{-\infty}^{\infty} R_{11}(r,0,0)\,dr = \frac{u^2}{\pi}L_{11}, \tag{8.4.17}$$

$$F_{22}(0) = \frac{1}{2\pi}\int_{-\infty}^{\infty} R_{22}(r,0,0)\,dr = \frac{u^2}{\pi}L_{22}. \tag{8.4.18}$$

Note that integral scales are defined as the integral of the correlation between zero and positive infinity, so that the factor $1/2\pi$ in front of the integrals becomes $1/\pi$ at the right-hand side of (8.4.17, 8.4.18). Evaluating $F_{11}(0)$ and $F_{22}(0)$ from (8.4.13, 8.4.14) and using $\alpha = 1.5$ and $\kappa_0\ell = (\frac{3}{2})^{3/2}$, we obtain

$$L_{11} = \ell/2, \quad L_{22} = \ell/4. \tag{8.4.19}$$

We recall that ℓ was defined by $\epsilon = u^3/\ell$. The relations between these length scales and the Lagrangian integral time scale are derived in the next section. Although (8.4.19) was derived for isotropic turbulence, it may be used also to obtain crude estimates in shear flows.

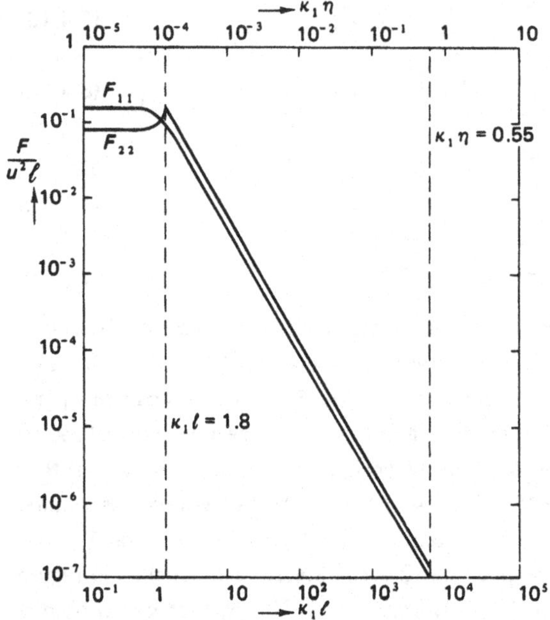

Figure 8.9. Crude approximations for the one-dimensional spectra at $R_\ell = 2 \times 10^5$ ($\ell/\eta = 10^4$).

8.5
Time spectra

So far we have discussed only space spectra, which are Fourier transforms of autocorrelations taken with a spatial separation and zero time delay. We now want to consider time spectra, which are obtained from correlations taken at the same point with varying time delay. If the point of measurement is a fixed point in a coordinate system chosen such that the mean velocity is zero, we obtain an *Eulerian time spectrum*; if the point of measurement is a wandering material point, we obtain a *Lagrangian time spectrum*. The measurements needed to obtain these spectra are quite difficult and time-consuming; very few experimental data are available.

Because time is a one-dimensional variable, time spectra are one-dimensional. We can have time spectra of u_1, u_2, or u_3; however, we are mainly interested in spectra which integrate to the total energy $\frac{1}{2}\overline{u_i u_i} = \frac{3}{2}u^2$. Let us define the Eulerian time spectrum $\psi_{ij}(\omega)$ by

$$\overline{u_i(\mathbf{x}, t)u_j(\mathbf{x}, t+\tau)} = R_{ij}(\tau) = \int_{-\infty}^{\infty} \exp(i\omega\tau)\, \psi_{ij}(\omega)\, d\omega,$$

$$\psi_{ij}(\omega) = \frac{1}{2\pi} \int_{-\infty}^{\infty} \exp(-i\omega\tau)\, R_{ij}(\tau)\, d\tau. \tag{8.5.1}$$

The Lagrangian time spectrum $\chi_{ij}(\omega)$ can be defined by

$$\overline{v_i(\mathbf{a}, t)v_j(\mathbf{a}, t+\tau)} = \mathscr{R}_{ij}(\tau) = \int_{-\infty}^{\infty} \exp(i\omega\tau)\, \chi_{ij}(\omega)\, d\omega,$$

$$\chi_{ij}(\omega) = \frac{1}{2\pi} \int_{-\infty}^{\infty} \exp(-i\omega\tau)\, \mathscr{R}_{ij}(\tau)\, d\tau. \tag{8.5.2}$$

In homogeneous turbulence, $\overline{u_i u_i} = \overline{v_i v_i}$ (see Section 7.2), so that

$$\int_{-\infty}^{\infty} \psi_{ii}(\omega)\, d\omega = \int_{-\infty}^{\infty} \chi_{ii}(\omega)\, d\omega = \overline{u_i u_i} = 3u^2. \tag{8.5.3}$$

If the turbulence is also isotropic, $\psi_{11} = \psi_{22} = \psi_{33}$ and $\chi_{11} = \chi_{22} = \chi_{33}$. These spectra do not vanish at the origin; instead, their values at $\omega = 0$ define integral time scales:

$$\psi_{ii}(0) = \frac{1}{2\pi} \int_{-\infty}^{\infty} R_{ii}(\tau)\, d\tau = \frac{\overline{u_i u_i}}{\pi} T = \frac{3u^2}{\pi} T, \tag{8.5.4}$$

$$\chi_{ii}(0) = \frac{1}{2\pi} \int_{-\infty}^{\infty} \mathscr{R}_{ii}(\tau)\, d\tau = \frac{\overline{u_i u_i}}{\pi} \mathscr{T} = \frac{3u^2}{\pi} \mathscr{T}. \tag{8.5.5}$$

Here, T is the Eulerian integral time scale and $\mathscr{T}$ is the Lagrangian integral time scale. It would be necessary to define more than one T and more than one $\mathscr{T}$ if the turbulence were not homogeneous and isotropic.

In order to understand what these time spectra mean, we have to use the energy cascade concept and extensions of the Kolmogorov scaling laws presented in Section 8.3. We begin with the equilibrium range for large wave numbers. The wave-number spectrum in the equilibrium range, (8.3.1), is normalized with the velocity and size of the dissipative eddies. If we accept the premise that large wave numbers correspond to high frequencies, the time spectra in the equilibrium range should be normalized with the velocity and frequency of the dissipative eddies.

The Lagrangian time spectrum relates to the temporal evolution seen by an observer moving with the turbulent velocity fluctuations. The highest dynamic frequency in turbulence is of order $(\epsilon/\nu)^{1/2} = \upsilon/\eta$ (υ is the Kolmogorov velocity, η is the Kolmogorov length); this frequency is a measure for the rate at which the dissipative eddies evolve. If the Lagrangian time spectrum has an equilibrium range at high frequencies, that range should be normalized with υ and $(\epsilon/\nu)^{1/2} = \upsilon/\eta$. The result is

$$\chi_{ii}(\omega) = \upsilon^2 (\nu/\epsilon)^{1/2} f\{\omega(\nu/\epsilon)^{1/2}\} = \upsilon\eta f(\omega\eta/\upsilon) . \quad (8.5.6)$$

The idea that equilibrium-range scaling can be applied to the Lagrangian time spectrum was first suggested by Inoue (1951).

The Eulerian time spectrum relates to the temporal changes seen by an observer located at a fixed point with respect to a frame of reference in which the mean velocity at the observation point is zero. Concentrating on a possible equilibrium range at high frequencies, such an observer sees small eddies being swept past him by larger eddies. The highest frequency seen this way occurs when the smallest eddies are advected past the observer by the most energetic eddies. This frequency is of order u/η, which is larger than the highest Lagrangian frequency by a factor proportional to $u/\upsilon \sim R_\ell^{1/4}$ (1.5.16). The high-frequency end of the Eulerian time spectrum thus is quite different from its Lagrangian counterpart; advection effects mask the Lagrangian temporal evolution. If the Eulerian time spectrum has an equilibrium range at high frequencies, the frequency in that range should be normalized as $\omega\eta/u$, and there results

$$\psi_{ii}(\omega) = \upsilon^2 \frac{\eta}{u} f(\omega\eta/u) . \quad (8.5.7)$$

This equilibrium range is unlike the others encountered so far, because the parameters ϵ and ν are not sufficient to define its form (Tennekes, 1975).

In the energy-containing range, the two time spectra should scale with u and ℓ. There is no significant spectral broadening in the large-scale end of the Eulerian time spectrum because the Lagrangian frequency u/ℓ that relates to the temporal evolution of large eddies is comparable to the frequency seen when a large eddy advects itself past a fixed observer. Therefore, we should write, for $\omega\eta/u < \omega\eta/\upsilon \ll 1$,

$$\chi_{ii}(\omega) = u\ell \mathscr{F}(\omega\ell/u), \quad \psi_{ii}(\omega) = u\ell F(\omega\ell/u) . \quad (8.5.8)$$

The shapes of F and $\mathscr{F}$ are different because the Eulerian spectrum is affected by advection while the Lagrangian spectrum is not.

The inertial subrange If the Reynolds number is so large that the high-frequency end of (8.5.8) overlaps with the low-frequency ends of (8.5.6) and (8.5.7), there should be inertial subranges in the two time spectra. A matching procedure similar to that used in Section 8.3 yields for the inertial subrange in the Lagrangian time spectrum (Inoue, 1951; Corrsin, 1963a)

$$\chi_{ii}(\omega) = \beta\epsilon\omega^{-2}, \tag{8.5.9}$$

and for the "inertial-advective" subrange in the Eulerian time spectrum (Tennekes, 1975)

$$\psi_{ii} = B\epsilon^{2/3}\mathscr{u}^{2/3}\omega^{-5/3}. \tag{8.5.10}$$

Both of these expressions can be obtained also from simple dimensional analysis. An eddy of wave number κ in the inertial subrange has a kinetic energy of order $\epsilon^{2/3}\kappa^{-2/3}$ and a Lagrangian frequency of order $\epsilon^{1/3}\kappa^{2/3}$. The kinetic energy at that Lagrangian frequency ω thus can be written as $\epsilon\omega^{-1}$, and χ_{ii} (which has the dimensions of kinetic energy per unit frequency) becomes as given in (8.5.9). The Eulerian frequency associated with the same wave number, however, is $\mathscr{u}\kappa$. The kinetic energy $\epsilon^{2/3}\kappa^{-2/3}$ at the Eulerian frequency $\omega = u\kappa$ thus can be written as $\epsilon^{2/3}\mathscr{u}^{2/3}\omega^{-2/3}$; this leads to (8.5.10).

The constants β and B are unknown: both are presumed to be of order one. At present, there is no way to estimate B, but β can be estimated if we accept the premise that the Lagrangian time spectrum is a simple rearrangement of $E(\kappa)$. If the energy at wave number κ is $\kappa E(\kappa)$ and if the angular frequency corresponding to κ is $\omega = \alpha^{1/2}\epsilon^{1/3}\kappa^{2/3}$ (8.4.3), we have

$$\kappa E = \omega\chi_{ii}, \tag{8.5.11}$$

$$\omega = \alpha^{1/2}\epsilon^{1/3}\kappa^{2/3}. \tag{8.5.12}$$

Eliminating κ between (8.5.11) and (8.5.12), and using $E(\kappa) = \alpha\epsilon^{2/3}\kappa^{-5/3}$, we obtain

$$\chi_{ii} = \alpha^{3/2}\epsilon\omega^{-2}. \tag{8.5.13}$$

With $\alpha = 1.5$, we find $\beta = 1.8$.

The Lagrangian integral time scale The form of the Lagrangian time spectrum at low frequencies is more difficult to predict. The Lagrangian spectrum is of interest because it could be used to find the Lagrangian integral scale, which plays such an important role in turbulent transport (Chapters 2 and 7). However, if we understood Lagrangian dynamics well enough to derive a spectrum, we would probably not need to estimate the Lagrangian integral scale in this way. Lacking any information other than that $\chi_{ii}(0)$ should be finite, the best we can do is to guess that χ_{ii} is constant for all frequencies below some ω_0 and that it follows (8.5.13) above ω_0. Thus, let us assume that

$$\chi_{ii} = \beta\epsilon\omega_0^{-2} \text{ for } 0 \leqslant \omega < \omega_0, \tag{8.5.14}$$

$$\chi_{ii} = \beta\epsilon\omega^{-2} \text{ for } \omega \geqslant \omega_0. \tag{8.5.15}$$

The value of ω_0 can be determined by requiring that the integral of $\chi_{ii}(\omega)$ over all ω be equal to $\overline{u_i u_i} = 3u^2$ (8.5.3). If the Reynolds number is so large that the viscous cutoff can be ignored, there results

$$\omega_0 \ell / u = 4\beta/3 \cong 2.4. \tag{8.5.16}$$

The resulting value of $\mathscr{T}$ is approximately $\ell/3u$. Using the Eulerian length scales L_{11} and L_{22} obtained in (8.4.19), we find

$$\mathscr{T} \cong \frac{1}{3}\frac{\ell}{u} \cong \frac{2}{3}\frac{L_{11}}{u} \cong \frac{4}{3}\frac{L_{22}}{u}. \tag{8.5.17}$$

This method of estimating the Lagrangian integral scale was first suggested by S. Corrsin (1963a); his result was somewhat different because he used $\chi_{ii} = 0$ for $0 < \omega < \omega_0$. Considering the crudeness of the assumptions involved in obtaining (8.5.17), we should not take the values of the coefficients too seriously. In effect, (8.5.17) gives barely more than the dimensional statement $\mathscr{T} \sim L_{11}/u \sim L_{22}/u$. In any case, in the absence of better estimates, (8.5.17) is useful for the purpose of obtaining Lagrangian integral scales from Eulerian correlation data.

An approximate Lagrangian spectrum Equations (8.5.14–8.5.16) define a crude approximation to the Lagrangian time spectrum, in the spirit of the one-dimensional spectra presented in Figure 8.9. One detail yet needs to be resolved: the spectrum has to be truncated at some frequency in the dissipation range. Because the Lagrangian dynamics of dissipation cannot be formulated in simple terms, we can do no more than compute the maximum value of $\omega = 2\pi s(\kappa) = (\kappa^3 E)^{1/2}$ from Corrsin's form (8.4.6) of the Eulerian space spectrum. If $\alpha = 1.5$, this maximum occurs at $\kappa\eta = 0.5$; its value is

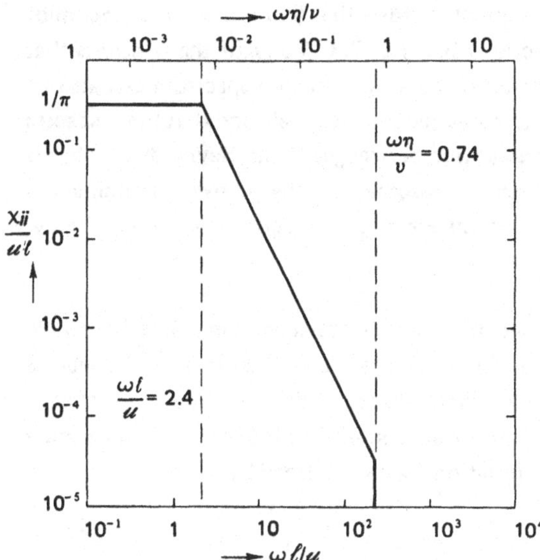

Figure 8.10. An approximate Lagrangian spectrum for $R_\ell = 10^5$.

$$\omega_d = 0.74(\epsilon/\nu)^{1/2} = 0.74 v/\eta. \tag{8.5.18}$$

This is in agreement with the ideas developed in Chapter 3; that is, the maximum frequency (vorticity, strain rate) is of order $(\epsilon/\nu)^{1/2}$. The ratio between ω_d and ω_0 is, from (8.5.16) and (8.5.18),

$$\omega_d/\omega_0 = 0.31\,(\epsilon/\nu)^{1/2}\,\ell/u = 0.31\,R_\ell^{1/2}. \tag{8.5.19}$$

The approximate Lagrangian spectrum for $R_\ell = 10^5$, $\omega_d/\omega_0 \sim 10^2$ is sketched in Figure 8.10. The separation of scales in the time spectrum is much less than that in the space spectrum; indeed, ω_d/ω_0 is a much better measure of the extent of the inertial subrange because vorticities and strain rates, which are frequencies, dominate the dynamics of turbulence.

8.6
Spectra of passive scalar contaminants

When a dynamically passive contaminant is mixed by a turbulent flow, a spectrum of contaminant fluctuations is produced. The scales of contaminant fluctuations range from the scale of the energy-containing eddies to a smallest scale that depends on the ratio of diffusivities (Prandtl number, Schmidt number), as we found in Sections 3.4 and 7.3. Many of the concepts that were used to elucidate the form of the kinetic energy spectrum can also be applied to the spectra of scalar contaminants; we shall find that these spectra have simple forms in various wave-number ranges if the Reynolds number is large. To simplify the discussion, we assume that the passive contaminant is heat. If the temperature fluctuations are small enough, the associated buoyancy is dynamically unimportant.

One- and three-dimensional spectra Passive scalar contaminants have one- and three-dimensional spectra. These spectra are defined in a similar way as velocity spectra, but they are simpler because there is only one variable, rather than three components, to be accounted for. If $\theta(\mathbf{x},t)$ is a temperature fluctuation, the spatial autocorrelation $R_\theta(\mathbf{r})$ is defined by

$$R_\theta(\mathbf{r}) = \overline{\theta(\mathbf{x}, t)\theta(\mathbf{x} + \mathbf{r}, t)}. \tag{8.6.1}$$

The Fourier transform of $R_\theta(\mathbf{r})$ is the spatial spectrum $\phi(\boldsymbol{\kappa})$:

$$R_\theta(\mathbf{r}) = \iiint_{-\infty}^{\infty} \exp(i\boldsymbol{\kappa}\cdot\mathbf{r})\, \phi_\theta(\boldsymbol{\kappa})\, d\boldsymbol{\kappa},$$

$$\phi_\theta(\boldsymbol{\kappa}) = \frac{1}{(2\pi)^3} \iiint_{-\infty}^{\infty} \exp(-i\boldsymbol{\kappa}\cdot\mathbf{r})\, R_\theta(\mathbf{r})\, d\mathbf{r}. \tag{8.6.2}$$

Just as for the velocity field, $\phi_\theta(\boldsymbol{\kappa})$ is the "energy" of waves of wave-number vector $\boldsymbol{\kappa}$. A *one-dimensional* spectrum $F_\theta(\kappa_1)$ may be defined by

$$R_\theta(r, 0, 0) \equiv \int_{-\infty}^{\infty} \exp(i\kappa_1 r)\, F_\theta(\kappa_1)\, d\kappa_1. \tag{8.6.3}$$

The relation between $\phi_\theta(\boldsymbol{\kappa})$ and $F_\theta(\kappa_1)$ is given by

$$\begin{aligned} R_\theta(r, 0, 0) &= \iiint_{-\infty}^{\infty} \exp(i\kappa_1 r)\, \phi_\theta(\boldsymbol{\kappa})\, d\boldsymbol{\kappa} \\ &= \int_{-\infty}^{\infty} \exp(i\kappa_1 r) \left(\iint_{-\infty}^{\infty} \phi_\theta\, d\kappa_2\, d\kappa_3 \right) d\kappa_1, \end{aligned} \tag{8.6.4}$$

$$F_\theta(\kappa_1) = \iint_{-\infty}^{\infty} \phi_\theta\, d\kappa_2\, d\kappa_3. \tag{8.6.5}$$

Clearly, $F_\theta(\kappa_1)$ suffers from the same aliasing problem as the one-dimensional spectra for the velocity (see (8.1.9)).

The *three-dimensional* spectrum $E_\theta(\kappa)$ is defined as the spectral density of waves which have the same wave-number magnitude κ ($\kappa^2 = \boldsymbol{\kappa}\cdot\boldsymbol{\kappa} = \kappa_i\kappa_i$), regardless of direction. This is obtained by integrating $\phi_\theta(\boldsymbol{\kappa})$ over a sphere with radius κ (see 8.1.4):

$$E_\theta(\kappa) = \oiint \tfrac{1}{2}\,\phi_\theta\, d\sigma, \tag{8.6.6}$$

$$\tfrac{1}{2}\,\overline{\theta^2} = \int_0^\infty E_\theta(\kappa)\, d\kappa. \tag{8.6.7}$$

Because the integral of E_θ is $\frac{1}{2}\overline{\theta^2}$, we call E_θ the spectrum of temperature variance.

In an isotropic field, there is only one one-dimensional spectrum because the direction of the spatial separation $\mathbf{r}$ in $R_\theta(\mathbf{r})$ is immaterial. For the same reason, $\phi_\theta(\boldsymbol{\kappa})$ depends only on the wave-number modulus κ under isotropic conditions. The isotropic relations between $F_\theta(\kappa_1)$, $\phi_\theta(\kappa)$, and $E_\theta(\kappa)$ are (Hinze, 1959)

$$E_\theta(\kappa) = 2\pi\kappa^2\phi_\theta(\kappa), \tag{8.6.8}$$

$$F_\theta(\kappa_1) = \int_{\kappa_1}^\infty \kappa^{-1}E_\theta(\kappa)\, d\kappa, \quad E_\theta(\kappa) = -\kappa\frac{d}{d\kappa}[F_\theta(\kappa)]. \tag{8.6.9}$$

If the temperature field has a finite integral scale, $F_\theta(0)$ is finite and $E_\theta(\kappa)$ begins parabolically upward from $\kappa = 0$ (recall that the kinetic energy spectrum starts quartically from $\kappa = 0$). This statement is valid even if the field is not isotropic.

The cascade in the temperature spectrum In the development of a model for the spectral transfer of temperature fluctuations, we use E and E_θ, because they represent the "energy" at a given wave number without effects due to aliasing. If the Reynolds number is large enough for an equilibrium range to exist in the kinetic energy spectrum, there is also an equilibrium range, exhibiting local isotropy, in the spectrum of temperature variance, because it is the turbulent motion that is mixing the temperature field.

The cascade in the temperature spectrum is similar to that in the velocity spectrum. The temperature gradient associated with an eddy of wave number

κ_1 is of order $[\kappa_1^3 E_\theta(\kappa_1)]^{1/2}$. The velocity fluctuations of the eddies of the next smaller scale ($\kappa_2 > \kappa_1$, say), distort this gradient, thus producing temperature fluctuations of smaller scale. This is like the production of temperature variance from a mean temperature gradient, which we discussed in Section 3.4. There, temperature variance was produced at a rate $-\overline{\theta u_j}\, \partial\Theta/\partial x_j$; perhaps we can reason by analogy here. The spatial heat flux θu_j was estimated as $u\ell\partial\Theta/\partial x_j$ in Chapter 2; the spectral flux of temperature variance thus should be the spectral equivalent of $u\ell(\partial\Theta/\partial x_j)^2$. Now, u and ℓ have to be substituted by the velocity $[\kappa_2 E(\kappa_2)]^{1/2}$ and the size $1/\kappa_2$ of the smaller eddies that distort the temperature gradients of the larger eddies. The spectral flux of temperature variance may then be estimated as

$$T_\theta = C\kappa_2^{-1}[\kappa_2 E(\kappa_2)]^{1/2}\kappa_1^3 E_\theta(\kappa_1). \tag{8.6.10}$$

If we ignore the difference between κ_1 and κ_2 because they are fairly close together, we obtain

$$T_\theta(\kappa) = C\kappa^2 E_\theta(\kappa E)^{1/2}. \tag{8.6.11}$$

This local estimate of T_θ is, of course, as crude as the cascade model developed for the kinetic energy spectrum. In particular, the significance at large and small wave numbers of scales associated with such quantities as kinematic viscosity ν, the thermal diffusivity γ, and the mean strain rate S is ignored.

Spectra in the equilibrium range Within the equilibrium range, $E_\theta(\kappa)$ should scale with the parameters governing the velocity field, which are ϵ and ν, and the corresponding parameters for the temperature field. The dissipation of temperature variance will be called N; it is defined by

$$N \equiv \gamma\, \overline{\frac{\partial\theta}{\partial x_i}\frac{\partial\theta}{\partial x_i}}. \tag{8.6.12}$$

Thus, we expect $E_\theta = E_\theta(\kappa, \epsilon, \nu, \gamma, N)$ in the equilibrium range. A convenient combination of variables is

$$E_\theta(\kappa) = N\epsilon^{-1/3}\kappa^{-5/3} f(\kappa\eta, \sigma). \tag{8.6.13}$$

Here, $\sigma = \nu/\gamma$ is the Prandtl number. Because of the presence of σ, the nondimensional spectrum f is different in different fluids.

If the Reynolds number is so large that the energy spectrum has an inertial subrange and if γ is small enough, so that there is an appreciable part of the spectrum where the thermal diffusivity is unimportant, we obtain an *inertial-convective subrange,* that is, an inertial range in which temperature fluctuations are simply convected. In this range, the spectrum should be independent of ν and γ, so that we have $E_\theta = E_\theta(\kappa, N, \epsilon)$, which can have only one form:

$$E_\theta(\kappa) = \beta N \epsilon^{-1/3} \kappa^{-5/3}. \tag{8.6.14}$$

This was first suggested independently by Corrsin (1951) and by Oboukhov (1949). Recent measurements give $\beta = 0.5$ approximately. If we substitute (8.6.14) and $E(\kappa) = \alpha\epsilon^{2/3}\kappa^{-5/3}$ into the estimate for T_θ given in (8.6.11), we find that the spectral transfer of temperature variance in the inertial-convective subrange is constant, as it should be. Conversely, if we assume that $T_\theta(\kappa) = N$, we obtain (8.6.14) from (8.6.11). In other words, the "mixing-length" model for T_θ given by (8.6.11) is consistent with (8.6.14).

If we want to take the effects of γ and ν into account, we have to distinguish between fluids with small Prandtl numbers and those with large Prandtl numbers. If $\sigma < 1$, so that $\gamma < \nu$, the thermal diffusivity becomes important within the inertial subrange, where the viscosity does not yet influence the spectrum. As in Section 3.4, we denote the Kolmogorov microscale for the temperature field by η_θ; recall that $\eta_\theta > \eta$ if $\gamma > \nu$ and that $\eta_\theta < \eta$ if $\gamma < \nu$. If $\gamma >> \nu$, there is a range of wave numbers where $\kappa\eta_\theta \geqslant 1$, but $\kappa\eta << 1$. This is called an *inertial-diffusive subrange*; it occurs in mercury, for example.

On the other hand, in water and most other liquids the Prandtl number is large, so that viscosity becomes important at wave numbers where the thermal diffusivity does not yet affect the temperature spectrum. The range of wave numbers where $\kappa\eta \geqslant 1$, but $\kappa\eta_\theta << 1$, is called a *viscous-convective subrange.* Of course, there is also a range where $\kappa\eta >> \kappa\eta_\theta \geqslant 1$; this is called the *viscous-diffusive subrange.*

The inertial-diffusive subrange In fluids with low Prandtl numbers, an inertial-diffusive subrange exists for $\kappa\eta_\theta \geqslant 1$, $\kappa\eta << 1$. In this range, the spectral flux of kinetic energy is constant and equal to ϵ. However, the spectral flux of temperature variance, which is equal to N in the inertial-convective subrange, decreases in the inertial diffusive subrange because of local dissipation of temperature variance:

$$dT_\theta/d\kappa = -2\gamma\kappa^2 E_\theta. \tag{8.6.15}$$

This equation can be solved only if we adopt the cascade model (8.6.11) of the spectral transfer T_θ. In the inertial-diffusive subrange, $E(\kappa) = \alpha\epsilon^{2/3}\kappa^{-5/3}$, so that (8.6.11) becomes

$$T_\theta = C\alpha^{1/2}\epsilon^{1/3}\kappa^{5/3}E_\theta. \tag{8.6.16}$$

Comparing this with (8.6.14), we find that (8.6.16) amounts to replacing N by T_θ in (8.6.14). Hence, we are using inertial scaling, as in (8.6.14), but based on the local value of T_θ. This can be a fair approximation only if T_θ changes slowly with wave number. The comparison of (8.6.14) and (8.6.16) also shows that we should take $C\alpha^{1/2} = \beta^{-1}$.

The solution of (8.6.15, 8.6.16), with $C\alpha^{1/2} = \beta^{-1}$, is (Corrsin, 1964)

$$E_\theta(\kappa) = \beta N\epsilon^{-1/3}\kappa^{-5/3}\exp[-\tfrac{3}{2}\beta(\kappa\eta_\theta)^{4/3}]. \tag{8.6.17}$$

Here, the temperature microscale η_θ is defined by

$$\eta_\theta \equiv (\gamma^3/\epsilon)^{1/4}. \tag{8.6.18}$$

This scale was discovered by Corrsin (1951). At present, no measurements exist with which (8.6.17) can be compared. The spectrum is well behaved at large wave numbers, but it cannot be valid far into the inertial-diffusive subrange because the assumptions on which it is based are not valid there.

Because (8.6.18) is identical in shape with (8.4.6), the peak in the spectrum of dissipation of temperature variance occurs at $\kappa\eta_\theta = 0.2$. Also, if we want to truncate the spectrum at the high wave-number end, we have to put one cutoff point at $\kappa\eta_\theta = 0.55$.

The viscous-convective subrange A viscous-convective subrange occurs at wave numbers such that $\kappa\eta \geqslant 1$, $\kappa\eta_\theta \ll 1$, in fluids with a large Prandtl number. In this range, the scales of temperature fluctuations are progressively reduced by the strain-rate field (see Sections 3.4 and 7.3), but the thermal diffusivity is not yet effective. Temperature fluctuations at wave numbers beyond $\kappa\eta = 1$ experience strain-rate fields of magnitude $(\epsilon/\nu)^{1/2}$ and size η. Because the energy spectrum drops off so sharply near $\kappa\eta = 1$, the extent of the strain-rate fields appears to be infinite to small temperature eddies at $\kappa\eta \gg 1$. Therefore, only $(\epsilon/\nu)^{1/2}$ should be important, but not η. In the viscous-convective subrange, we thus expect that $E_\theta = E_\theta(\kappa, N, (\epsilon/\nu)^{1/2})$.

This must have the form

$$E_\theta(\kappa) = cN(\nu/\epsilon)^{1/2}\kappa^{-1}. \tag{8.6.19}$$

This spectrum was first predicted by Batchelor (1959); measurements by Gibson and Schwartz (1963) have confirmed its existence. It should be noted that (8.6.19) can also be obtained (but less rigorously) from the "mixing-length" estimate (8.6.11) of the spectral transfer T_θ. This is done by replacing the strain rate $(\kappa^3 E)^{1/2}$ in (8.6.11) by $(\epsilon/\nu)^{1/2}$ and putting $T_\theta = N$, because the effects of γ are presumed to be small.

The viscous-diffusive subrange At very large wave numbers, molecular diffusion of temperature fluctuations becomes effective. The viscous-convective subrange ends when the scale of the temperature fluctuations has become so small that diffusion becomes significant for time scales of the order of the period $(\nu/\epsilon)^{1/2}$ of the strain-rate fluctuations. Diffusion spreads hot spots of size ℓ at a rate determined by $\ell^2 \sim \gamma t$; the smallest scale (η_θ) is obtained if t is replaced by $(\nu/\epsilon)^{1/2}$. This yields

$$\eta_\theta/\eta = (\gamma/\nu)^{1/2}. \tag{8.6.20}$$

This estimate, which is valid only for $\gamma/\nu \ll 1$, was obtained earlier in Sections 3.4 and 7.3.

The shape of the spectrum near $\kappa\eta_\theta = 1$ can be estimated in the now familiar way by adopting almost-inertial scaling for T_θ. As the viscous-diffusive subrange is approached, T_θ begins to decrease slowly. As long as T_θ is not too different from N, we may generalize (8.6.19) as

$$E_\theta(\kappa) = cT_\theta\,(\nu/\epsilon)^{1/2}\kappa^{-1}. \tag{8.6.21}$$

This states that T_θ is proportional to the amount of temperature variance in eddies of scale κ, which is κE_θ, and to the strain rate $(\epsilon/\nu)^{1/2}$. Substituting (8.6.21) into (8.6.15), we obtain

$$E_\theta(\kappa) = cN(\nu/\epsilon)^{1/2}\kappa^{-1}\exp[-c(\kappa\eta_\theta)^2], \tag{8.6.22}$$

in which η_θ is given by (8.6.20). The location of the exponential cutoff obtained this way agrees with the estimate (8.6.20). Again, although (8.6.22) is well behaved, it is certainly not valid for $\kappa\eta_\theta \gg 1$, because it is based on (8.6.21), which certainly is not valid there.

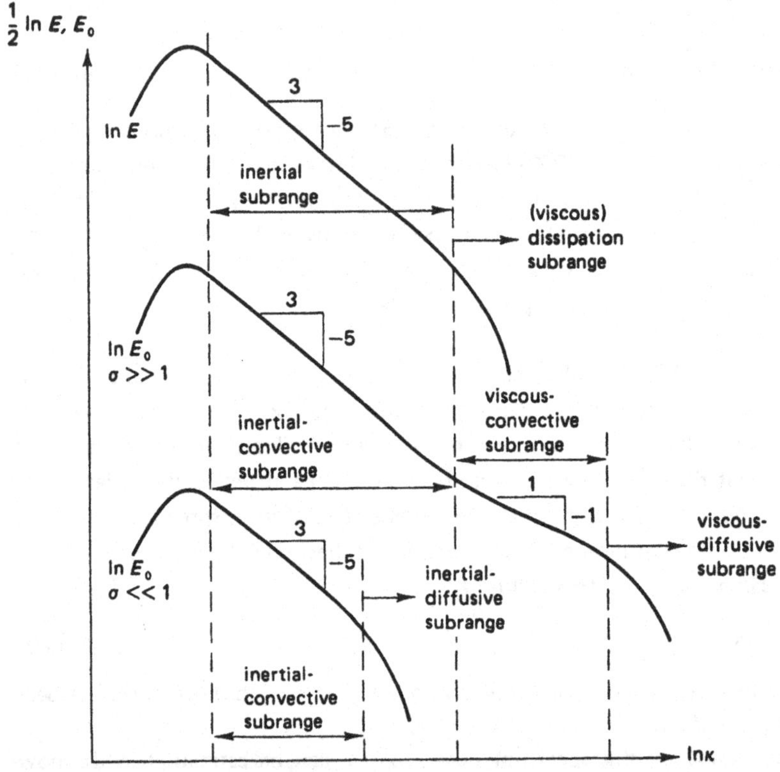

Figure 8.11. Spectra of temperature variance in liquids with large and small Prandtl numbers.

Summary The various subranges in the spectrum of temperature variance, for liquids with large and small Prandtl number, are sketched in Figure 8.11. The Reynolds number, of course, is assumed to be large.

Problems

8.1 What is the shape of the correlation function $\overline{u(x)u(x+r)}$ in a range of values of r which corresponds to the inertial subrange?

8.2 In the spectral energy transfer model of Corrsin, the energy transfer across given wave number is approximated by a mixing-length expression that is not corrected for viscous effects as $\kappa\eta = 1$ is approached. Make a similar model for the momentum transfer in the inner layer of a turbulent boundary layer in zero pressure gradient. Integrate the resulting equation of motion and show that, unlike its spectral counterpart, this model does not give an accurate representation of the mean flow in the inner layer.

8.3 Derive expressions for the evolution of the kinetic energy and of the integral scale of isotropic turbulence in the initial period of decay (see Section 3.2). Do this by calculating the evolution of an approximate energy spectrum that consists of an inertial subrange at high wave numbers and a spectrum of the type $E(\kappa) \sim \kappa^4$ (8.1.16) at low wave numbers. Assume that the constant C in (8.1.16) is independent of time (this is called the "permanence of the largest eddies" (Batchelor, 1953)). Show that the Reynolds number of isotropic turbulence decreases in time during the initial period of decay, in contradiction with the result given in Section 3.2.

8.4 In the final period of decay of isotropic turbulence, the Reynolds number is so small that no energy exchange between wave numbers takes place. Calculate the rate of decay of the kinetic energy, assuming that the spectrum at the beginning of the final period of decay is given by (8.1.16), with C independent of time (see also Problem 8.3).

8.5 A small, heavy particle rapidly falls through a field of isotropic turbulence. Because the terminal velocity of the particle is large, its path is nearly straight, so that the particle, in first approximation, experiences a frequency spectrum corresponding to the one-dimensional Eulerian space spectrum. If the terminal velocity is V_T, the relation between frequency and wave number is $\omega = \kappa V_T$. Under certain conditions, the equation for the horizontal particle velocity v may be approximated by $T\,dv/dt + v = u$, where $T = V_T/g$ is the particle time constant and u is the horizontal fluid velocity experienced by the particle. Calculate the horizontal dispersion of the particle and compare it with the Lagrangian dispersion experienced by a particle with vanishingly small V_T.

BIBLIOGRAPHY AND REFERENCES

A comprehensive up-to-date bibliography may be found in Monin and Yaglom, 1971.

Abramovich, G. N., 1963.
The theory of turbulent jets. The M.I.T. Press, Cambridge, Mass.
Bakewell, H. P., and Lumley, J. L., 1967.
Viscous sublayer and adjacent wall region in turbulent pipe flow. *Physics of Fluids* **10**, 1880.
Batchelor, G. K., 1953.
The theory of homogeneous turbulence. Cambridge University Press, London.
———, 1959.
Small-scale variation of convected quantities like temperature in a turbulent fluid, part 1. *Journal of Fluid Mechanics* **5**, 113.
———, 1967.
An introduction to fluid mechanics. Cambridge University Press, London.
Batchelor, G. K., and Moffat, H. K. (eds.), 1970.
Proceedings of the Boeing Symposium on Turbulence. *Journal of Fluid Mechanics* **41**, parts 1 and 2.
Batchelor, G. K., and Townsend, A. A., 1956.
Turbulent diffusion. *Surveys in mechanics*, pp. 353–399. Cambridge University Press, London.
Blackadar, A. K., and Tennekes, H., 1968.
Asymptotic similarity in neutral barotropic planetary boundary layers. *Journal of the Atmospheric Sciences* **25**, 1015.
Bowden, K. F., Frenkiel, F. N., and Tani, I. (eds.), 1967.
Boundary layers and turbulence. *Physics of Fluids* **10**, S1–S321 (supplement issue, September 1967).
Bradshaw, P., 1971.
An introduction to turbulence and its measurement. Pergamon Press, London.
Clauser, F. H., 1956.
The turbulent boundary layer. *Advances in Applied Mechanics* **IV**, 1–51.
Cole, J. D., 1968.
Perturbation methods in applied mathematics. Blaisdell Publishing Co., Waltham, Mass.

Coles, D., 1956.
The law of the wake in the turbulent boundary layer. *Journal of Fluid Mechanics* **1**, 191.
Comte-Bellot, G., and Corrsin, S., 1965.
The use of a contraction to improve the isotropy of grid-generated turbulence. *Journal of Fluid Mechanics* **25**, 657.
Corrsin, S., 1951.
On the spectrum of isotropic temperature fluctuations in isotropic turbulence. *Journal of Applied Physics* **22**, 469.
———, 1953.
Remarks on turbulent heat transfer. *Proceedings of the Iowa Thermodynamics Symposium*, pp. 5–30. State University of Iowa, Iowa City.
———, 1959.
Outline of some topics in homogeneous turbulent flow. *Journal of Geophysical Research* **64**, 2134.
———, 1963a.
Estimates of the relations between Eulerian and Lagrangian scales in large Reynolds number turbulence. *Journal of the Atmospheric Sciences* **20**, 115.
———, 1963b.
Turbulence: Experimental methods. *Handbuch der Physik* **VII/2**, pp. 524–590. Springer-Verlag, Berlin.
———, 1964.
Further generalizations of Onsager's cascade model for turbulent spectra. *Physics of Fluids* **7**, 1156.
Corrsin, S., and Kistler, A. L., 1954.
The free-stream boundaries of turbulent flows. National Advisory Committee for Aeronautics, Technical Note NACA TN 3133.
Csanady, G. T., 1973.
Turbulent diffusion in the environment. D. Reidel Publishing Co., Dordrecht, the Netherlands.
Davies, T. J., 1972.
Turbulence phenomena. Academic Press, New York.
Favre, A. (ed.), 1962.
Mécanique de la turbulence. Colloques Internationaux du Centre National de la Récherche Scientifique. Éditions CNRS, Paris. English edition: *The mechanics of turbulence*, Gordon and Breach, New York, 1964.

Frenkiel, F. N. (ed.), 1962.
Fundamental problems in turbulence and their relation to geophysics. American Geophysical Union, New York.
Friedlander, S. K., and Topper, L. (eds.), 1962.
Turbulence: Classical papers on statistical theory. Interscience, New York.
Gibson C. H., and Schwartz, W. H., 1963.
The universal equilibrium spectra of turbulent velocity and scalar fields. *Journal of Fluid Mechanics* **16**, 365.
Hinze, J. O., 1959.
Turbulence. McGraw-Hill, New York.
Inoue, E., 1951.
On the Lagrangian correlation coefficient for turbulent diffusion and its application to atmospheric diffusion phenomena. Report, Geophysics Institute, University of Tokyo.
Jeans, J., 1940.
An introduction to the kinetic theory of gases. Cambridge University Press, London.
Kadomtsev, B. B., 1965.
Plasma turbulence. Academic Press, New York.
Kármán, Th. von, 1930.
Mechanische Ähnlichkeit und Turbulenz. *Nachrichten der Akademie der Wissenschaften Göttingen*, Math.-Phys. Klasse, 58.
Leslie, D. C., 1973
Developments in the theory of turbulence. Clarendon Press, Oxford.
Lin, C. C., 1961.
Statistical theories of turbulence. Princeton University Press, Princeton, N.J.
Ludwieg, H., and Tillman, W., 1949.
Untersuchungen über die Wandschubspannung in turbulenten Reibungsschichten. *Ingenieur Archiv* **17**, 288. English edition: National Advisory Committee for Aeronautics, Report NACA TM 1285, Investigations of the wall shearing stress in turbulent boundary layers.
Lumley, J. L., 1965.
On the interpretation of time spectra measured in high intensity shear flows. *Physics of Fluids* **8**, 1056.
———, 1970.
Stochastic tools in turbulence. Academic Press, New York.

———, 1972a.
Application of central limit theorems to turbulence problems. In *Statistical models and turbulence*. Lecture notes in mathematics. Springer-Verlag, New York.
———, 1972b.
On the solution of equations describing small scale deformation. In *Symposia Matematica: Convegno sulla Teoria della Turbolenza al Istituto Nazionale di Alta Matematica.* Academic Press, New York.
Lumley, J. L., and Panofsky, H. A., 1964.
The structure of atmospheric turbulence. Interscience, New York.
Millikan, C. B., 1939.
A critical discussion of turbulent flow in channels and circular tubes. *Proceedings of the Fifth International Congress on Applied Mechanics* (Cambridge, Mass., 1938), pp. 386–392. Wiley, New York.
Monin, A. S., and Yaglom, A. M., 1971.
Statistical fluid mechanics. The M.I.T. Press, Cambridge, Mass.
Morkovin, M. V., 1956.
Fluctuations and hot-wire anemometry in compressible flows. AGARDograph 24, NATO, Paris.
Oboukhov, A. M., 1949.
Structure of the temperature field in turbulent flows. *Izvestiya Akademii Nauk SSSR, Geogr. and Geophys. Ser.* **13**, 58.
Pao, Y. H., 1965.
Structure of turbulent velocity and scalar fields at large wave numbers. *Physics of Fluids* **8**, 1063.
Pao, Y. H., and Goldberg, A. (eds.), 1969.
Clear-air turbulence and its detection. Plenum Press, New York.
Pasquill, F., 1962.
Atmospheric diffusion. Van Nostrand, New York.
Phillips, O. M., 1966.
The dynamics of the upper ocean. Cambridge University Press, London.
Plate, E. J., 1971.
Aerodynamic characteristics of atmospheric boundary layers. National Technical Information Service, Springfield, Virginia.
Priestley, C. H. B., 1959.
Turbulent transfer in the lower atmosphere. University of Chicago Press, Chicago.

Reynolds, A. J., 1974.
Turbulent flows in Engineering. Wiley-Interscience, New York.

Reynolds, O., 1895.
On the dynamical theory of incompressible viscous fluids and the determination of the criterion. *Philosophical Transactions of the Royal Society of London*, Series A, **186**, 123.

Rotta, J. C., 1972.
Turbulente Strömungen. B. G. Teubner Verlag, Stuttgart, Germany.

Rouse, H., Yih, C. S., and Humphreys, H. W., 1952.
Gravitational convection from a boundary source. *Tellus* **4**, 201.

Saffman, P. G., 1963.
On the fine-scale structure of vector fields convected by a turbulent fluid. *Journal of Fluid Mechanics* **16**, **545**.

Schlichting, H., 1960.
Boundary-layer theory. McGraw-Hill, New York (4th ed.).

Schubauer, G. B., and Tchen, C. M., 1961.
Turbulent flow. Princeton University Press, Princeton, N.J.

Spitzer, L., Jr., 1968.
Diffuse matter in space. Interscience, New York.

Stewart, R. W., 1969.
Turbulence. (Motion picture film). Educational Services, Inc., Cambridge, Mass.

Stratford. B. S., 1959.
An experimental flow with zero skin friction throughout its region of pressure rise. *Journal of Fluid Mechanics* **5**, 17.

Tatarski, V. I., 1961.
Wave propagation in a turbulent medium. McGraw-Hill, New York.

Taylor, G. I., 1915.
Eddy motion in the atmosphere. *Philosophical Transactions of the Royal Society of London*, Series A, **215**, 1.

———, 1921.
Diffusion by continuous movements. *Proceedings of the London Mathematical Society*, Series 2, **20**, 196.

———, 1932.
The transport of vorticity and heat through fluids in turbulent motion. *Proceedings of the Royal Society of London*, Series A, **135**, **685**.

———, 1935.
Statistical theory of turbulence. *Proceedings of the Royal Society of London*, Series A, **151**, 421.

———, 1938.
The spectrum of turbulence. *Proceedings of the Royal Society of London*, Series A, **164**, 476.

Tennekes, H., 1965.
Similarity laws for turbulent boundary layers with suction or injection. *Journal of Fluid Mechanics* **21**, 689.

———, 1968.
Outline of a second-order theory for turbulent pipe flow. *AIAA Journal* **6**, 1735.

Tennekes, H., 1975.
Eulerian and Lagrangian time microscales in isotropic turbulence. *Journal of Fluid Mechanics* (in press).

Townsend, A. A., 1956.
The structure of turbulent shear flow. Cambridge University Press, London.

Uberoi, M. S., and Kovasznay, L. S. G., 1953.
On mapping and measurement of random fields. *Quarterly of Applied Mathematics* **10**, 375.

Vinnichenko, N. K., Pinus, N. Z., Shmeter, S. M., and Shur, G. N., 1973.
Turbulence in the free atmosphere. Consultants Bureau, New York.

Yaglom, A. M., and Tatarski, V. I. (eds.), 1967.
Atmospheric turbulence and radio-wave propagation. Publishing House Nauka, Moscow.

Zel'dovich, Ya. B., 1937.
Limiting laws for turbulent flows in free convection. *Zhurnal Eksperimental'noi Teoreticheskoi Fiziki* (Journal of Experimental and Theoretical Physics) **7**, No. 12, 1463.

INDEX

Printed in the United States
By Bookmasters